Structural Mechanics

Pearson Education

We work with leading authors to develop the strongest educational materials in construction and engineering, bringing cutting-edge thinking and best learning practice to a global market.

Under a range of well-known imprints, including Prentice Hall, we craft high quality print and electronic publications which help readers to understand and apply their content, whether studying or at work.

To find out more about the complete range of our publishing please visit us on the World Wide Web at: www.pearsoneduc.com

Structural Mechanics

LOADS, ANALYSIS, DESIGN AND MATERIALS

Sixth edition

H AL NAGEIM, F DURKA, W MORGAN
AND D T WILLIAMS

Prentice
Hall

An imprint of **Pearson Education**

Harlow, England · London · New York · Reading, Massachusetts · San Francisco
Toronto · Don Mills, Ontario · Sydney · Tokyo · Singapore · Hong Kong · Seoul
Taipei · Cape Town · Madrid · Mexico City · Amsterdam · Munich · Paris · Milan

Pearson Education Limited
Edinburgh Gate
Harlow
Essex CM20 2JE
England

and Associated Companies throughout the world

Visit us on the World Wide Web at:
www.pearsoneduc.com

First published in Great Britain by Pitman Publishing Limited 1980
Fourth edition published by Longman Group UK Limited 1989
Fifth edition published by Addison Wesley Longman Limited 1996
Sixth edition published 2003

ISBN 0 582 43165 4

British Library Cataloguing-in-Publication Data
A catalogue record for this book is available from the British Library

Library of Congress Cataloging-in-Publication Data
Structural mechanics : loads, analysis, design, and materials /
H. Al Nageim ... [et al.].—6th ed.
   p. cm.
Includes bibliographical references and index.
ISBN 0–582–43165–4
1. Structural analysis (Engineering) I. Al Nageim, H. (Hassan)

TA645 .S757 2002
624.1′71—dc21

2002025831

10 9 8 7 6 5 4 3 2 1
06 05 04 03 02

Typeset by 68 in 10/11 pt EhrhardtMT
Printed and bound in Malaysia

Contents

Part 1 Behaviour of structures

1 Introduction   3

2 Loads on buildings and structures   18

3 Concurrent coplanar forces   36

Part 2 Understanding structural design

List of tables

Preface

Understanding structural mechanics requires knowledge of many interlinked factors, such as the loads and load actions on the structure, the strength and properties of the materials from which structural elements are made, the ways by which the loads and load actions are transferred via the structure to the foundations, the interaction between the foundations and the supporting ground, structural stability, durability and environmental conditions. To address the above effectively, this edition includes three new chapters, along with a general update of the core content, especially the inclusion of British and Eurocode information.

This book covers discussions of conceptual design, including design methods, design loads and the design strength of construction materials. Also included is the presentation of principles for the various methods of analysis, as well as detailed analysis and preliminary design for selected structural elements. As part of the learning process, students are given selected self-assessment exercises at the end of each chapter. These generally consist of making an analysis, explaining important issues related to structural mechanics, design, material durability and/or structural safety.

The main aim of the book is to make understanding structures simple and to help students foster an interest in structural analysis, design and evaluation.

In the preface to the first edition of this book the late W. Morgan and D.T. Williams expressed the opinion that structural mechanics, the elementary stage of the theory of structures, 'is often rendered unnecessarily difficult by a highly mathematical approach and, therefore, the mathematical treatment here has been kept as simple as is consistent with accuracy. In all cases a full understanding of the basic principles has been considered more important than mere mathematical agility'.

Whilst the original authors' simplicity of the mathematical treatment has been maintained throughout subsequent editions, more emphasis has been placed on the relevance of structural mechanics to the process of structural design, material design strength and properties, design loads and load action on materials as outlined in Chapters 1 and 2. This approach makes the text suitable for students taking degree and diploma courses in architecture, building and surveying, and those in the early stages of civil and structural engineering courses. The text also covers the syllabus of the BTEC Higher National Unit 8162E Structural Mechanics.

The initial chapters of this book deal with the concept of structural design, design methods, material design strength, loads on structures, suitability of materials for constructions, forces in their various aspects – concurrence, non-concurrence, moments, etc. – and their effects on structural materials and elements in terms of stress and strain. The significance of the shape of the cross-section of structural elements is considered in Chapter 9. The remaining chapters are devoted to the design of simple structural elements (beams, columns, main elements of portal frames and dome frames, and gravity retaining walls) based on fully updated relevant data as given in the current editions of British Standards and the European codes of practice.

As Mr Frank Durka emphasized in his preface to the fifth edition, the preparations of this and the earlier editions of this book have not been confined to a single person, as the title page may suggest. In this work, I have been helped by many, both directly and indirectly, and my thanks are due and gratefully given to them all and in particular to Dr Ralf. M. Romaya, Principal Lecturer, Liverpool Polytechnic, Mr Patrick Dunlea-Jones, Senior Lecturer, Liverpool John Moores University, Mrs Alison Cotgrave, Principal Lecturer, Liverpool John Moores University, Mr Paul Hodgkinson and Emma Green, Liverpool John Moores University. None of this, however, would have been possible without the generous co-operation of my wife and sons Haydar and Yassier. I am, therefore, extremely grateful for their continuous support and help.

I hope that many students will find this current edition as helpful in their studies as the first edition was to myself and successive editions to my students. Then all the effort put into this work will have been well worthwhile.

Hassan Al Nageim
Liverpool, July 2002

Acknowledgements

We are grateful to the following for permission to reproduce copyright material:

British Steel: Sections, Plates and Commercial Steels for our Tables 11.3 and 14.4 from *Structural Sections to BS 4: Part 1 and BS 4848: Part 2*, available from British Steel, Steel House, Redcar, Cleveland TS10 5QW.

Extracts from British Standards reproduced with the permission of BSI under licence number 2002/SK0068. British Standards are available from BSI Customer Services, Telephone: 020 8996 9001.

The layout of the book and what it covers

From a 'static' point of view, structural mechanics is the science that describes and predicts the conditions under which structures remain or should remain at rest (static) under the action of various loads or forces. This means the structure must not collapse.

This textbook has been prepared to help you to understand structural mechanics. It is divided into 18 chapters, which are grouped into 2 parts.

Part 1 contains nine chapters and explains:

- The principles and philosophy of structural mechanics and structural design using the limit state design approach (Chapter 1).
- How applied loads or forces on an individual part of a structure:
  (a) can be calculated (Chapter 2), and
  (b) give rise to damaging effects and how those effects can be calculated and/or assessed (Chapters 3 to 6).
  From a 'static' point of view, the most efficient structure is one that allows the applied loads or forces to reach the supporting structure or the ground with no damaging effects. Chapters 3 to 8 give guidance on how to calculate the most obvious damaging effects of forces that an applied load or a combination of loads may exert on an individual part of a structure, or the structure as a whole. The methods for calculating the damaging effects occurring at any intermediate position of an individual part of the structure and the position where that part tends to disintegrate if it is loaded excessively are detailed. The damaging effects of an applied load on a given position of the structural element or the structure as a whole may depend on the magnitude of that load *only* or on both the magnitude of the load and the distance from the applied load. This is dealt with in Chapter 8.
- The behaviour of the structure or structural members in their efforts to resist the actions of applied loads. In other words, the possible structural consequences if a load or combination of loads is applied on a structural member is described in Chapters 7 and 8. This involves the study of the strength properties and behaviour of a range of construction materials, including steel, concrete and timber, to the latest British and European codes of practice.

All materials alter in shape when they are stressed. For a state of stress to exist there must be action and reaction. Engineers and architects use the term stress to indicate intensity of stress, which is another way of saying that each unit of cross-sectional area of a material transmits a certain load or that the load per unit area is a certain value. That value should be less than the ultimate strength of the material. Designers would be **unwise** to stress construction materials up to their yield point at which permanent deformation would take place. In real structures and buildings this could create all sort of problems such as cracking of plaster and brickwork, sagging of floors, etc. BS 8110: 1997 *Structural Use of Concrete* states that 'with an appropriate degree of safety, structures should sustain all the loads and deformations of normal construction and use and have adequate durability and resistance to the effects of misuse and fire'. Chapter 7 will help you to learn how to calculate the stress, strength and modulus of elasticity, and the criteria upon which designers decide the selection of specific materials for the design of structural members.

Part 2

In Part 2, the design of the structural member itself is covered, and the way in which the structural member 'fights back' to resist the damaging effects of the applied loads and reactions will be explained in detail (Chapters 10 to 18).

The load carrying capacity of a structural member depends on many factors. For example, the bending resistance of a beam depends on two factors: material design strength, which depends solely on the material you or the designer propose to use, and the section modulus, which is a measure of the efficiency of a section in terms of its shape (see Chapter 9). On the other hand, the load carrying capacity of a rectangular column is a function of the slenderness of the column (which is a measure of both the length of the column and its lateral dimension), the compressive strength of the material from which the column is made and the eccentricity of the applied loads.

The design principles, behaviour, and assessment of connections in steelwork; additions of direct and bending stress; portal frames and arches; and gravity retaining walls are covered in Chapters 15 to 18.

The book has been designed to be easily adapted to a modular teaching programme culture if this culture is required in the teaching strategy.

Part 1, which consists of nine chapters, covers the behaviour of a structure in its efforts to resist the action of applied loads or forces. Part 2, which also consists of nine chapters, covers an understanding of structural design and evaluation. Each chapter demands active participation from the student and demands correct responses to the exercises at the end of each chapter. The following flow diagram shows the inter-relationship between Part 1 and Part 2.

Flow diagram Showing inter-relationship between Part 1 and Part 2

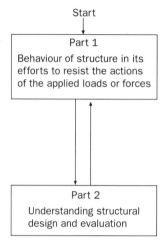

Start

Part 1

Behaviour of structure in its efforts to resist the actions of the applied loads or forces

Part 2

Understanding structural design and evaluation

1. Introduction to structural mechanics and structural design
2. Loads on buildings and structures
3. Concurrent coplanar forces
4. Non-concurrent coplanar forces
5. Moments of forces
6. Framed structures
7. Construction materials
8. Shear force and bending moment
9. Properties of sections

10. Introduction to structural stability, durability and environmental conditions
11. Simple beam design
12. Beams of two materials
13. Deflection of beams
14. Axially loaded columns
15. Connections
16. Addition of direct and bending stress
17. Portal frames and arches
18. Gravity retaining walls

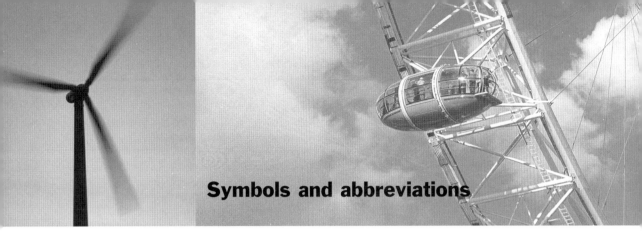

Symbols and abbreviations

γ_f	=	partial factor of safety applied to loads
γ_m	=	factor of safety applied to the strength of materials
π	=	22/7
k	=	stiffness of a spring
A	=	cross–sectional area
A_c	=	cross–sectional area of concrete excluding any finishing material and reinforcing steel bars
A_g	=	cross–sectional area of rolled steel beam
A_{gross}	=	gross area
A_{sc}	=	cross–sectional area of longitudinal steel bars
BMD	=	bending moment diagram
BS	=	British Standard
BV	=	bearing value of a rivet
C30	=	concrete grade 30
DSV	=	double shear value
E	=	modulus of elasticity of a material
e_x	=	eccentricity about the X–X axis
e_y	=	eccentricity about the Y–Y axis
f	=	stress
f_{cu}	=	characteristic strength of concrete
F_k	=	characteristic loads
f_m	=	mean strength of materials
F_v	=	shear force
f_y	=	characteristic strength of reinforcement steel bars
GL	=	ground level
G_k	=	dead loads
s	=	gravitational acceleration = 9.8 m/s^2
GS	=	general structural grade
HSFG	=	high strength friction grip
I	=	moment of inertia
I_{xx}	=	moment of inertia about the principal axis X–X
I_{yy}	=	moment of inertia about the principal axis Y–Y
kN	=	kilonewton = 1000 N = gravitational force, taken to be 10 m/s^2 for this purpose, due to a 100 kg weight = 100 kg × 10 m/s^2
kN m	=	kilonewton metre
L or l	=	original length
L_e or l_e	=	effective length
M	=	moment
m	=	metre
m	=	modular ratio of two materials

M_b	=	buckling moment resistance
M_c	=	moment capacity
MGS	=	machine general structural grade
M_{max}	=	maximum bending moment
M_r	=	moment of resistance
M_{rc}	=	moment of resistance in compression
M_{rt}	=	moment of resistance in tension
M_{uc}	=	ultimate moment resistance of concrete
M_{ut}	=	ultimate moment resistance of steel reinforcement
N	=	newton = the force that when applied to an object having a mass of one kilogram gives it an acceleration of one metre per second per second, i.e. N = kg $\times$ m/s^2
NA	=	neutral axis
ϕ	=	angle in radians
OPC	=	Ordinary Portland Cement
p_{bb}	=	bearing strength of a bolt obtained from Table 31 of BS 5950: Part1
p_{bs}	=	bearing strength of connected parts obtained from Table 32 of BS 5950: Part 1
p_c	=	compressive strength
p_{cc}	=	permissible stress for concrete in compression
p_{sc}	=	permissible compression stress for column steel bars
p_{SL}	=	slip resistance
P_v	=	section shear capacity
p_w	=	fillet weld design strength
p_y	=	design strength of steel
Q_k	=	imposed loads
Q_s	=	dynamic pressure
r or i	=	least radius of gyration
r_{yy}	=	radius of gyration about the Y–Y axis
s	=	second
S	=	standard deviation
SFD	=	shear force diagram
SS	=	special structural grade
SSV	=	single shear value
S_x	=	plastic modulus of a section at the X–X axis = first moment of area about the X–X axis
u	=	buckling parameter
UB	=	universal beam
V	=	vertical force
v	=	slenderness factor, see Table 19 of BS 5950: Part 1
V_e	=	effective wind speed
W/C	=	water/cement ratio
W_k	=	wind loads
χ	=	torsional index
Z	=	section modulus
Z_{xx}	=	section modulus about the principal axis X–X
β	=	M_{small}/M_{large}
δl	=	small change in length
λ	=	slenderness ratio
μ	=	coefficient of friction
ρ	=	air density

Part one Behaviour of structures

Chapter one Introduction

From a static point of view, structural mechanics is the science that describes and predicts the conditions under which structures remain or should remain at rest (static) under the action of various loads or forces. This means the structure must not collapse.

Structural design

Structural design is the process of choosing a suitable system or framing arrangement to support a shape or a form and prevent it from collapsing. The support system is called structure. A structure is made from one part only or from individual parts called structural elements. Structural elements are the parts of the frame that help to support the structure (see Fig. 1.1).

Structural classification

Structures are classified according their use and need, as in Table 1.1.

Table 1.1 Structural classification

Classification	Use/need
Domestic and residential	Dwelling houses, hotels, motels and guest houses
Offices and commercial	Banks, shopping centres and department stores
Institutional and exhibitions	Hospitals, colleges and universities, museums, art galleries
Industrial	Warehouses, factories, power stations
Other structures	Bridges, towers, water towers, electric towers, offshore structures, telescopes, etc.

Fig. 1.1 Structural elements are the parts of the frame that help to support the structure: (a) multi-storey steel building; (b) framed building; (c) details A; (d) details B; (e) details C

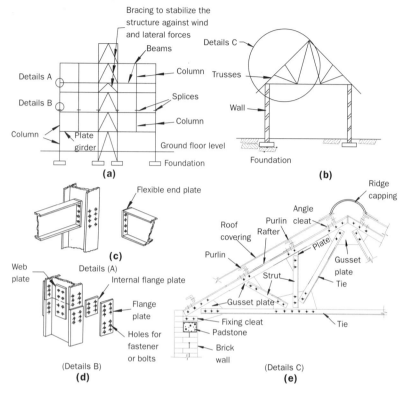

The purpose of a structure

The fundamental purpose of a structure is to transmit loads from the point of application to the point of support and, ultimately, through the foundations to the ground. That purpose must be fulfilled within the constraints imposed by the client's brief, which will inevitably insist on low initial costs and low maintenance costs and may vaguely stipulate the functional needs of the project. Consequently, the process of structural design begins with the appraisal of the client's requirements through collaboration with other members of the project design team.

Project design team

The project design team normally consists of the following:

- Architectural engineers are involved in producing structural plans, models and/or computer simulations to meet the client's requirements; controlling the project and employee consultant engineers until completion of the project.
- Consultants carry out the detailed design; prepare tender documents and construction drawings; make decisions about materials, structural types and forms, and design methods to be used; supervise, inspect and

approve materials, fabrication and construction activities; and inspect construction activities. They include structural, electrical, mechanical, heating and ventilation engineers, as well as quantity surveyors and building surveyors. In some cases, the builders or contractors may also be consulted at this stage.

- Builders or contractors carry out all the construction activities, such as earthworks, foundations, fabrication of structural elements and building frameworks, walls, bracing, finishes, installation of equipment and services (heating and ventilation systems, etc).

The design process

The structural design itself is a combination of art and science in that it is the creation of a structural form that will accommodate the often con flicting aspects of cost, function, services and aesthetics, and be capable of being quantified to produce dimensioned details for the purpose of erection. There are, therefore, two distinct stages in the design process as follows:

- Stage 1: the conceptual design and planning stage. Structural engineers draw on their experience, intuition and knowledge to make an imaginative choice of a preliminary scheme in terms of layout, materials and erection methods, based on the most economical structural form, construction and erection methods, and materials which could be used. They will examine the structural site and ensure that it is suitable for the project. The behaviour of the supporting ground under the load applied by the structure should be fully investigated. Knowledge of the interaction of soil and structure is required so that the most suitable structural form and type of foundations can be chosen.
- Stage 2: the detailed design stage. In this stage, the chosen scheme is subjected to detailed analysis based on the principles of structural mechanics. The resulting scheme must be consistent with the engineer's basic aim to provide a structure that satisfies the criteria of safety and serviceability at reasonable cost. This stage of design includes:

  (a) idealization of the structure and any complex parts of the structure by means of mathematical models for the analysis and design process;
  (b) estimation of all relevant loads, and any realistic combination of loads, that produce the most critical effects on the individual elements of the structure and on the structure as a whole;
  (c) design of all the structural elements, such as foundations, walls, structural frames and connections;
  (d) preparation of final detailed drawings, materials lists, specifications for both fabrication and construction activities, bills of quantities and tender documents.

The fabrication and construction of the structure are the processes of making the individual elements of precast concrete or steel frame buildings.

Structural design aims include technical, architectural and financial concepts. The primary technical aim is that the structural engineer must ensure

Fig. 1.2 A typical breakdown of the construction cost of a steel frame building

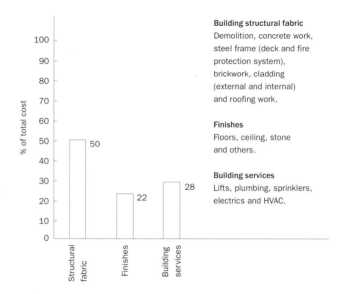

Building structural fabric
Demolition, concrete work, steel frame (deck and fire protection system), brickwork, cladding (external and internal) and roofing work.

Finishes
Floors, ceiling, stone and others.

Building services
Lifts, plumbing, sprinklers, electrics and HVAC.

that the structure as a whole and its individual elements are strong enough to withstand the most critical action of the applied loads or combination of loads to which the structure will be subjected throughout its design life. There should be no possibility of progressive collapse of the structure or its individual elements under normal and/or accidental loadings. Structural engineers need to ensure that they are familiar with issues such as the strength of material used and generate a greater consideration of serviceability, including dynamic structural response, as part of the development of a safe structural form. The design engineer also aims to increase the efficiency of the structure, and to ensure adequate resistance to fire, overturning, corrosion, etc. Financial aims look to the achievement of a minimum overall cost of the structure by striking a balance between the costs of materials, labour and the method of erection. Figure 1.2 shows a typical breakdown of construction costs for a steel frame building.

Design methods and design philosophies

Elastic or permissible stress method

A safe structure may not be easy to define but it must, at least, not collapse under an applied load. The standards of safety have changed considerably since the Code of Hammurabi (*c*. 2000 B.C.) declared the life of the builder forfeit should the built house collapse and kill the owner. It was not until the nineteenth century, however, that the concept of a factor of safety was first formulated in terms of the ratio of ultimate or, in certain cases, yield stress to the working stress. The method of design is called the elastic or permissible stress method. In this design method, the design stresses of the materials are calculated by dividing their ultimate strength by a factor of safety. The design or permissible stress is given by:

Fig. 1.3 Short-term design
stress–strain curves for steel

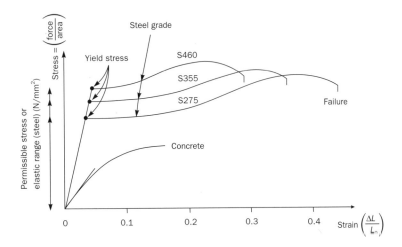

$$\text{permissible stress} = \frac{\text{ultimate strength}}{\text{factor of safety}}$$

The size of the structural elements is selected on the basis that the actual stress the material will experience, under the applied loads, during its design life, must be kept within the elastic region, i.e. the yield stress of the material must not be reached during the life span of the structural element (see Fig. 1.3). Such an idea formed the basis of BS 449.

Example 1.1

Calculate the cross-sectional area and the diameter of a circular mild steel bar that is required to safely support a dead load of 5 kN and an imposed load of 2 kN (see Fig. 1.4). The yield stress (ultimate strength) of mild steel is 250 N/mm². Use a factor of safety of 1.8 for the material strength.

Solution

$$\text{permissible stress} = \frac{\text{total load}}{\text{cross-sectional area}}$$

$$\text{cross-sectional area, } A = \frac{\text{total load}}{\text{permissible stress}}$$

$$\text{permissible stress} = \frac{250}{1.8} = 138.89 \text{ N/mm}^2$$

$$\text{cross-sectional area} = \frac{(5+2) \times 1000}{138.89} = 50.4 \text{ mm}^2$$

$$\text{The cross-sectional area of the circular bar} = \frac{\pi D^2}{4}$$

$$\text{therefore the bar diameter, } D = \sqrt{\left(\frac{50.4 \times 4}{\pi}\right)}$$

$$= 8 \text{ mm}$$

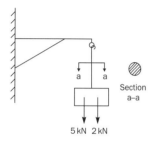

Fig. 1.4 Example 1.1

Because the elastic theory or the elastic design method is based on an elastic stress distribution, the method is not applicable to a semi-plastic material such as concrete, (see Fig. 1.3). It is also not suitable when the deformation of a structural element is not proportional to the load, as in the case of a slender column. Furthermore, its use is found to be unsafe when dealing with the stability of a steel structure subjected to overturning.

Load factor method or plastic theory

Some decades later an alternative approach was introduced in terms of the load factor based on the ratio of loads rather than stresses, i.e.

$$\text{load factor} = \frac{\text{ultimate load}}{\text{working load}}$$

Using factors of safety in the elastic method ensures a satisfactory performance under working loads, but only assumes a reasonable margin against failure, while load factors ensure a definite margin against failure and assume a satisfactory performance under working loads.

In the load factor method, the working loads are multiplied by a factor of safety, with the ultimate strength of the materials being used, and the design load is defined as follows:

$$\text{design load} = \text{working loads} \times \text{factor of safety}$$

The load factor method or plastic theory is based on finding the load that causes the structure to fail. Then the working load is the collapse load or the load that causes plastic deformation (see Fig. 1.5). Because this method does not apply a factor of safety to the material strength, it cannot directly take into account the variability in the strength of material used and its effects on the load carrying capacity of the structure. The method also cannot be used to calculate the deflection at working loads. The load factor method is in fact permitted in BS 449 but is rarely used by practical engineers.

Example 1.2

For the material used in Example 1.1, calculate the cross-sectional area and the diameter of the steel bar, using the load factor method and a factor of safety of 1.8 for the total load.

Fig. 1.5 Stress–strain curve for the plastic method

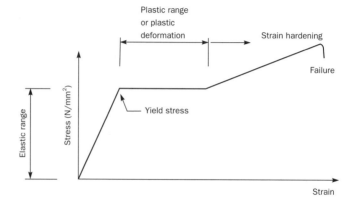

$$\text{cross-sectional area, } A = \frac{\text{total factored load}}{\text{yield stress}}$$

$$= \frac{(5 + 2) \times 1000 \times 1.8}{250} = 50.4 \text{ mm}^2$$

$$\text{and the bar diameter, } D = \sqrt{\frac{(50.4 \times 4)}{\pi}} = 8 \text{ mm}$$

Limit state design method (LSM)

In the two previous design approaches, both factors operate on the implicit assumption that the determined values, both for the loading which the structure is expected to carry, and for the strength of the materials of which the structure is to be made, remain constant, thus inferring a guarantee of absolute safety. Whilst this may be satisfactory in certain cases, it is generally recognized that a more realistic measure of the safety of a structure can be achieved through an assessment of the probability of its failure. In this method, called the limit state design, partial safety factors are used, separating the probability of failure due to overloading from that due to variability of strength of the materials.

Whereas safety deals with the structure's ability to carry its loads without unwarranted risks to human life and limb (i.e. within the ultimate limit states, ULS), the requirements of serviceability of the structure (i.e. the serviceability limit states, SLS) insist on its fitness for use without excessive deflections, disturbing vibrations, noticeable cracking or other local failures which may demand costly remedial work.

The limit state design method, recently introduced into the codes of practice, is briefly outlined in this chapter, and may be defined as that state of a structure at which it becomes unfit for the use for which it was designed.

To satisfy the object of this method of design, all relevant limit states should be considered in order to ensure an adequate degree of safety and serviceability. The codes of practice therefore distinguish between the ultimate limit states, which apply to the safety of the structure, and the serviceability limit states, which deal with factors such as deflection, vibration, local damage and cracking of concrete.

Whereas the permissible stress design and load factor methods relied on a single factor of safety, the limit state design codes introduce two partial safety factors, one applying to the strength of the materials, γ_m, and the other to loads, γ_f. This method, therefore, enables the designer to vary the degree of risk by choosing different partial safety factors. In this way the limit state design ensures an acceptable probability that the limit states will not be reached, and so provides a safe and serviceable structure economically.

As mentioned above, the current codes of practice, e.g. BS 5950, BS 8110, distinguish between the ultimate limit state, which concerns the safety of the structure, and the serviceability limit states, which deal with deflection, cracking of concrete, vibration, corrosion and durability. The former relates to the safety of the structure against collapse, i.e. the structure's stability and load carrying capacity, taking into account structural properties, such as strength (including general yielding, rupture, buckling and transformation into a mechanism), overturning, sway, and brittle fracture.

Part 1 of BS 5950: 2000 and BS 8110: 1997, *Structural Use of Steelwork in Building*, and *Structural Use of Concrete* respectively, give some examples of limit states and factors of safety for material strength, γ_m, loading, γ_f, and structural performance, γ_p.

Characteristic strength and design strength of materials

The limit state design method refers to the following design terms: the characteristic strength of materials, design strength of materials, characteristic dead loads, characteristic imposed loads and design loads. The first two are described in detail below whereas the characteristic dead loads, characteristic imposed loads, and the design loads are covered in full detail in Chapter 2.

Characteristic strength of materials

The characteristic strength of materials, f_k, is defined as the value of material strength below which results are unlikely to fall, or that value below which it is unlikely that more than x per cent of results will fall. When $x = 5$ per cent, statistically f_k is given by;

$$f_k = f_m - 1.64S \tag{1}$$

where

$$f_m \text{ (mean strength)} = \frac{\sum x}{n}$$

$$S \text{ (standard deviation)} = \sqrt{\frac{\sum (x - f_m)^2}{n}}$$

x = individual test results
n = number of tested specimens or samples

The value of f_k from equation (1) accounts for variations in the results of tested specimens, which are affected by the quality control during the making of the specimens and the quality of the materials used. Where the value of f_k is required for a particular material, this may be found by carrying out strength tests, such as tensile tests on steel or compression tests on concrete cubes. If a large number of tests are carried out on a particular material then a normal frequency distribution curve such as the one shown in Fig. 1.6 can be obtained and f_k can be found by applying equation (1).

The value of $1.64S$ is taken since, for a normal distribution, this means that only 5 per cent of the area under the frequency distribution curve lies beyond this range, so there is only a 5 per cent probability that the strength of a specimen would be below f_k.

Example 1.3

Calculation of f_k, the characteristic strength of a material
Table 1.2 shows the results of 14 compressive strength tests of steel bars made under the same conditions.

Fig. 1.6 Normal frequency distribution curves

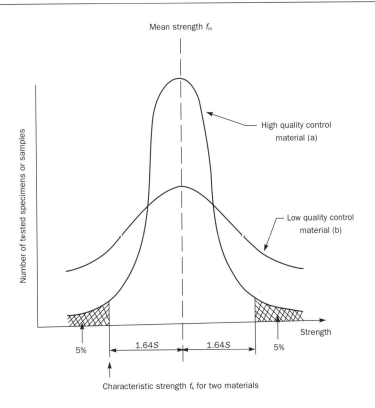

1 Determine the characteristic strength of the material tested.
2 Determine the compressive strength above which only 5 per cent of the results are likely to fall.
3 Draw the normal distribution curve for the results.

Table 1.2 Steel bars – compressive strength test results

Test values, x (N/mm²)						
278.0	280.0	274.0	274.5	282.0	285.0	290.0
286.0	285.5	278.5	277.0	282.0	276.0	281.5

Sum of test values $= 3930\ \text{N/mm}^2$

therefore

$$\text{mean value}(f_{\text{m}}) = \frac{\Sigma x}{n} = 280.714\ \text{N/mm}^2$$

and the values for $(x - f_{\text{m}})^2$ are as shown in Table 1.3.

Table 1.3 Values for $(x - f_{\text{m}})^2$ calculation

$(x - f_{\text{m}})^2$ for test values (N/mm²)						
7.365	0.509	45.077	38.613	1.653	18.369	86.229
27.941	22.905	4.901	13.793	1.653	22.221	0.617

Therefore

$$\text{sum of } (x - f_m)^2 = 291.846 \text{ N/mm}^2$$

$$S \text{ (standard deviation)} = \sqrt{\frac{\sum (x - f_m)^2}{n}}$$

$$= \sqrt{\frac{291.846}{14}}$$

$$= 4.565 \text{ N/mm}^2$$

Solution

1   $f_k = 280.714 - 1.64S = 273.227 \text{ N/mm}^2$
2   f_k (above which) 5 per cent of results are likely to fall $= 280.714 + 1.64S$
$$= 288.200 \text{ N/mm}^2$$

3   The curve is shown in Fig. 1.7.

Table 1.4 Compressive strengths

Range of compressive strength (N/mm²)	Number of tested samples
270–274	1
274–278	4
278–282	5
282–286	3
286–290	1

Fig. 1.7 Normal distribution curve for the tested samples. This has been drawn based on the range of compressive strength (N/mm²) and number of tested samples shown in Table 1.4

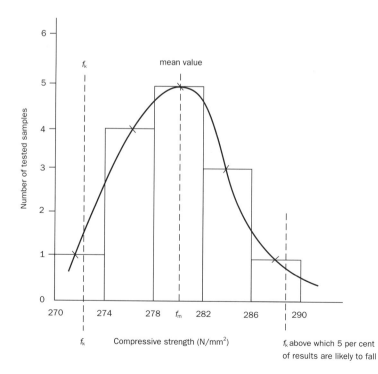

Design strength of materials

Design strength of a material $= f_k/\gamma_m$, where f_k = characteristic strength of the material, and γ_m = the partial factor of safety. The γ_m value takes into account possible variation such as construction tolerances and quality control in both manufacturing and construction. For example, quality control is required for steel both in the selection of its raw materials and/or contents and during the rolling and construction process.

Mathematical models

To ensure the safety of a structure it is necessary to provide it with sufficient strength and, to safeguard its serviceability, it must have adequate stability. In order to assess the sufficiency of its strength and the adequacy of its stability, the structure needs to be presented in quantitative terms before it can be analysed and, since it does not yet exist, its behaviour has to be simulated. A scaled-down version of the proposed structure may seem the obvious choice here, but now even fairly complex structures can be analysed by means of mathematical models.

Consider, for example, the behaviour of the steel spring shown in Fig. 1.8. As the load P is increased, the extension χ in the length of the spring will also increase. From several pairs of measurements of P and χ it will be found that χ is proportional to P. Therefore, the relationship between the two quantities can be formulated by means of the equation $P = k \times \chi$, where k is a constant representing, in this case, the stiffness of the spring. This equation serves as the mathematical model for the behaviour of the loaded spring since it allows the extension in the length of the spring to be determined for any given load.

The behaviour of a structure is clearly more complex than that of a steel spring. It calls, therefore, for more complex models. In fact, even a simple structure may contain too many unknown factors to be analysed completely. In such cases, it becomes necessary to replace the real structure by a simplified, idealized version of it.

An example of this simplification can be shown in the case of a steel roof truss. Figure 1.9(a) represents the real structure. Note that the roof loads are applied to the truss at the joints; the members of the truss are made up of standard structural steel sections and are welded or bolted together. Figure 1.9(b) shows an idealized truss. Here the roof loads are indicated by arrows directed vertically downwards, and the truss members are shown as simple bars with pinned connections.

Fig. 1.8 Loading a spring

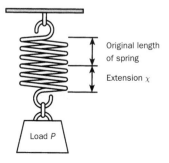

Original length of spring

Extension χ

Load P

Fig. 1.9 Steel roof truss:
(a) illustrative representation;
(b) idealized representation

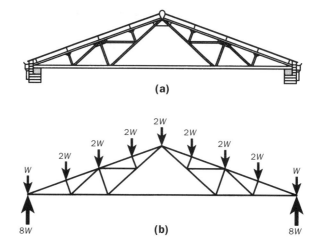

(a)

(b)

These two structures differ in several aspects, the most significant being the substitution of pinned connections for the welded ones. A welded connection gives some degree of rigidity, which complicates the behaviour of the truss members. A pinned connection, on the other hand, is assumed to be frictionless allowing the members free movement relative to each other, thus simplifying their analysis. This idealization is made on the assumption that the truss would function satisfactorily if the connections were in fact pinned, and that the difference in the behaviour of the members of the real and the idealized trusses is sufficiently small. This has been shown by experiment and experience to be the case for the ordinary range of trusses.

It is, of course, of vital importance that the structural engineer keeps these assumptions within strict limits in the knowledge that each assumption tends to reduce the accuracy of the subsequent analysis. The engineer must be aware, for example, that the application of the model for the behaviour of the steel spring, discussed earlier, is restricted by the fact that the stiffness of the spring remains constant only within a certain range of values for P. The validity of the model, after all, depends on how closely it represents the real behaviour. The engineer is required, therefore, to make grave and far-reaching choices which must be founded on sound experience, supported by an intuitive understanding and fostered by a broad knowledge of structural behaviour. In the search for the appropriate solution, the engineer may find assistance in the codes of practice, which are based on the collective opinion and corporate judgement of experienced designers. They are intended to provide a set of guidelines within which design decisions can be formulated and the analyses carried out in accordance with the principles of structural mechanics.

Forces

The term mechanics, according to a dictionary definition, refers to 'that branch of applied science which deals with the action of forces in producing motion or equilibrium'. In the context of structural mechanics, forces represent the loads that the structure is expected to carry. These are generally classified into three groups: dead loads, imposed (live) loads, and wind loads.

Dead loads comprise the permanent loads due to the static weight of the structure itself, the cladding, floor finishes, and any other fixtures which form the fabric of the building.

Imposed (live) loads are those produced by the intended occupancy of the building, i.e. loads due to the weight of plant and equipment, furniture, the people who use the building, etc. They also include snow loads, impact and dynamic loads arising from machinery, cranes and other plant, and such irregular loads as those caused by earthquakes, explosions and changes in temperature.

Wind loads are classified separately owing to their transitory nature and the complexity of their effects. Loads and loads combinations are explained in detail in Chapter 2.

The structure may be subjected to the action of almost any combination of these loads and, in order to produce equilibrium, it must provide an equal and opposite **reaction**. The necessary reaction is generated by the **stress** caused by the action of the loads within the material, and by the ensuing **strain** in the elements of the structure.

The two concepts (stress and strain) are discussed in detail in Chapter 7. They are the direct outcome of the action of **forces** and the **deformations** they produce.

The definition of a force is generally derived from Newton's first law of motion as that influence which causes change in an object's uniform motion in a straight line or, as is more appropriate to structural mechanics, its state of rest.

Most of the loads supported by a structure exert forces on it by virtue of their mass, which is subjected to the gravitational pull of the earth. This pull, as was demonstrated in the case of the steel spring in Fig. 1.8, acts downwards, thus indicating that the force has a direction as well as a magnitude and is, therefore, a vector quantity.

As such, a force can be represented by a straight line of a given length indicating the magnitude of the force. The direction of this straight line is drawn parallel to the line of action of the force and denoted by an arrow. Such a line is called a **vector**.

The SI unit of force is the newton, denoted by the capital letter N, and is that force which, when applied to an object having a mass of one kilogram, gives it an acceleration of one metre per second per second, i.e. $N = kg \times m/s^2$.

Attention must be drawn here to the ambiguity that exists in the use of the term *weight*. In this textbook, weight is considered as the force exerted on a body by the gravitational pull of the earth and is, therefore, stated in force units (newtons). However, in accordance with the Weights and Measures Act 1963, the weights of structural materials are specified in mass units (kilograms). Hence, in order to calculate the forces due to dead loads, the kilograms have to be converted to newtons by multiplying them by the gravitational acceleration. The value of this acceleration is 9.81 m/s^2, but it is generally accepted that, to simplify the conversion for the purpose of structural calculations, the figure may be taken to be 10 m/s^2. For example the gravitational force due to a 100 kg weight is

$$100 \text{ kg} \times 10 \text{ m/s}^2 = 1000 \text{ N or } 1 \text{ kN}$$

To counteract the action of the gravitational (and other) forces, the structure must provide an equal and opposite reaction in order to remain in a state of rest, i.e. in equilibrium. The action of the forces may manifest itself in any of the following ways:

Fig. 1.10 Action of forces
(a) tension; (b) compression;
(c) shear; (d) bending; (e) torsion

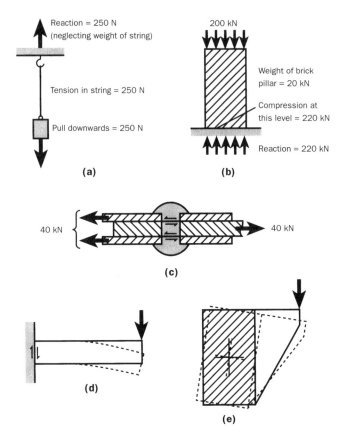

- *Tension*   The forces acting on the object act away from each other; the object (string in Fig. 1.10(a)) is being stretched.
- *Compression*   The forces acting on the object act towards each other; the object (brick pillar in Fig. 1.10(b)) is being squeezed.
- *Shear*   The forces acting on the object cause parts of the object to slide in relation to each other (Fig. 1.10(c)).
- *Bending*   The forces acting on the object produce a bending effect as shown in Fig. 1.10(d). Note that this action also causes sliding (shear).
- *Torsion*   The forces acting on the object cause it to be twisted (Fig. 1.10(e)).

Force, however, cannot be observed directly. It is recognized and, therefore, measured by the deformations it causes. The deformation which is of particular interest to the structural engineer is that resulting from bending, i.e. deflection.

The control of deflection is of great importance in safeguarding the stability of the structure. The deflection must, therefore, be quantified. To this end the study of the deflected forms of structural elements or frameworks provides a useful basis for the appreciation of structural behaviour and it is advisable to acquire this helpful ability.

A number of simple examples are illustrated in Fig. 1.11(a)–(g).

Fig. 1.11 Deflection:
(a) straight cantilever;
(b) propped cantilever;
(c) encastré beam;
(d) simply supported beam;
(e) continuous beam;
(f) three-hinged arch;
(g) portal frame

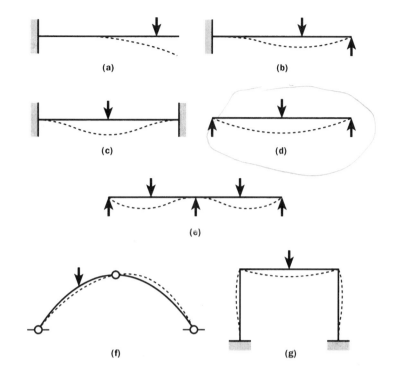

Exercises

1 Explain what is meant by the following:

- permissible stress, load factor, and limit state design methods
- characteristic strength
- characteristic loads

2 Explain why the partial factors of safety for materials are incorporated when the limit state design method is used.

3 Determine the cross-sectional area of the mild steel bar in Example 1.1 if the bar is supporting a dead load of 10 kN and an imposed load of 6 kN. Use all the three design methods and critically comment on your results.

4 The results shown in Table 1.Q4 are the compressive strengths of 20 steel bars made under the same conditions.

1. Determine the characteristic strength of the materials.
2. Determine the compressive strength above which only 5 per cent of the tested results are likely to fall.
3. Draw the normal frequency distribution curve of compressive strength, mark f_k and f_m on the distribution curve.

5 Write a computer program to solve question 4 above.

Table 1.Q4 Compressive strength of steel samples (N/mm²)

273.0	275.1	280.3	273.1	276.0	270.9	271.8	277.1	281.8	281.9
271.7	271.8	277.1	281.8	290.6	289.1	296.3	284.1	288.1	272.1

Chapter two Loads on buildings and structures

Understanding structural mechanics and structural design requires knowledge of many inter-linked factors. These include the loads and load actions on the structure, the strength and properties of the materials from which structural elements are made, the ways by which the loads and load actions are transferred via the structure to the foundations, the interaction between the foundations and the supporting ground, structural stability, durability and environmental conditions.

It is therefore important to estimate the loads that a structure has to withstand during its intended useful life accurately, in order to achieve safety and economy in design.

The behaviour of structures under loads depends on the strength properties of the materials of construction and the interaction between the components and parts of the structural frame and between the structural frame, its foundations and the supporting ground. Designers in their structural analyses try to predict this behaviour of the structure and identify the model to be used in the structural analyses. If they succeed then designs will usually be safe and economic.

At present, existing knowledge of the loads on structures, properties of the materials of construction and analysis of structural frames is well advanced so that structural design can usually be considered to be economic with regard to these aspects. However, future research on understanding the actions of loads on structures will help to reduce a number of the existing uncertainties and hence result in safer and more economic designs.

Load types

In design, the loads on buildings and structures are classified into different types based on their frequency of occurrence and method of assessment. These are:

1 dead loads
2 imposed loads
3 wind loads
4 earth and liquid pressures
5 other load effects such as thermal effects; ground movement; shrinkage and creep in concrete; and vibration.

For each type of load, there will be a characteristic value and a design value. These will be explained later in this chapter. The design of any particular element of the frame of the structure or of the structure as a whole has to be based on the design load or design load combination that is likely to produce the most adverse effect on that element or the structure as a whole in terms of compression, tension, bending, moment, shear, deflection, torsion and overturning.

Dead loads

BS 6399–1: 1996 *Loading for buildings, Part 1: Code of practice for dead and imposed loads.*

Dead load is the weight of structural components, such as floors, walls and finishes, and includes all other permanent attachments to structures like pipes, electrical conduits, air conditioning, heating ducts and all items intended to remain in place throughout the life of the structure. It is calculated from the unit weights given in BS 648: 1964 *Schedule of weights of building materials* or from the actual known weights of the materials used.

In the analysis process, although the dead load of structural parts or members can be calculated accurately, it is usual practice to simplify complicated load distributions to reduce the analysis and design time, for example, in the design of beams an approximate uniformly distributed load is usually used instead of the actual stepped-type loading.

In the design process, the assessment of the dead load of most load bearing structural parts has to be done in practice by a method of trial and error to determine the approximate dimensions required for such parts. However, for most of the common types of structural elements, for example, slabs, beams and columns, there are some simple rules for assessing the approximate dimensions required. These rules are explained in the relevant code of practice, for example, for reinforced concrete and steel structures see BS 8110: Part 1: 1997 and BS 5950: 2000 respectively.

Imposed loads

BS 6399–1: 1996 *Loading for buildings, Part 1: Code of practice for dead and imposed loads.*

Imposed loads are sometimes called live loads or superimposed loads. They are gravity loads varying in magnitude and location. They are assumed to be produced by the intended occupancy or use of the structure. They include distributed, concentrated, impact and snow loads but exclude wind loads. Such loads are usually caused by human occupancy, furniture and storage of materials or their combinations. Because of the unknown nature of the magnitude, location and distribution of imposed load items, realistic values are difficult to determine. These values are prescribed by both government and local building codes.

BS 6399–1: 1996 *Loading for buildings, Part 1: Code of practice for dead and imposed loads* gives imposed loads for various occupancy and functional requirements of buildings, such as

- domestic and residential (dwelling houses, flats, hotels, guest houses)
- institutional and exhibitions (schools, colleges and universities)

- industrial (warehouses, factories, power stations)
- bridges (pedestrian, highway and railway)
- shopping areas
- warehousing and storage areas.

Even with this classification there is still broad variation in the imposed loads, for example, within the high school building some space is used in classrooms and laboratories. The imposed loads for these various buildings are different and hence different values should be specified for design.

In structures such as highway bridges, it is necessary to consider traffic loads in terms of both a concentrated load and a varying uniformly distributed load. In addition, the effect of impact forces due to traffic loading must be accounted for.

Reduction in total imposed floor loads

The code of practice allows for the reduction of imposed loads in the design of certain structural components and should be consulted for full details. Briefly the main reductions are as follows:

Beams and girders. Where a single span of a beam or girder supports not less than 46 m² of floor at one general level, the imposed load may in the design of the beam or girder be reduced by 5 per cent for each 46 m² supported subject to a maximum reduction of 25 per cent. No reduction, however, shall be made for any plant or machinery for which specific provision has been made nor for buildings for storage purposes, warehouses, garages and those office areas that are used for storage and filing purposes.

Columns, piers, walls, their supports and foundations. The imposed floor loads contributing to the total loads for the design of such structural elements may be reduced in accordance with Table 2.1.

This reduction is allowed because of the reduced probability that the full imposed loads will occur at all the floors simultaneously.

Table 2.1 Reduction in total distributed imposed floor loads

Number of floors, including the roof, carried by member under consideration	Reduction in total distributed imposed load on all floors carried by the member under construction (%)
1	0
2	10
3	20
4	30
5 to 10	40
Over 10	50

Dynamic loads

Dynamic loads are those that produce dynamic effects from machinery, run-ways, cranes and other plant supported by or connected to the structure. Allowance is made for these dynamic effects, including impact, in the design of the relevant structural parts.

To allow for such effects in practical design, it is common practice in most cases to increase the dead-weight value of machinery or plant by an adequate amount to cater for the additional dynamic effect, and a static analysis is then carried out for these increased loads and the computed load effects used in the design. The appropriate dynamic increase for all affected members is ascertained as accurately as possible and must comply with the relevant code of practice.

Load from partitions

Clause 5.1.4 of BS 6399–1: 1996.

Dead loads from permanent partitions. Where permanent partitions are shown in the construction plans their actual weights shall be included in the dead load. For floors of offices, this additional uniformly distributed partition load should be not less than 1.0 kN/m^2.

Imposed loads from demountable partitions. To provide for demount-able partitions it is normal practice to consider an equivalent uniformly dis-tributed load of not less than one third of the per metre run of the finished partitions and treat it as an imposed load in design.

Wind loads on structures

BS 6399–2: 1997 *Loading for buildings. Part 2: Code of practice for wind loads.*

Wind loads depend on the wind environment and on the aerodynamic and aeroelastic behaviour of the building. Wind loads on structures are dynamic loads due to changes in wind speed. When the wind flow meets an obstruction, such as a building or a structure, it has to change speed and direction to keep flowing around the building and over it. In this process of change in direction it exerts pressures of varying magnitudes on the face, sides and roof of the building. In structural analysis and design it is neces-sary to consider the design wind loads due to these pressures in combina-tion with other applied imposed and dead loads. For convenience in design it is usual practice to consider the wind loads as static loads. However, for some light tall structures, such as metal chimneys, the dynamic effects of the wind, such as induced oscillations, have to be considered in design.

Owing to the change in direction when wind flow encounters stable struc-tures, the induced wind pressure can vary in direction such that the resultant wind loads are horizontal and vertical. Furthermore, since the wind direction varies with time the wind loads on structures have to be considered as of possible application from all directions.

In view of the complexity of the assessment of wind loads on structures it is not possible to give the subject full treatment here and the reader is advised to consult one of the references at the end of the book.

Fig. 2.1 Wind speed versus time

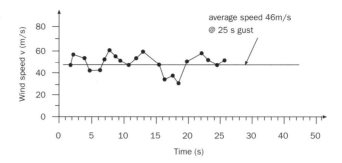

The effective wind loads on structures are dependent on the wind speed, geographical location of structure or building, size, shape and height.

The wind normally blows in gusts of varying speed, and its direction depends on the wind environment. Figure 2.1 shows a typical graph of speed versus time during a gale.

The wind pressure, which is caused by changes of wind speed from V_e in m/s (metre/second) to zero, as occurs when the wind meets a building and has to change direction, is given by q_s:

$$\text{dynamic pressure } q_s = \tfrac{1}{2}\rho V_e^2 \text{ (in pascals, Pa (N/m}^2\text{))}$$

the air density $\rho = 1.226 \text{ kg/m}^3$

V_e = effective wind speed from section 2.2.3 of BS 6399: 1997 *Loading for buildings – Part 2: Code of practice for wind loads.*

Therefore

$$q_s = 0.613 \ V_e^2 \tag{1}$$

The wind speed to be used in equation (1) is not the maximum recorded value. It should be calculated from the relevant section of the code of practice. For example from section 2.2.3 of BS 6399: 1997 *Loading for buildings – Part 2: Code of practice for wind loads.*

If the shape of the structure is streamlined, then the change in wind speed is reduced and hence the dynamic wind pressure will also be reduced (see the relevant code of practice).

Loads on structures – summary

• **Dead loads**
They are the self-weight of structures or buildings, and are caused by the effect of gravity, and so act downward. Dead loads are calculated from the actual known weights of the materials used (see Table 2.2). Where there is doubt as to the permanency of dead loads, such loads should be considered as imposed loads. Dead loads are the unit weight multiplied by the volume. For more information, see the relevant code of practice or, in the UK, see BS 6399–1: 1996 and BS 648: 1964.

Table 2.2 Weights of building materials. (See also BS 648: 1964 *Schedule of weight of building materials*.) From PP 7312: 1998 *Extracts from British Standards for Students of Structural Design*, British Standards Institute

Material	Weight	Material	Weight
Ashphalt		**Plaster**	
Roofing 2 layers, 19 mm thick	42 kg/m^2	Two coats gypsum, 13 mm thick	22 kg/m^2
Damp-proofing, 19 mm thick	41 kg/m^2	**Plastic sheeting (corrugated)**	4.5 kg/m^2
Road and footpaths, 19 mm thick	44 kg/m^2	**Plywood**	
Bitumen roofing felts		per mm thick	0.7 kg/m^2
Mineral surfaced bitumen	3.5 kg/m^2	**Reinforced concrete**	2400 kg/m^3
Blockwork		**Rendering**	
Solid per 25 mm thick, stone aggregate	55 kg/m^2	Cement : sand (1 : 3), 13 mm thick	30 kg/m^2
Aerated per 25 mm thick	15 kg/m^2	**Screeding**	
Board		Cement : sand (1 : 3), 13 mm thick	30 kg/m^2
Blackboard per 25 mm thick	12.5 kg/m^2	**Slate tiles**	
Brickwork		(depending upon thickness and source)	24–78 kg/m^2
Clay, solid per 25 mm thick medium density	55 kg/m^2	**Steel**	
		Solid (mild)	7850 kg/m^3
Concrete, solid per 25 mm thick	59 kg/m^2	Corrugated roofing sheets, per mm thick	10 kg/m^2
Cast stone	2250 kg/m^3	**Tarmacadam**	
Concrete		25 mm thick	60 kg/m^2
Natural aggregates	2400 kg/m^3	**Terrazzo**	
Lightweight aggregates (structural)	$1760 \begin{smallmatrix} + 240 \\ - 160 \end{smallmatrix}$ kg/m^3	25 mm thick	54 kg/m^2
Flagstones		**Tiling, roof**	
Concrete, 50 mm thick	120 kg/m^2	Clay	70 kg/m^2
Glass fibre		**Timber**	
Slab, per 25 mm thick	2.0–5.0 kg/m^2	Softwood	590 kg/m^2
Gypsum panels and partitions		Hardwood	1250 kg/m^3
Building panels 75 mm thick	44 kg/m^2	**Water**	1000 kg/m^2
Lead		**Woodwool**	
Sheet, 2.5 mm thick	30 kg/m^2	Slabs, 25 mm thick	15 kg/m^2
Linoleum			
3 mm thick	6 kg/m^2		

- **Imposed loads**

They are gravity loads which vary in magnitude and location and are appropriate to the types of activity or occupancy for which a floor area will be used in service; see the appropriate code of practice or Table 1 of BS 6399–1: 1996.

Moveable imposed loads. Such as furniture, stored material, people, etc. Caused by gravity, act downward. Considered in structural design and analysis as static loads. Also called superimposed loads or live loads.

Moving imposed loads. Such as vehicles, cranes, trains, etc. Their dynamic effects should be considered in addition to their static effects.

- **Wind loads**

Due to dynamic wind movements, these depend on the wind environment and on the aerodynamic and aeroelastic behaviour of the structure or building. Variable in intensity and direction. Depend on:

1 shape of structure/building
2 height of structure/building above its base
3 location of structure/building, directional and topographic effects.

See the relevant national code of practice or BS 6399: 1997 – *Part 2: Code of practice for wind loads.*

• **Others**

Soil pressure, hydraulic pressure, thermal effects, ground movement, shrinkage and creep in concrete, and vibration are determined by special methods found in specialist literature.

Characteristic load

Characteristic load, F_k, is a statistically determined load value above which not more than x per cent of the measured values fall. Using the principles of probability and standard deviation, and when $x = 5$ per cent, characteristic loads can be defined as:

$$\text{charateristic load} = \text{mean load} \pm 1.64S$$
$$S = \text{standard deviation for load} \qquad (2)$$

The plus sign is 'commonly' used, since in most cases the characteristic load is the maximum load on a critical structural member.

However, for stability or the behaviour of continuous members, readers are referred to the relevant code of practice.

At the present state of knowledge, the characteristic load is that obtained from the relevant national codes of practice, such as, in the UK, BS 6399: Parts 1–3: 1996 and 1997 for dead, imposed and wind loads and BS 2573 for crane loads.

Design loads and partial factors of safety

The design load is calculated by multiplying the characteristic load F_k by the appropriate partial safety, i.e.

$$\text{design load} = F_k \times \gamma_f$$

where γ_f = the partial factor of safety for loads, which is introduced to take into account the effects of errors in design assumptions, minor inaccuracies in calculation, unusual increases in loads and construction inaccuracies. The partial factor of safety also takes into account the importance of the sense of the limit state under consideration, and the probability of particular load combinations occurring. BS 5950: 2000 and BS 8110: 1997 give recommendations for practical partial factors of safety for loads.

Load combinations

A structure is usually exposed to the action of several types of loads, such as dead loads, imposed loads and wind loads. They should be considered separately and in such realistic combinations as to take account of the most critical effects on the structural elements and on the structure as a whole. For the ultimate limit state, the loads should be multiplied by the appropriate factor of safety given in the relevant table of the code of practice. The factored loads should be applied in the most unfavourable realistic combination to the part of the structure or the effect under consideration. Different load combinations are recommended by the codes of practice. For example, see BS 5950: Part 1: 2000, Table 2, *Partial factors for loads* γ_f. Some examples on load combinations are as follows:

1 **Dead and imposed load**

   (a) design dead load $= 1.4G_k$ or $1.0G_k$
   (b) design imposed load $= 1.6Q_k$
   (c) design earth and water load $= 1.4E_n$

where $Q_k =$ imposed load, $G_k =$ dead load and $E_n =$ design earth and water load.

For example, in the design of a simply supported beam the following load combination is commonly used:

$$\text{design load} = 1.4G_k + 1.6Q_k \text{ (vertical load)}$$

2 **Dead and wind loads**

   (a) design dead load $= 1.4G_k$ or $1.0G_k$
   (b) design wind load $= 1.4W_k$

where $G_k =$ dead load (vertical load), and $W_k =$ wind load.

3 **Dead, imposed and wind loads**

   design loads $= 1.2G_k$
   $\quad\quad\quad\quad\quad + 1.2Q_k + 1.2W_k$

where $1.2G_k + 1.2Q_k =$ vertical load, and $1.2W_k =$ wind load.

Design comments

1 The criterion for any load combination is that it is likely to produce the worst effect on a structure or structural element for design and/or analysis purposes. Obviously only possible design load combinations should be considered.
2 In the design of a continuous beam, the worst load combination should be associated with the design dead load of $1.0G_k$ or $0.9G_k$ acting on some parts of the structure to give the most severe condition; see Fig. 2.2 (case 3, more load combinations are possible in this case).

Fig. 2.2 Load combinations

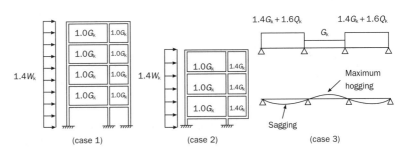

3 In Fig. 2.2, for case 1, γ_f (dead loads) = 1.0 and for case 2, γ_f (dead loads) = 1.0 and 1.4 for load resisting uplift or overturning.
4 Other realistic combinations that give the most critical effects on the individual structural elements or the structure as a whole are shown in the relevant code of practice, for example, see Table 2 of BS 5950 – Part 1: 2000.

It is important that the design loads are assessed accurately. If the design loads are wrongly assessed at the beginning then all the subsequent structural design and/or analysis calculations will also be wrong.

Example 2.1

Figure 2.3 shows a 3 m long reinforced concrete beam and a 914 mm deep × 419 mm wide universal steel beam that is 6 m long.
  Calculate the following:

(a) the weight of each beam per unit length (the uniformly distributed loads per unit length)
(b) the total weight of each beam
(c) the design dead load for each beam.

Fig 2.3 Example 2.1 beams; (a) reinforced concrete; (b) steel

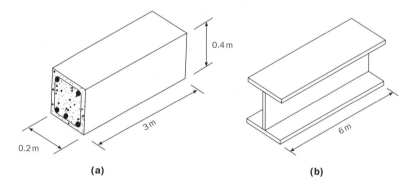

(a)   (b)

Solution

1 Reinforced concrete beam (see Fig. 2.4)

Cross-sectional area = 0.2 × 0.4 = 0.08 m²

From Table 2.2, unit weight of concrete = 24 kN/m³
Therefore the unit weight per unit length = 0.08 m² × 24 kN/m³
                                          = 1.92 kN/m

Total weight of beam = 1.92 kN/m × 3 m = 5.76 kN

Design dead load of the beam = $1.4G_k$ = 1.4 × 5.76 = 8.064 kN

Fig. 2.4 Example 2.1 loads on reinforced concrete beam

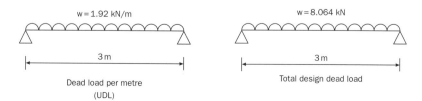

2 Steel beam (see Fig. 2.5)

Cross-sectional area $= 49\,400$ mm^2 (from Table 11.3, p. 224)

From Table 2.2, unit weight of steel (mild steel) $= 78.5$ kN/m^3

The weight per unit length $= (49\,400/10^6)$ m$^2 \times 78.5$ kN/m$^3 = 3.88$ kN/m (i.e. mass per metre of the beam $= 388$ kg/m, since 1 kN is equivalent to a mass of 100 kg)

Total weight of beam $= 3.88$ kN/m $\times 6$ m $= 23.287$ kN

Design dead load of the beam $= 1.4G_k = 1.4 \times 23.287 = 32.602$ kN

Fig. 2.5 Example 2.1 loads on steel beam

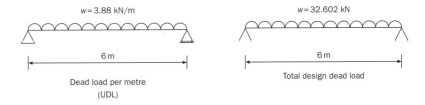

Dead load per metre
(UDL)

Total design dead load

Example 2.2

Figure 2.6 shows plan and roof details of a flat roof single-storey extension to an existing house. Calculate the design loads on the reinforced concrete beam A (including self-weight), which is 300 mm wide and 600 mm deep. Access is to be provided to the roof, therefore use an imposed load of 1.5 kN/m^2. Unit weight of concrete $= 24$ kN/m^3.

Roof construction:
asphalt (two layers) 19 mm thick	42 kg/m^2
25 mm timber boards, softwood	590 kg/m^3
50 $\times$ 175 mm timber joists spaced at	
400 mm centre to centre	590 kg/m^3
plaster board and skim (plaster finish)	15 kg/m^2

Fig. 2.6 Example 2.2: plan and roof details of single-storey extension to an existing house

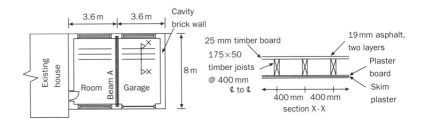

Solution

(See Fig. 2.7)

Design loads $= 1.4G_k + 1.6Q_k$

Area carried by the beam $= 3.6 \times 8 = 28.8$ m^2

Dead load:
asphalt $= 42 \times 10 = 420$ N/m^2
             $= 0.42$ kN/m^2

timber boards	$= 590 \times 10 \times 0.025$
	$= 147.5 \text{ N/m}^2$
	$= 0.147 \text{ kN/m}^2$
timber joists	$= 590 \times 10 \times 0.050 \times 0.175 \times 1/0.4$
	(number of joists in 1 m width)
	$= 129.06 \text{ N/m}^2$
	$= 0.129 \text{ kN/m}^2$
plaster board and skim	$= 15 \times 10 = 150 \text{ N/m}^2$
	$= 0.15 \text{ kN/m}^2$

Total dead loads (excluding self-weight of the beam) $= 0.847 \text{ kN/m}^2$

Beam self-weight $\qquad = 24 \times 0.300 \times 0.600 \times 8 = 34.56 \text{ kN}$

Design loads $\qquad = (1.4 \times (0.847 \times 28.8))$
$\qquad\qquad + (1.4 \times 34.56) + (1.6 \times 1.5 \times 28.8)$
$\qquad\qquad = 34.151 + 48.384 + 69.12 = 151.655 \text{ kN}$

Fig. 2.7 Example 2.2 solution

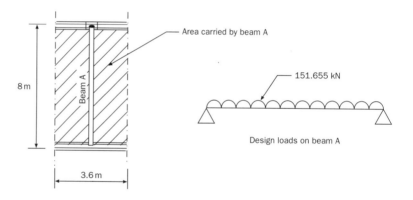

Example 2.3

Construction of a roof beam using a rolled steel joist
Figure 2.8 shows part of a roof plan for a small steel building. The flat roof consists of felt, steel decking, insulation boards and a suspended ceiling below the rolled steel joists. Calculate the design loads acting on one steel beam (joist).

Fig. 2.8 Part of a roof in a steel building

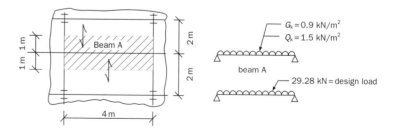

Solution

Dead load $= 0.9 \times 4 \times 2 = 7.2 \text{ kN}$

Imposed load $= 1.5 \times 4 \times 2 = 12 \text{ kN}$

Design loads $= (1.4 \times 7.2) + (1.6 \times 12) = 29.28 \text{ kN}$

General notes on the calculation of loading from roofs or slabs onto supporting beams

A roof or slab can be designed and detailed to span one way so that load is physically transmitted only to supporting beams and not directly to the beams running at right angles to them. If this is not the case, two-way action of the slab or roof must be taken into account.

One-way span

1 Where the ratio $L_1/L_2 \geqslant 2$ (see Fig. 2.9(a)), *and normally when the slab is made from concrete which is cast in situ*. The load on the roof ABCD can be assumed to be carried equally between beams AB and DC.

2 When the roof or the slab is constructed of pre-cast concrete units with the ratio $L_1/L_2 \geqslant 2$ (see Fig. 2.9(b)). The loads on the roof ABCD may be assumed to be carried equally between beams AB and DC.

Two-way span

When the above rules are not satisfied then the loads on the roof or the floor can be distributed as shown in Fig. 2.9(c) to take account of two-way action.

Fig. 2.9 Example 2.3: (a) and (b) one-way action of the slab or floor; (c) two-way action of the slab

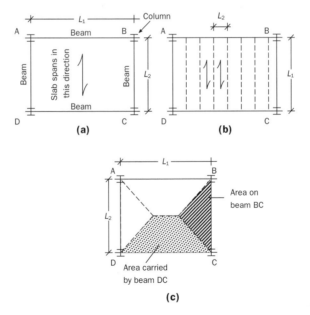

Figure 2.10 shows the relevant details at side walls and columns for a fully braced (in both directions) four-storey steel office building comprising a steel frame with reinforced concrete slabs on profiled metal decking. External cladding is brick/breeze block and double-glazing. Calculate the design loads acting on roof beam C2–D2, floor beam C2–D2, and inner column, lower length 2–5–8 (for column reference numbers, see Fig. 2.14).

Fig. 2.10 Fully braced steel building: (a) plan; (b) side elevation; (c) end elevation; (d) building details at parapet, side walls, side column and internal column

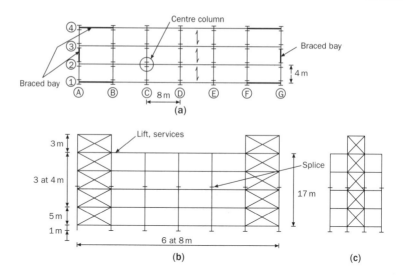

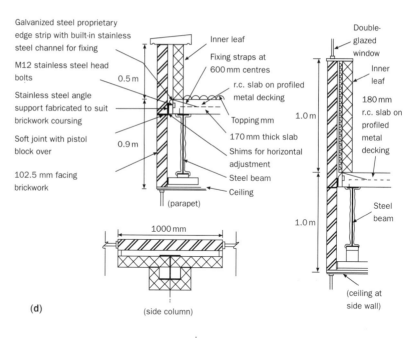

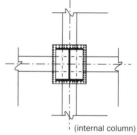

Loading details

Imposed loads (BS 6339 Part 1)
On roof $= 1.5$ kN/m^2
On floors $= 3.5$ kN/m^2
Reduce the imposed loads in accordance with number of stories as recommended in the practical code of practice (see Table 2.1, p. 20).

Dead loads on the flat roof (kN/m^2)		Dead loads on the floor (kN/m^2)	
topping materials (asphalt and screed)	1.0	tile screed	0.6
concrete slab, 170 mm thick	4.1	concrete slab, 180 mm thick	4.32
steel	0.2	steel	0.30
ceiling	0.5	partitions and ceiling (0.58)	1.58
services	0.2	services	0.2
Total dead load	**6.0 kN/m^2**		**7.0 kN/m^2**
Parapet wall (cavity wall)	**4.8 kN/m^2**	**external wall** 1 m high	**5.0 kN/m^2**

Column and casing
internal column, dead loads $= 1.5$ kN/m
external column, dead load $= 6.3$ kN/m

External steel beam at roof level
cavity wall dead load $= 4.8$ kN/m^2
concrete slab $= 4.1$ kN/m^2
steel $= 0.5$ kN/m
ceiling $= 0.5$ kN/m^2

External side wall at floor level
cavity wall 5.0 kN/m^2
concrete slab 4.32 kN/m^2
steel 0.68 kN/m
glazing (double) 0.6 kN/m
ceiling 0.5 kN/m^2

Solution

Roof Beam C2–D2 (see Fig. 2.11)

Dead loads	$= 6$ kN/m^2
Imposed loads	$= 1.5$ kN/m^2
Weight per unit length	$= (6 \times (4 \times 1 \text{ m (long)}))$ $+ (1.5 \times (4 \times 1 \text{ m (long)}))$
	$= 24 + 6 = 30$ kN/m

Design loads per unit length

$$= (1.4 \times 6 \times (4 \times 1 \text{ m}))$$
$$+ (1.6 \times 1.5 \times (4 \times 1 \text{ m}))$$
$$= 33.6 + 9.6 = 43.2 \text{ kN/m}$$

Total design loads

$$= (1.4 \times 6 \times (4 \times 8))$$
$$+ (1.6 \times 1.5 \times (4 \times 8))$$
$$= 268.8 + 76.8 = 345.6 \text{ kN}$$

Fig. 2.11 Design loads on beam C2–D2 at roof level

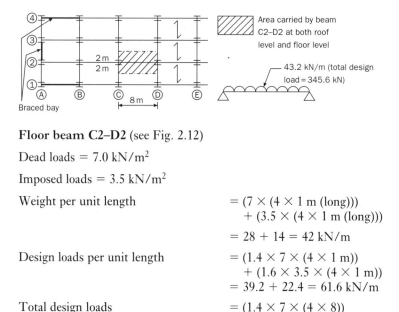

Floor beam C2–D2 (see Fig. 2.12)

Dead loads $= 7.0 \text{ kN/m}^2$

Imposed loads $= 3.5 \text{ kN/m}^2$

Weight per unit length

$$= (7 \times (4 \times 1 \text{ m (long)}))$$
$$+ (3.5 \times (4 \times 1 \text{ m (long)}))$$
$$= 28 + 14 = 42 \text{ kN/m}$$

Design loads per unit length

$$= (1.4 \times 7 \times (4 \times 1 \text{ m}))$$
$$+ (1.6 \times 3.5 \times (4 \times 1 \text{ m}))$$
$$= 39.2 + 22.4 = 61.6 \text{ kN/m}$$

Total design loads

$$= (1.4 \times 7 \times (4 \times 8))$$
$$+ (1.6 \times 3.5 \times (4 \times 8))$$
$$= 313.6 + 179.2 = 492.8 \text{ kN}$$

Fig. 2.12 Design loads on beam C2–D2 at floor level

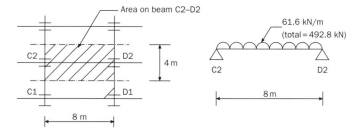

External steel beam C1–D1 at roof level (see Fig. 2.13)

Loads from:

cavity wall	$= 4.8 \times 0.5 \text{ m (high)} = 2.4 \text{ kN/m}$
steel	$= 0.5 \times 1 \text{ m (one unit length)} = 0.5 \text{ kN/m}$
concrete slab	$= 4.1 \times 2 \text{ m (wide strip)} = 8.2 \text{ kN/m}$
ceiling	$= 0.5 \times 2 \text{ m (wide strip)} = 1.0 \text{ kN/m}$
Total dead load	$= 12.1 \text{ kN/m}$
Total design dead load	$= 1.4 \times 12.1 \times 8 = 135.52 \text{ kN}$
Imposed design load	$= 1.6 \times 1.5 \times 2 \times 8 = 38.4 \text{ kN}$
Total design loads	$= 135.52 + 38.4 = 173.92 \text{ kN}$

Fig. 2.13 Design loads on external roof beam C1–D1

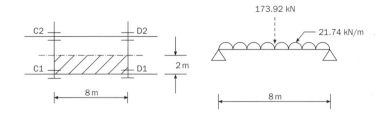

Centre column, lower length 2–5–8 (see Fig. 2.14)

Load on column above joint 5
Design dead loads for

Roof	$= 1.4 \times 6 \times (4 \times 8) = 268.8$ kN
2 floors	$= 1.4 \times (2 \times 7) \times (4 \times 8) = 627.2$ kN
3 columns (4 m high) and casing	$= 1.4 \times (3 \times 4 \times 1.5) = 25.2$ kN

Design imposed loads for:

Roof	$= 1.6 \times 1.5 \times (8 \times 4) = 76.8$ kN
2 floors	$= 1.6 \times 2 \times 3.5 \times (8 \times 4) \times 0.8$
	(20% reduction, see the code)
	$= 286.72$ kN

Fig. 2.14 Column reference numbers

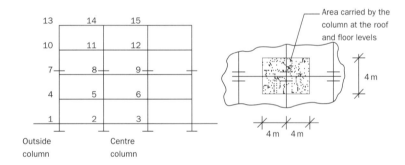

Loads from beam 4–5 and beam 5–6

Design dead loads	$= 1.4 \times 7 \times (4 \times 4)$ [from beam 4–5]
	$+ 1.4 \times 7 \times (4 \times 4)$ [from beam 5–6]
	$= 313.6$ kN
Design imposed loads	$= 1.6 \times 3.5 \times (4 \times 4)$ [from beam 4–5]
	$+ 1.4 \times 3.5 \times (4 \times 4)$ [from beam 5–6]
	$= 179.2$ kN
Total design loads on the column below joint 5	$= 1777.52$ kN

Note: For the calculation of the moment acting on the column at joint 5, calculate the loads on the beams as follows:

Load on beam 5–4	$= 1.4 \times 7 \times (4 \times 4) +$ zero imposed loads
	$= 156.8$ kN

Load on beam 5–6 (usually the longest beam of 5–4 or 5–6)

$$= (1.4 \times 7 \times (4 \times 4)) \\ + (1.6 \times 3.5 \times (4 \times 4)) \\ = 246.4 \text{ kN}$$

The design loads and bending moments are shown in Fig. 2.15.

Fig. 2.15 Design loads and moments

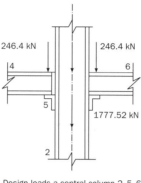

246.4 kN   246.4 kN

4   6

5

1777.52 kN

2

Design loads a central column 2–5–6

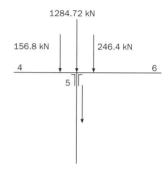

1284.72 kN

156.8 kN   246.4 kN

4   6

5

For moment at 5 use the above loads

Exercises

1 'In the design and/or assessment of an individual structural element or a structure as a whole, all relevant loads should be considered separately and in such realistic combinations as to comprise the most critical effects on the elements and the structure as a whole. The magnitude and frequency of fluctuating loads should also be considered.' Discuss the above statements. Use annotated sketches to support your answers.

2 Figure 2.Q2 shows plan and roof details of a flat roof single-storey extension to an existing house. Calculate and sketch the total design loads on one timber joist, steel beam A and steel beam B.

Access is to be provided to the roof, therefore use imposed load of 1.5 kN/m².

Roof construction:

asphalt (two layers) 19 mm thick	42 kg/m²
25 mm timber boards	590 kg/m³
50 × 175 mm timber joists spaced at 400 mm centre to centre	590 kg/m³
plaster board and skim (plaster finish)	15 kg/m²

3 A reinforced concrete bridge between two buildings spans 8 m. The cross-section of the bridge is shown in Fig. 2.Q3. The brickwork weighs 20 kN/m³, the

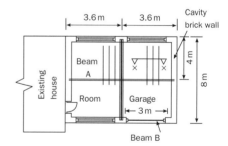

3.6 m   3.6 m   Cavity brick wall

Existing house

Beam A

Room   Garage

4 m   8 m

3 m

Beam B

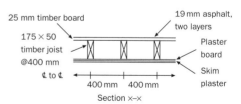

25 mm timber board

175 × 50 timber joist @400 mm

₵ to ₵

19 mm asphalt, two layers

Plaster board

Skim plaster

400 mm   400 mm

Section x–x

Fig. 2.Q2 Exercise 2: plan and roof details of a single-storey extension to an existing house

concrete weighs 24 kN/m³ and the coping weighs 0.5 kN/m. The floor of the bridge has to carry a uniformly distributed imposed load of 1.5 kN/m² in addition to its own weight. For beam A shown in

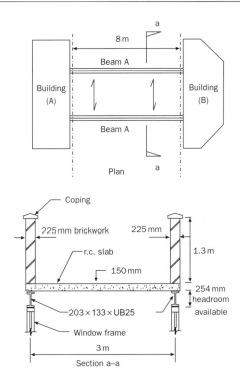

Fig. 2.Q3 Reinforced concrete bridge

Fig. 2.Q3, calculate and sketch the total uniformly distributed design loads in kN/m and the total design loads in kN. The beam is simply supported at both ends.

4 Figure 2.10 (Example 2.4) shows the relevant details at side walls and columns for a fully braced (in both directions) four-storey steel office building that is made of a steel frame with a reinforced concrete slab on profiled metal decking. External cladding is brick/breezeblock and double-glazing. Calculate and sketch the design loads acting on roof beam A1–A2, floor beam A1–A2, floor beam B1–B2 and outer column, lower length 1–4–7 (see Fig. 2.14).

All the problems in this chapter are concerned with forces which:

- meet at a point, i.e. they are concurrent
- act in one plane, i.e. they are coplanar.

The chapter deals with the graphical methods of resolving such forces into convenient components, and then determining their resultant effect, for the purpose of establishing the equilibrium of the system of forces.

Triangle of forces

AB in Fig. 3.1(a) represents a horizontal wooden beam containing a number of hooks. By means of string, a small ring, and two spring balances a weight of 35 N is suspended as shown. It is assumed in this example that the readings on the balances are 20 N and 25 N respectively. The ring is in equilibrium (at rest) under the action of three forces, i.e. the vertical pull downwards of the weight and the pulls exerted by the two strings. The condition can be represented on drawing paper by Fig. 3.1(b) which is called the **free-body diagram** with respect to point O.

The three lines are called the **lines of action** of the forces and the arrows represent the **direction** in which the forces are acting on the ring. In order to plot on paper the lines of action of the forces, the angles between the strings can be measured by means of a circular protractor. An alternative and simpler method of transferring to paper the lines of action is to hold a sheet of paper or cardboard behind the strings and, by means of a pin, to prick two or three points along each string. Pencil lines can then be drawn to connect these points and thus to fix on paper the free-body diagram.

The magnitudes of the forces can be represented on paper by vectors, i.e. lines drawn to scale. For example, if it is decided to let 1 mm represent 1 N then the force of 35 N will be represented by a line 35 mm long, the force of 25 N by a line 25 mm long, etc.

Fig. 3.1 Triangle of forces: (a) load diagram; (b) free-body diagram at point O; (c) force diagram

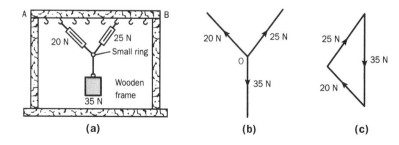

Now, if the free-body diagram is drawn on paper to represent the lines of action and directions of the forces, and three lines are drawn, as shown in Fig. 3.1(c), parallel to these forces to represent to scale their magnitudes, it will be found that a **force diagram** will be formed in the shape of a triangle. In an actual experiment the result may not be quite as accurate as indicated in Fig. 3.1(c) because of possible lack of sensitivity of the spring balances and the difficulty of transferring accurately to paper the free-body diagram. The fact is, however:

If three forces meeting at a point are in equilibrium, they may be represented in magnitude and direction by the three sides of a triangle drawn to scale, as shown in Fig. 3.1(c).

This is the **principle** or **law of the triangle of forces**.

By placing the spring balances on different hooks, e.g. as in Fig. 3.2, further illustrations of the principle of the triangle of forces can be obtained.

Fig. 3.2 Second example of a triangle of forces: (a) load diagram; (b) free-body diagram; (c) force diagram

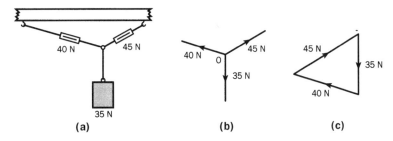

Action and reaction are equal and opposite, yet in Fig. 3.2 it appears that 85 N are required to support a downward force of 35 N. This is because the strings are inclined in opposite directions and react on each other as well as on the force of 35 N. For example, the string which has a tension of 40 N is pulling towards the left in addition to pulling up. This pull to the left is resisted by the other string pulling to the right, so that only a part of the force in each string is effective in holding up the weight. This will be demonstrated more formally on page 40.

It is advisable to carry out a few experiments similar to those of Figs. 3.1 and 3.2 in order to verify the law of the triangle of forces. The law can then be used to determine unknown forces and reactions.

Example 3.1

Determine the tensions in the two strings L and R of Fig. 3.3(a).

Fig. 3.3 Example 3.1: (a) load diagram; (b) free-body diagram; (c) force diagram

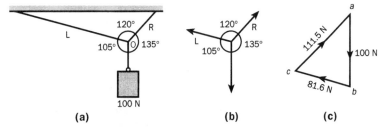

Solution

Point O is in equilibrium under the action of three forces as shown in the free-body diagram of Fig. 3.3(b), therefore the triangle of forces law must apply.

Draw a line *ab* parallel to the force of 100 N to a scale of so many newtons to the millimetre. From *b* draw a line parallel to string L. From *a* draw a line parallel to string R. The two lines intersect at point *c*. The length of line *bc* gives the tension in string L (i.e. 81.6 N) and the length *ca* gives the tension in string R (i.e. 111.5 N). (A student with a knowledge of trigonometry can obtain the magnitudes of the forces by calculation.)

Example 3.2

Two ropes are attached to a hook in the ceiling and a man pulls on each rope as indicated in Fig. 3.4(a). Determine the direction and magnitude of the reaction at the ceiling (neglect the weights of the ropes).

Solution

Since the hook is in equilibrium under the action of the forces in the ropes and the reaction at the ceiling, the triangle of forces principle applies. Draw *ab* parallel to the force of 250 N to a suitable scale (Fig. 3.4(c)). From *b* draw

Fig. 3.4 Example 3.2: (a) load diagram; (b) free-body diagram; (c) force diagram

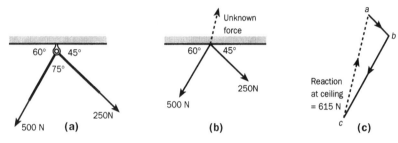

bc parallel to the other rope to represent 500 N. The closing line *ca* of the force triangle represents to scale the reaction at the ceiling and is nearly 615 N, acting at approximately 7° to the vertical. This force can be called the **equilibrant** of the two forces in the ropes, since the ceiling must supply a force of 615 N in order to maintain equilibrium.

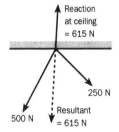

Fig. 3.5 Resultant force

In the last problem, the reaction at the ceiling due to the pulls in the two ropes equalled 615 N acting as shown in Fig. 3.5. If we substitute one rope for the two ropes and pull with a force of 615 N in the direction indicated, the effect on the ceiling will be exactly as before, i.e. the result of the two forces is equivalent to one force of 615 N. This force is, therefore, called the **resultant**. Note also that this resultant is equal in value to the equilibrant and acts in the same straight line but in the opposite direction. The resultant of a given number of forces is therefore the single force which has the same effect on the equilibrium of the body as the combined effects of the given forces.

Parallelogram of forces

The law of the parallelogram of forces is in essentials the same as the law of the triangle of forces. Any problem which can be solved by the parallelogram of forces law can be solved by the triangle of forces law, although it may sometimes be slightly more convenient to use the former method.

Example 3.3

Referring to Example 3.2 determine the resultant of the pulls in the two ropes by using the parallelogram of forces.

Solution

Measure a distance OA along line *a* in Fig. 3.6 to represent 500 N and measure a distance OB along line *b* to represent 250 N. From A draw a line parallel to OB and from B draw a line parallel to OA thus forming a parallelogram. The length of line OC, which is the diagonal of the

Fig. 3.6 Example 3.3:
parallelogram of forces

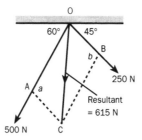

parallelogram (from the point where the two forces meet), gives the value of
the resultant of the two forces (615 N) and also its direction. Note that the
triangle OBC is identical to the triangle of forces *abc* in Fig. 3.4(c).

Example 3.4

Two ropes pull on an eye-bolt as indicated in Fig. 3.7(a). Determine the
resultant pull on the bolt.

Fig. 3.7 Example 3.4:
(a) load diagram; (b) resultant
force diagram

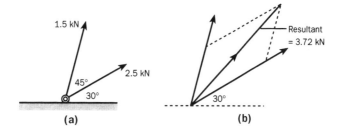

The solution is indicated in Fig. 3.7(b).

   Note that when the parallelogram of forces law is used to determine the
resultant of two forces, the two forces must be drawn so that the arrows representing their directions both point towards the meeting point or both point
away from it. In other words, a thrust and a pull must be converted into two
thrusts or into two pulls.

Example 3.5

Determine the resultant force on the peg due to the thrust of 2.5 kN and the
pull of 1.5 kN (Fig. 3.8(a)).

Figure. 3.8 Example 3.5: (a) load
diagram; (b) and (c) resultant
force diagrams

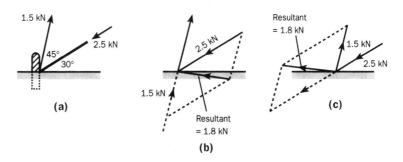

Solution

The resultant is found either as shown in Fig. 3.8(b) or as shown in Fig.
3.8(c). Note that the effect on the peg is the same whether it is pushed or
pulled with a force of 1.8 kN.

Example 3.6

A rigid rod is hinged to a vertical support and held at 60° to the horizontal by means of a string when a weight of 250 N is suspended as shown in Fig. 3.9(a). Determine the tension in the string and the compression in the rod, ignoring the weight of the rod.

Fig. 3.9 Example 3.6: (a) load diagram; (b) free-body diagram; (c) force diagram

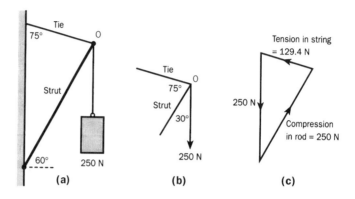

Solution

Point O is in equilibrium under the action of three forces which meet at the point, i.e. the weight of 250 N, the tension in the string and the compression in the rod. The solution is given in Fig. 3.9(c). Note that a tension member is called a **tie** and a compression member is called a **strut**.

Rectangular components

Figure 3.10(a) shows two rods of wood connected by a hinge at the top and supported by two concrete blocks. If a gradually increasing pull is applied to the

Fig. 3.10 Rectangular components: (a) load diagram; (b) force diagram of forces meeting at hinges; (c) force at each compression member

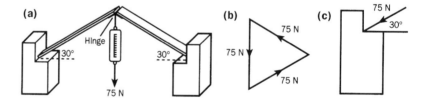

spring balance, the vertical force necessary to cause lifting of the blocks can be recorded. Assume this downward force is 75 N when the angles are as given and when the blocks just begin to lift. The compression acting along each member is 75 N as obtained by the force diagram, Fig. 3.10(b). Now consider what happens at the bottom end of one compression member. The member is pushing on the block with a force of 75 N (Fig. 3.10(c)). Since this force is inclined it has a twofold effect. It is tending to cause crushing of the concrete as well as overturning. If the force were vertical (and axial), there would be a crushing effect only and no overturning effect. If the force were horizontal, there would be an overturning effect and no crushing effect. In order to evaluate the crushing and overturning effects of the inclined force it is necessary to **resolve** such a force into its **horizontal component** and its **vertical component**.

It was demonstrated on page 39 that the resultant of two forces meeting at a point is given by the diagonal of a parallelogram. By a reverse process, one force can be replaced by (or resolved into) two forces. By constructing a rectangle (parallelogram) (Fig. 3.11) with the force in the rod (75 N) as the diagonal, the horizontal and vertical components are given by the two sides of the rectangle. The components can be found either by drawing to scale or by calculation.

Fig. 3.11 Resolving horizontal and vertical force components

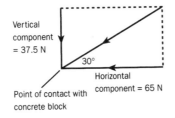

Example 3.7

Replace the tie-rope shown in Fig. 3.12(a) by two ropes, one vertical and one horizontal, which together will have the same effect on the eye-bolt.

Fig. 3.12 Example 3.7: (a) load diagram; (b) horizontal and vertical force components

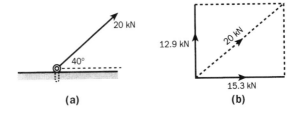

(a) (b)

Solution

By constructing a rectangle with the force of 20 kN, drawn to scale, as the diagonal (Fig. 3.12(b)) the vertical component (and therefore the force in the vertical rope) is found to be 12.9 kN. The horizontal component is 15.3 kN.

In the above problem, the components are called **rectangular components** because the angle between them is 90°. When a force has to be split into two components it is usually the rectangular components which are required. It is possible, however, to resolve a force into two components with any given angle between them.

Example 3.8

Replace the rope X which is pulled with a force of 500 N by two ropes L and R as indicated in Fig. 3.13.

Fig. 3.13 Example 3.8 resolving a single force into its components

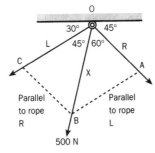

Solution

Measure a length along line OB to represent 500 N and assume this line to be the diagonal of a parallelogram by drawing lines from B parallel to ropes L and R respectively. The forces in the two ropes which together have the same effect on the bolt as the single rope can now be scaled off, i.e.

Force in rope L = 450 N (length OC)

Force in rope R = 365 N (length OA)

These two forces can be said to be components of the original force of 500 N.

Example 3.9

Determine the tensions in the two chains X and Y of Fig. 3.14(a). Then determine the horizontal and vertical components of these tensions. In addition, determine the horizontal and vertical components of the reactions at A and B. Neglect the weight of the chains.

Fig. 3.14 Example 3.9: (a) load diagram; (b) free-body diagram; (c) force diagram; (d) horizontal and vertical components within the chains; (e) reactions at points A and B

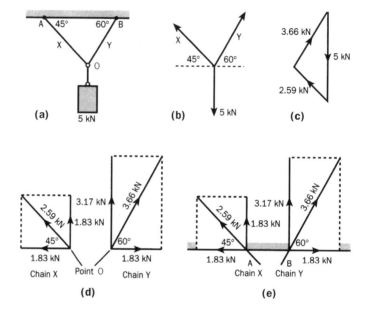

Solution

The free-body and force diagrams for the point are given in Fig. 3.14(b) and (c). The tensions in the chains are respectively 2.59 kN and 3.66 kN. The horizontal and vertical components of the tensions in the chains are given in Fig. 3.14(d).

Note that the sum of the two vertical components is equal to the value of the suspended load of 5 kN (action and reaction are equal and opposite). Note also that the horizontal component of the force in chain X is equal to the horizontal component of the force in chain Y and acts in the opposite direction (action and reaction are equal and opposite).

Now consider the reactions at points A and B, which equal the tensions in the chains (Fig. 3.14(e)). Again, neglecting the weights of the chains, the sum of the vertical components of the reaction is equal to 5 kN and the horizontal components are equal and opposite.

Referring back to Fig. 3.10(a) on page 40, it was stated that the concrete blocks were on the point of overturning when the compression acting along each member was 75 N. The horizontal component of this force is 65 N, and this component can be prevented from having an overturning effect by connecting the two inclined members by a horizontal tie member (Fig. 3.15(a)). It will now be impossible to cause overturning of the concrete blocks by pulling downwards at the apex. The tension in the tie member will be equal to the horizontal component (65 N) of the thrust in the inclined member.

Another way of finding the tension in the tie is by applying the triangle of forces law to the joint between the inclined and horizontal members (Fig. 3.15(b)). Neglecting the weights of the members, the vertical downward force of 75 N is balanced by the reactions of the blocks (37.5 N each block).

Fig. 3.15 Tied truss: (a) and (b) load diagrams; (c) free-body diagram; (d) force diagram

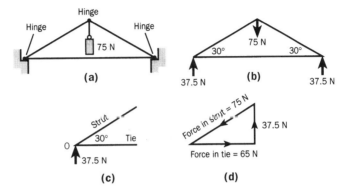

Consider the equilibrium of point O (Fig. 3.15(c)). Note that, when three forces meeting at a point are in equilibrium, the arrows in the force diagram follow one another around in the same direction (Fig. 3.15(d)). The strut is pushing towards point O and the tie is pulling away from the same point.

Polygon of forces

The same apparatus used in Fig. 3.1 can be employed to demonstrate the principle of the polygon of forces.

Referring to Fig. 3.16(a), the small ring is at rest (i.e. in equilibrium) as the result of the downward pull of the 35 N force and the pulls in the four strings. The forces in the strings are given by the readings on the spring balances and these forces are given for one particular arrangement of the strings.

Fig. 3.16 Polygon of forces: (a) load diagram; (b) free–body diagram; (c) force polygon

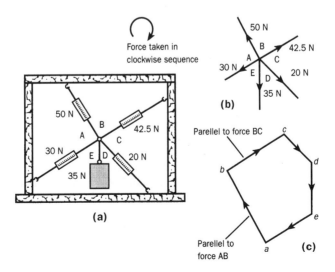

The lines of action of the forces can be transferred to a piece of paper or cardboard (in the manner described on page 36) to give the free-body diagram (Fig. 3.16(b)).

It will be noted that the *spaces* between the forces in the free-body diagram have been lettered A, B, C, etc., and each force is therefore denoted by two letters, e.g. force AB (which separates space A from space B), force BC, force CD, etc. (If the forces are taken in anticlockwise sequence they will be called force DC, force CB, force BA, etc., but it is more usual to take the forces in clockwise sequence.) This method of notation was devised by an engineer (R. H. Bow) about 1870 and is known as **Bow's notation**. It is not essential to use a notation for simple problems but some form of notation is indispensable when a large number of forces are involved, such as in roof trusses (see Chapter 6).

Having plotted the lines of action of the forces in the free-body diagram, the next step is to choose a scale of so many newtons to the millimetre in order to plot the force diagram. Starting with the force AB, a line *ab* is drawn parallel to it, the length *ab* representing to scale the magnitude of the force (50 N). From *b* a line *bc* is drawn in the direction indicated by the arrow on force BC (42.5 N).

The other forces are drawn in the same manner and, if the experiment has been performed accurately, it will be found that the last line drawn parallel to force EA (30 N) will finish at *a*, where line *ea* represents to scale the force EA.

It is not necessary to start the force diagram with force AB. Any one of the forces can be chosen as the starting force.

Further experiments can be performed by connecting the strings to the other hooks in the framework, and it will be found in every case (apart from small experimental errors) that, when a number of forces are in equilibrium, the force diagram will form a polygon, the sides of which represent to scale the magnitudes and directions of the forces.

Study carefully the force diagram or polygon in Fig. 3.16(c) and note that the arrows 'chase each other' in the same sense around the diagram, i.e. the tip of each arrow is pointing to the tail of the arrow in front of it. This will always be true for force diagrams when the forces are in equilibrium, and use can be made of this fact to discover the direction of unknown forces, as in the following example.

Example 3.10

A rod (the weight of which is negligible) is hinged to a support at S and is supported by a tie (Fig. 3.17(a)). From the point O three ropes are pulling with the forces indicated. Determine the forces in the rod and tie.

Solution

The point O is in equilibrium as the result of the action of five forces, i.e. the three given forces and the forces in the rod and tie.

The free-body diagram at point O is given in Fig. 3.17(b).

It is an advantage to start Bow's notation in such a manner that the two unknown forces are last. First, draw a line *ab* parallel to force AB to represent to scale 2.5 kN, then draw lines *bc* and *cd* to represent the other known forces. From *d* draw a line parallel to force DE. Since point O is in equilibrium the force diagram must form a closed polygon. Therefore from *a* draw a line parallel to force EA to intersect the line drawn parallel to force DE. The intersection point of these two lines is point *e*.

Fig. 3.17 Example 3.10:
(a) load diagram; (b) free-body
diagram; (c) force polygon

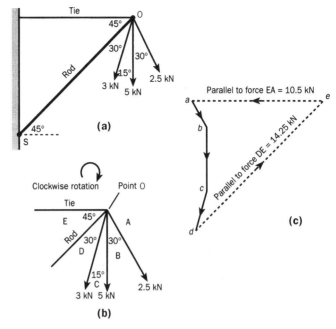

The force in the rod is given by the length of line *de* and is 14.3 kN approximately. Similarly, the force in the tie is given by the length of line *ea* and is 10.5 kN approximately. The arrows on lines *de* and *ea* must follow the general direction; therefore the tie is pulling away from point O and the rod (which is a strut) is pushing towards point O.

Note that the forces in the rod and tie have been stated to be 14.3 kN and 10.5 kN approximately. The forces can be obtained by calculation and may be slightly different from the values given, since calculation is more accurate than drawing. Graphical methods are, however, sufficiently accurate for most structural work and should be used when they are quicker and more easily applied than calculation methods.

The resultant of two forces meeting at a point can be found by the triangle or parallelogram of forces as described earlier. The resultant of any number of forces can be determined by the polygon of forces.

Example 3.11

Three ropes pull on a bolt as indicated in Fig. 3.18(a). Determine the resultant pull.

Fig. 3.18 Example 3.11:
(a) load diagram; (b) free-body
diagram; (c) force polygon

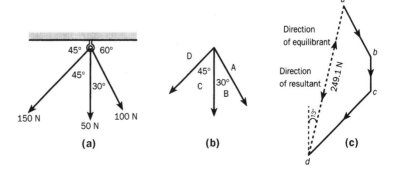

Solution   The force polygon is constructed from *a* to *d* by drawing lines to scale parallel to the three given forces (Fig. 3.18(c)). The closing line *da* of the polygon gives the magnitude and direction of the reaction (equilibrant) supplied by the support. The resultant of the pulls in the three ropes is, of course, equal to the equilibrant and is 249.1 N pulling downwards at an angle of approximately 13° to the vertical.

Summary

Triangle of forces   If three forces, the lines of action of which meet at one point, are in equilibrium, they can be represented in magnitude and direction by the sides of a triangle if these sides are drawn parallel to the forces.

Parallelogram of forces   If two forces meeting at a point are represented in magnitude and direction by the two sides of a parallelogram, their resultant is represented in magnitude and direction by the diagonal of the parallelogram which passes through the point where the two forces meet.

Polygon of forces   If any number of forces acting at a point are in equilibrium, they can be represented in magnitude and direction by the sides of a closed polygon taken in order.

Components of forces   A given force can be replaced by any two forces (components) which meet at the point of application of the given force. When the angle between the two forces is a right angle the components are called rectangular components.

Equilibrant   The equilibrant of a given number of forces which are acting on a body is the single force which keeps the other forces in equilibrium. Any one of the forces can be considered as being the equilibrant of the remainder of the forces.

Resultant   The resultant of a given number of forces is the single force which, when substituted for the given forces, has the same effect on the state of equilibrium of the body.

Exercises

Note: All pulleys in these exercises are **smooth** (i.e. frictionless).

1   Determine the tensions in the two chains L and R in the case of Fig. 3.Q1(a), (b) and (c).

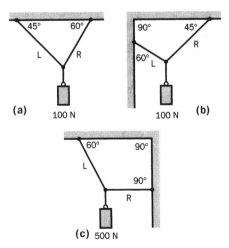

Fig. 3.Q1

2 The angles marked A in Fig. 3.Q2 are equal. Determine the minimum value of this angle if the tension in each rope must not exceed 1 kN.

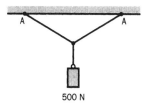

Fig. 3.Q2

3 The string marked L in Fig. 3.Q3 passes over a pulley. Determine the tensions in the two strings L and R by considering the equilibrium of point 1, then determine the reaction at the pulley by considering the equilibrium of point 2.

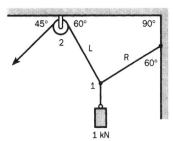

Fig. 3.Q3

4 The string marked L in Fig. 3.Q4 passes over a pulley and the system is in equilibrium. Determine the tensions in the strings L and R, then determine the magnitude of the weight W and the tension in the string M. (Consider first the equilibrium of point 1.)

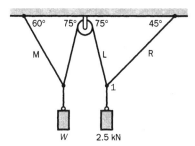

Fig. 3.Q4

5 Two struts thrust on a wall as indicated in Fig. 3.Q5. Determine the magnitude and direction of the resultant thrust.

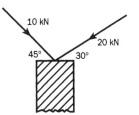

Fig. 3.Q5

6 The top of a pole resists the pulls from two wires as indicated in Fig. 3.Q6. Determine the magnitude and direction of the resultant pull.

Fig. 3.Q6

7 A bolt resists the pull from three wires as indicated in Fig. 3.Q7. Determine the resultant of the two pulls of 5 kN and 2.5 kN then combine this resultant with the remaining pull of 4 kN to determine the resultant pull on the bolt.

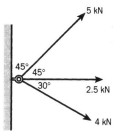

Fig. 3.Q7

8 Referring to Fig. 3.Q8, determine the value of the angle A so that the resultant force on the wall is vertical. What is the value of the resultant?

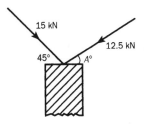

Fig. 3.Q8

9 The top of a pole sustains a pull and a thrust as shown in Fig. 3.Q9. Determine the resultant force.

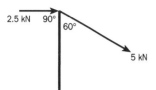

Fig. 3.Q9

10 Two men pull on ropes attached to a peg as shown in Fig. 3.Q10. Determine the resultant pull on the peg.

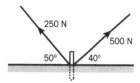

Fig. 3.Q10

11 In trying to move a block of stone resting on the ground, one man pushes and another man pulls as shown in plan in Fig. 3.Q11. Determine the resultant of the two forces.

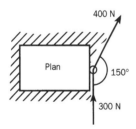

Fig. 3.Q11

12 Determine the resultant pull on the bolt in Fig. 3.Q12 and the reactions at the pulleys.

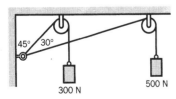

Fig. 3.Q12

13 Determine the forces in the tie and strut in the case of Fig. 3.Q13(a), (b) and (c), neglecting the weights of the members. All joints are hinged.

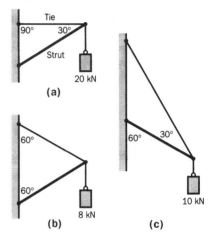

Fig. 3.Q13

14 Referring to Fig. 3.Q14 a weight of 5 kN is suspended from A by a flexible cable. The rope C is parallel to the member AB. Determine the force in the cable, the angle x and the forces in the strut and tie. Ignore the weights of the members.

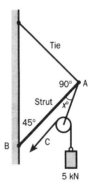

Fig. 3.Q14

15 Determine the forces in the two struts of Fig. 3.Q15 then determine the vertical and horizontal thrusts at each support.

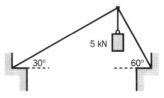

Fig. 3.Q15

16 Figure 3.Q16 shows a simple roof truss consisting of two struts and a tie. Consider the equilibrium of joint C to determine the forces in the two struts, then, using these values, apply the triangle of forces law to joints A and B respectively to determine the tension in the tie and the vertical reactions at the supports.

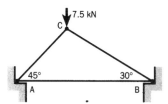

Fig. 3.Q16

17 Determine the forces in the tie and strut of Fig. 3.Q17, then determine the vertical and horizontal components of the reactions at A and B.

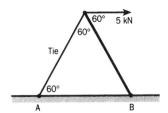

Fig. 3.Q17

18 A continuous string ABCDE has weights suspended from it as shown in Fig. 3.Q18. Determine the tensions in the portions AB and BC. In addition, determine the horizontal and vertical components of the reactions at A and E.

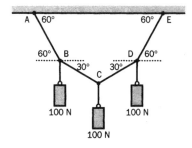

Fig. 3.Q18

19 In Fig. 3.Q19 the masts are vertical and hinged at their bases. Determine the tension in the rope supporting the weight, the tension in the stays and the compression in the masts.

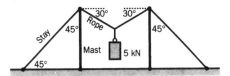

Fig. 3.Q19

20 In Fig. 3.Q20 the rope supporting the weight passes over pulleys at the tops of the masts. Determine the angle A that the masts make with the horizontal and determine the compression in the masts.

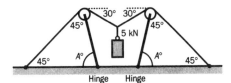

Fig. 3.Q20

21 The block of stone in Fig. 3.Q21 will begin to slide when the horizontal force is 2.5 kN. What is the maximum value of the thrust P without sliding occurring?

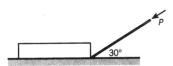

Fig. 3.Q21

22 Determine the force in the ropes A and B which will replace the rope C of Fig. 3.Q22.

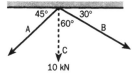

Fig. 3.Q22

23 Two posts support a weight of 1 kN as shown in Fig. 3.Q23. What horizontal force is each post exerting on the other?

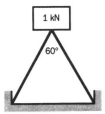

Fig. 3.Q23

24 A triangular framework of hinged rods is supported by two cables (Fig. 3.Q24). Determine the forces in the framework and in the cables, and determine the vertical components of the reactions at A and B. Neglect the weights of the members.

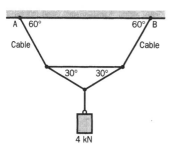

Fig. 3.Q24

25 A framework consisting of five rods hinged at their joints is supported as shown in Fig. 3.Q25. Neglecting the weights of the members, determine the forces in the members of the framework and the reaction at A.

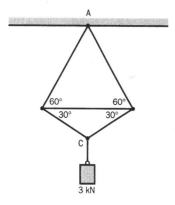

Fig. 3.Q25

26 Figure 3.Q26(a) to (f) shows the free-body diagrams for systems of concurrent forces which are in equilibrium. Determine the magnitude and direction of the unknown force (or forces) marked X and Y.

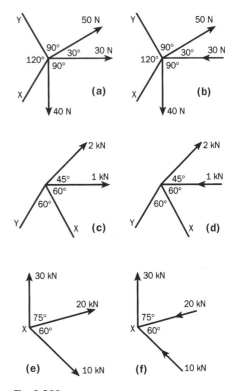

Fig. 3.Q26

27 Three ropes are attached to point O and are pulled with forces as indicated in Fig. 3.Q27. Determine the forces in the strut and tie.

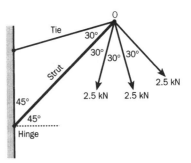

Fig. 3.Q27

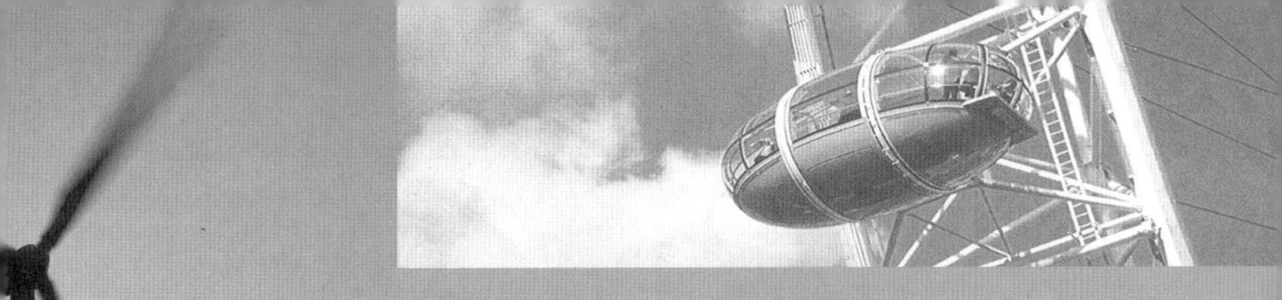

For the solution of the problems in this chapter it is necessary to know the following facts:

- The whole weight of a body can be assumed to act at one point for the purpose of determining supporting reactions, etc., in a system of forces which are in equilibrium. This point is called the **centre of gravity** of the body. This subject will be discussed in detail in Chapter 9. For the present, it will be sufficient to remember that the centre of gravity (c.g.) of a rod or beam which is uniform in cross-section is at the centre of its length; the c.g. of a thin rectangular plate is at the intersection of the diagonals of the rectangle; the c.g. of a thin circular disc is at the centre of the circle; and the c.g. of a cube or rectangular solid is at the intersection of its diagonals.
- If a body is supported by a perfectly smooth surface, the reaction of the surface upon the body acts at right angles to the surface (Fig. 4.1).

Fig. 4.1 Reactions at a smooth surface

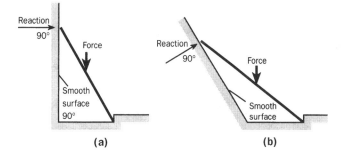

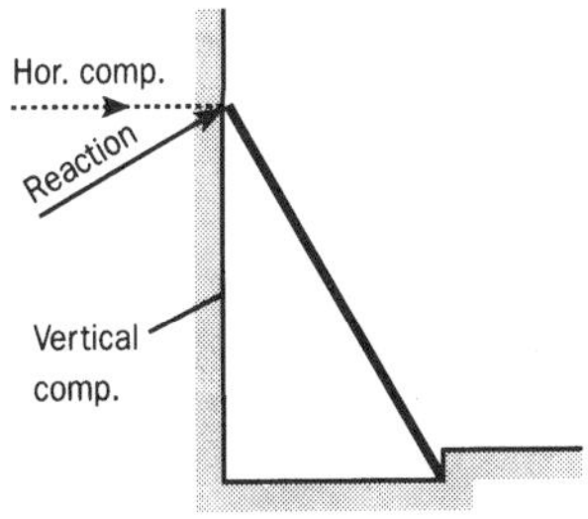

(a)   (b)

If the reaction were not at right angles to the surface, it would mean that there would be a component of the reaction acting along the wall surface as shown in Fig. 4.2. This shows there is frictional resistance of the wall opposing the downward trend of the ladder, and if the wall is smooth it can offer no frictional resistance.

- If two forces (which are not parallel) do not meet at their points of contact with the body, their lines of action can be produced until they do meet.
- If a body is in equilibrium under the action of three forces, and two of these forces are known to meet at a point, the line of action of the remaining force must pass through the same point.

Fig. 4.2 Reactions due to frictional resistance

Example 4.1

A thin rectangular plate weighing 100 N is supported by two strings as indicated in Fig. 4.3(a). Determine the tensions in the strings.

Fig. 4.3 Example 4.1: (a) load diagram; (b) free-body diagram at point O; (c) force diagram

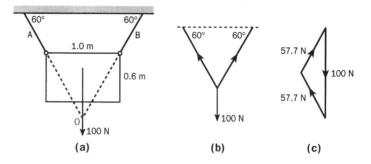

Solution

The whole weight of the plate can be assumed to act at its centre of gravity. The lines of action of the forces in the strings, when produced, meet the line of action of the weight of the plate at O. Consider the equilibrium of this point.

The tension in string A (and B) is 57.7 N. The principle involved in the solution of this problem can be demonstrated on the apparatus illustrated in Fig. 4.1.

Example 4.2

A ladder rests against a smooth wall and a man weighing 750 N stands on it as indicated in Fig. 4.4. Neglecting the weight of the ladder determine the reactions at the wall and at the ground.

Fig. 4.4 Example 4.2: (a) ladder resting against a smooth surface; (b) free-body diagram at point O; (c) force diagram

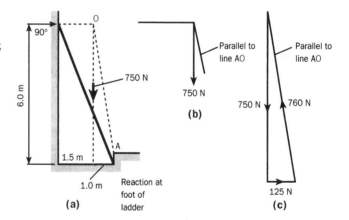

Solution

The ladder is in equilibrium as the result of the action of three forces: the weight of the man, the reaction at the top of the ladder, and the reaction at the foot. Since the wall is smooth the reaction at the top of the ladder is at right angles to the wall. Draw Fig. 4.4(a) to scale. The vertical line representing the weight of the man and the horizontal line representing the reaction at the top of the ladder meet at point O. The line of action of the reaction at the foot of the ladder must therefore pass through the same point.

We now have three forces meeting at a point and the solution is shown in Figs. 4.4(b) and (c) which give the reaction at the top of the ladder as 125 N, and the reaction at the foot as 760 N.

Example 4.3

The uniform rod 2 m long shown in Fig. 4.5(a) weighs 250 N. Determine the reactions at A and B.

Fig. 4.5 Example 4.3: (a) load diagram; (b) free-body diagram at point O; (c) force diagram

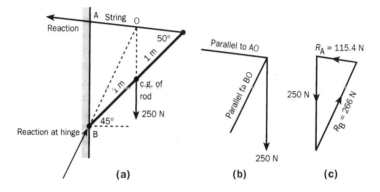

(a)   (b)   (c)

Solution

The reaction at A is in line with the string. The weight of the rod can be assumed to act at its centre of gravity and to act vertically downwards. Draw the diagram to scale. Produce the line of action of the force of 250 N to meet at O, the line of action of the reaction at A. The line of action of the reaction at B is given by joining B to O. Consider the equilibrium of point O as shown in Fig. 4.5(b). The reactions are 115.4 N at A and 266 N at B.

The link polygon

Non-concurrent forces are forces whose lines of action do not all meet at one point. (A system in which the forces are all parallel to one another is a special example of non-concurrent forces.)

Figure 4.6 shows a system of non-concurrent forces. Assume that the forces are acting on a block of stone, the plan view of which is a square of 0.8 m side, the forces being coplanar. If the resultant force on the stone is required, one method of solution based on work in the previous chapter is first to produce any two of the forces until they meet, and then to find the resultant of these two forces by the parallelogram of forces. This resultant can now be combined with one of the remaining forces and a new resultant found, and so on until the forces have been reduced to one resultant. This method, however, is cumbersome when the number of forces is large and a much neater solution is shown in Fig. 4.7.

The free-body diagram giving the lines of action of the forces is drawn accurately and the forces lettered with Bow's notation. These forces are treated as if they all meet at one point and a force polygon is drawn, the sides of which are parallel to the given forces.

The closing line fa (direction from f to a) of the force polygon will give the magnitude and direction of the equilibrant of the five forces and the same line

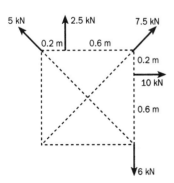

Fig. 4.6 System of non-concurrent forces

Fig. 4.7 Resultant force: (a) free-body diagram with link polygon; (b) force polygon with polar diagram

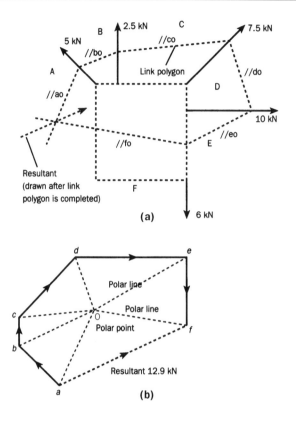

in the direction *a* to *f* will give the resultant. If the five forces met at one point, the answer to the problem would be complete since the resultant must pass through the same point. In this case, there is no common meeting point of the forces and the position of the resultant on the body must be determined. This is done by placing a point O (the polar point) *anywhere* inside or outside the force polygon and connecting it to the letters *a*, *b*, *c*, etc., by polar lines.

A line parallel to the polar line *ao* is now drawn on the free-body diagram *anywhere across space* A to cut force AB. Then a line parallel to *bo* is drawn across space B to cut force BC and so on. Where the last line (parallel to the polar line *fo*) intersects the first line (parallel to polar line *ao*) is one point on the line of action of the resultant (or equilibrant). The resultant can now be shown in its position on the free-body diagram by drawing a line parallel to the closing line *af* of the force polygon. In this construction the link lines form a polygon on the forces of the free-body diagram and it is known as a **link polygon**.

The graphical conditions of equilibrium for a system of non-concurrent forces are:

- the force polygon must be a closed one
- the link polygon must be a closed one.

The validity of the construction of the link polygon can be proved as follows.

Consider the two forces AB and BC of 5 kN and 2.5 kN respectively, which are shown in Fig. 4.8(a) to a larger scale than in Fig. 4.7.

Any two forces *m* and *n* can be substituted for the one force AB without altering the state of equilibrium of the body. The value of these two components

Fig. 4.8 Validity of the link polygon: (a) detail of Fig. 4.7; (b) force triangles

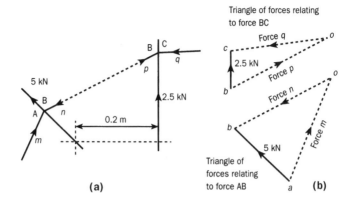

(a)   **(b)**

m and n which replace force AB can be obtained from a triangle of forces (Fig. 4.8). The force of 5 kN represented by the line ab is the resultant of the two forces. In a similar manner, force BC can be replaced by two forces p and q; the values of these two forces also being given by a triangle of forces.

If it is arranged that component n of force AB is equal to component p of force BC, the net result on the body of these two components is nil, since they act in the same straight line but in the opposite direction (see Fig. 4.8(b)). Similarly, by having a common polar point in the force diagram (given again in Fig. 4.9) component q of force BC will balance with an equal component of force CD and so on. It follows that, assuming the equilibrant acts through the point where the two link lines parallel to polar lines ao and fo meet, the net result on the body of all the forces, including the equilibrant, is nil. Therefore the construction is valid.

Fig. 4.9 Force diagram about a common polar point

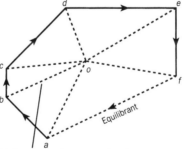

This line represents component n of force AB and also component p of force BC

Note: the arrow representing the direction of the equilibrant will be in the opposite direction to that given for the resultant in Fig. 4.7.

In brief, this method of dealing with non-concurrent forces consists in replacing each of the given forces by two forces or components and arranging that one component of each force is equal in value to a component of the succeeding force. This can be done by having a common polar point in the force diagram. If the polar point is at any other position different from that shown in Figs. 4.7 and 4.9, the construction is still valid, since a force can be replaced by two components in a multitude of ways. Although the polar point can theoretically be placed anywhere, a neater solution can be obtained by choosing its position so that the link polygon will fit nicely on the free-body diagram. A multitude of link polygons can be drawn for any one problem, by using different positions of the polar point, but the closing lines (such as lines drawn parallel to *ao* and *fo* in Fig. 4.7) will all intersect on points along the line of action of the equilibrant (or resultant).

Application of the link polygon

Examples 4.4–4.9 explore the use of the link polygon construction in a variety of situations.

Example 4.4

A vertical post is subjected to pulls from four cables as shown in Fig. 4.10. The magnitude, direction and position of the resultant pull are obtained as shown.

Note that in Fig. 4.10(a) line *ao* has to be produced beyond space A until it intersects line *eo*.

Example 4.5

A vertical post is subjected to parallel pulls from four cables as shown in Fig. 4.11. Determine the position of the resultant pull.

The resultant pull is 8 kN acting at a point 2.25 m from the foot of the post as shown in Fig. 4.11.

It is usually quicker and easier to obtain the resultant of parallel forces by calculation (see Chapter 5).

The link polygon may also be used to determine the reactions to a beam or truss. It must be realized, however, that the values of two unknown forces in a system of non-concurrent forces which are in equilibrium can only be determined when all the following conditions are satisfied:

- All the forces (apart from the two unknowns) must be known completely, that is, in magnitude, line of action and direction.
- The line of action of one of the unknown reactions (not necessarily its direction) must be known.
- At least one point on the line of action of the remaining reaction must be known.

Example 4.6

A horizontal beam is hinged at one end and supported on a smooth wall at the other end. It is loaded as indicated in Fig. 4.12(a) (the central load including the weight of the beam). Determine the reactions at the supports.

Solution

Since the wall is smooth, the reaction at the wall is at right angles to the surface and is therefore vertical. The line of action of one of the reactions is thus known and so is one point (the hinge) on the line of action of the remaining reaction. It is not necessary, for the solution, to know beforehand the direction of the arrow on the vertical reaction, but it should be obvious that in this example the reaction is acting in an upward direction.

The free-body diagram is drawn to scale and the three known forces are plotted on the force diagram, i.e. *a* to *b*, *b* to *c*, and *c* to *d*, as shown in Fig. 4.12. From *d* a vertical line is drawn parallel to the force DE (the unknown vertical reaction). The point *e* will be somewhere on this line, and its position cannot be fixed until the link polygon is drawn.

Fig. 4.10 Example 4.4: (a) free-body diagram with link polygon; (b) force polygon with polar diagram

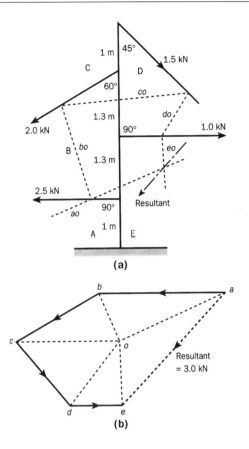

(a)

(b)

Fig. 4.11 Example 4.5: (a) free-body diagram with link polygon; (b) force and polar diagrams

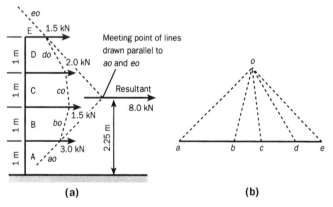

(a)   **(b)**

Choose a polar point O and draw the polar lines. *From the hinge* draw a line across the A space parallel to polar line *ao*. Where this line cuts the force AB, draw a line across the B space parallel to polar line *bo* and so on, the last line being drawn parallel to the polar line *do* to cut the vertical reaction (force DE). From this point on the force DE draw a line to the hinge. This line closes the link polygon and is known as the **closer**. Draw a line on the force diagram from the polar point O and parallel to the closer. Where this line cuts the vertical line from *d* is the position of *e*. Join *e* to *a*. The length

Fig. 4.12 Example 4.6: (a) load diagram; (b) free-body diagram with link polygon; (c) force and polar diagrams

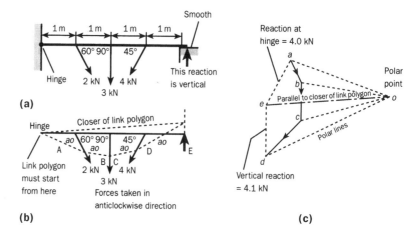

of line *de* gives the magnitude of the vertical reaction at the right-hand support (4.1 kN) and the length of line *ea* gives the magnitude and direction of the reaction at the hinge (4.0 kN).

It is absolutely essential that the link polygon is started by drawing a line from the *only known point* on the reaction at the hinge. Referring to Fig. 4.8 and the procedure necessary to obtain the equilibrant of a system of non-concurrent forces, it was shown that lines drawn parallel to the first and last polar lines intersect on one point of the line of action of the equilibrant. In the procedure just described for obtaining two unknown forces, advantage has been taken of the fact that one point on the unknown reaction is known, and therefore it can be arranged that lines drawn parallel to the first and last polar lines meet at this point (the last polar line being the closer).

Example 4.7

A beam as shown in Fig. 4.13(a) is supported by two walls and loaded as shown, the central load including an allowance for the weight of the beam. Determine the value of the reactions.

Solution

Solving a problem of this type (where all the forces are vertical) by means of the link polygon is equivalent to using a sledgehammer to drive in a tin-tack. Solution by calculation is much quicker and easier (Chapter 6). The graphical solution is shown in Fig. 4.13. The reaction at the right support is 156 kN (given by *de* on the force diagram) and the reaction at the left support is 109 kN (given by *ea*).

Fig. 4.13 Example 4.7: (a) load diagram; (b) free-body diagram with link polygon; (c) force and polar diagrams

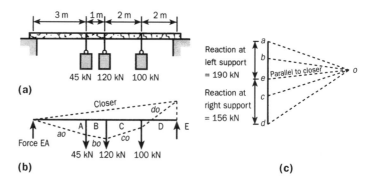

Example 4.8

Figure 4.14(a) represents a roof truss, one end of which is supported by a roller bearing. Determine the reactions. The vertical forces are due to the weight of the roof covering, etc., and the inclined forces (which are at right angles to the slope of the roof) are due to wind loads.

Solution

The manner in which the reactions are determined is shown in Fig. 4.14. It should be noted that the right-hand reaction is vertical because the rollers, as indicated, cannot resist horizontal forces, and although both reactions can be determined without combining the pairs of forces by the parallelogram law, the solution given here is probably the most straightforward.

Procedure

The free body diagram is drawn and the parallelogram of forces law used to combine the vertical and inclined forces. The known loads AB, BC, etc., are plotted in the force diagram, the last load to be plotted being force GH (represented by *gh* on the force diagram). Since the reaction at the right support (force HJ) is vertical, the point *j* will be somewhere on the vertical line drawn from *h* and the link polygon will be the means of fixing this point *j*.

It will be observed that two polar lines (*ao* and *ho*) appear not to be used for constructing the link polygon, and this fact needs explanation. The link polygon must start from the hinge which is the only known point on the line of action of the reaction JA. Going clockwise, the next force to the reaction is the force AB. Therefore from the hinge a line must be drawn parallel to *ao* across the A space

Fig. 4.14 Example 4.8: (a) load diagram; (b) free-body diagram with link polygon; (c) force and polar diagrams

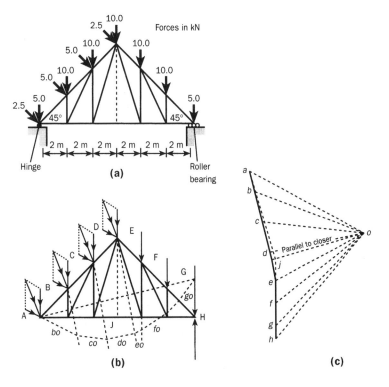

to cut force AB. But force AB passes through the hinge. Therefore a line parallel to *ao* must be drawn *from* the hinge *to* the hinge. A line drawn parallel to polar line *ao* may therefore be considered as contained in the hinge.

The next force in order is BC, so a line is drawn from the hinge (which is a point on force AB as well as a point on the reaction) parallel to polar line *bo* to cut force BC and so on, finishing with a line drawn parallel to polar line *go* to cut force GH.

The next force to GH is the vertical reaction HJ so a line parallel to polar line *ho* must be drawn from force GH to cut force HJ. But these two forces are in the same straight line, so, in effect, the line drawn parallel to polar line *ho* from force HG to force HJ is of zero length.

When the point *j* is fixed by drawing a line parallel to the closer of the link polygon, the line *hj* gives the reaction at the right support (34 kN approximately) and the line *ja* gives the direction and magnitude (38 kN approximately) of the reaction at the left support.

Many roof trusses, particularly those of small spans, are equally supported or hinged at both ends so that neither of the reactions is actually vertical. Even so, for the purposes of determining the reactions and the forces in the members of the truss, it is frequently assumed that one reaction is vertical. An alternative assumption with respect to the reactions is that, if the ends of the truss are similarly supported, the horizontal components of the two reactions are equal. This implies that each support resists equally the tendency to horizontal movement due to any inclined loads on the truss.

Example 4.9

A truss as shown in Fig. 4.15 is subjected to wind forces and dead loads. Determine the magnitudes and directions of the reactions at the supports assuming that the total horizontal component of the applied loads is shared equally by the two supports.

Solution

After combining by the parallelogram of forces, the loads are plotted, finishing with *de* on the force diagram. The perpendicular line *ax* gives the sum of the vertical components of all the loads on the truss, and the horizontal line *xe* gives the sum of the horizontal components of the loads. The total horizontal load on the truss is therefore given by *xe* and, since this is resisted equally by the two supports, the horizontal component of each reaction is given by half the length of line *xe*. Point *f* can now be fixed, *ef* representing the horizontal component EF of the reaction at the right support.

The next force in order is the vertical component FG, so from *f* a vertical line is drawn. The position of *g* on this line is not known but it is known that the vertical component FG plus the vertical component GH must equal the total vertical component of the applied loads. Point *h* must therefore be in a horizontal line from *a* and the force polygon is completed by the line *ha* which represents the horizontal component HA of the reaction at the left support.

The problem is now reduced to determining the vertical components of the two reactions – in other words, to fixing the position of point *g* on the vertical line *fh*.

Starting the link polygon at the left hinge, the first effective link line is the one drawn parallel to polar line *bo* across the B space to cut force BC. (The

Fig. 4.15 Example 4.9:
(a) free-body diagram with link
polygon; (b) force and polar
diagrams

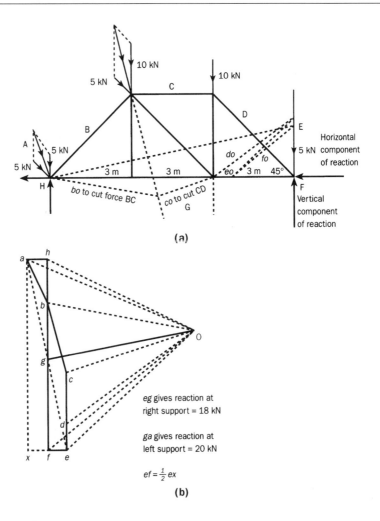

(a)

(b)

eg gives reaction at
right support = 18 kN

ga gives reaction at
left support = 20 kN

$ef = \frac{1}{2}ex$

first imaginary link line is *from* the hinge on the unknown vertical component
GH to cut the horizontal component HA *at* the hinge, and the second link
line is *from* the hinge on force HA to cut the force AB *at* the hinge.) Link
lines parallel to *co* and *do* are now drawn.

The next force in order is EF which is horizontal. Therefore the link line
parallel to *eo* is drawn across the E space to cut the horizontal force EF. The
last link line is drawn parallel to *fo* across the F space to cut the unknown vertical component FG.

Finally, the closer connecting the two unknown vertical components of
the reactions is drawn, and a line parallel to the closer is drawn on the force
diagram from O. Where this line cuts the vertical line *fh* is the position of *g*.
Then *fg* gives, to scale, the vertical component of the reaction at the right
support and *gh* gives the vertical component of the reaction at the left support. The actual magnitudes and directions of the reactions are given by lines
eg and *ga* respectively.

Exercises

1 Neglecting the weight of the pole, determine the reactions at A and B, Fig. 4.Q1.

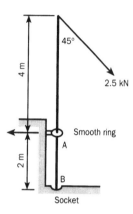

Fig. 4.Q1

2 Neglecting the weights of the members, determine the reactions at A and B of Fig. 4.Q2. (Point A is supported by a horizontal cable.) In addition, determine the forces in the three members of the framework, and the horizontal and vertical components of the reaction at B.

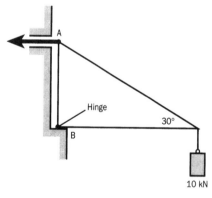

Fig. 4.Q2

3 A gate weighing 500 N is supported by two hinges so that the lower hinge takes the whole weight of the gate (Fig. 4.Q3). Find the reactions at the hinges, and the vertical and horizontal components of the reactions at the bottom hinge.

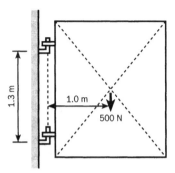

Fig. 4.Q3

4 Determine the tension in the tie-cable and the reaction at the hinge (Fig. 4.Q4).

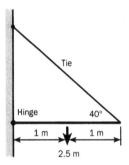

Fig. 4.Q4

5 Determine the tension in the cable and the reaction at the hinge (Fig. 4.Q5).

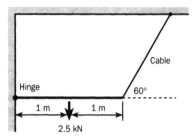

Fig. 4.Q5

6 A ladder rests against a smooth sloping wall, as shown in Fig. 4.Q6. The vertical load includes the weight of the ladder. Determine the reactions at A and B.

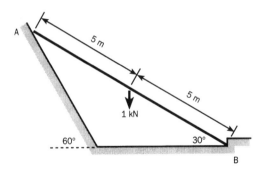

Fig. 4.Q6

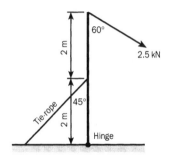

Fig. 4.Q9

7   The structure shown in Fig. 4.Q7 is supported by a smooth ring bolt at A and by a socket at B. Determine the reactions at A and B. Determine also the horizontal and vertical components of the reaction at B.

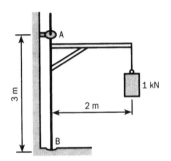

Fig. 4.Q7

8   Determine the tension in the string and the reaction at the hinge (Fig. 4.Q8). Neglect the weight of the members.

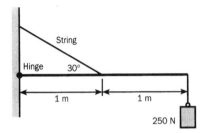

Fig. 4.Q8

9   Determine the tension in the tie-rope and the reaction at the hinge (Fig. 4.Q9).

10   A uniform beam weighing 500 N is supported horizontally as shown in Fig. 4.Q10. Determine the reactions at the smooth wall and hinge.

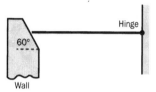

Fig. 4.Q10

11   The horizontal cable of Fig. 4.Q11 is halfway up the rod, the weight of which may be neglected. Determine the tension in the cable and the reaction at the hinge.

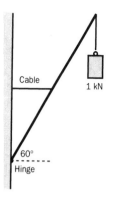

Fig. 4.Q11

12   A uniform rod is hinged halfway along its length as shown in Fig. 4.Q12. Determine the tension in the tie-rope and the reaction at the hinge, neglecting the weight of the rod.

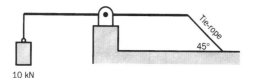

10 kN

Fig. 4.Q12

13 A uniform rod is hinged halfway along its length as shown in Fig. 4.Q13. Determine the tension in the rope and the reaction at the hinge, neglecting the weight of the rod.

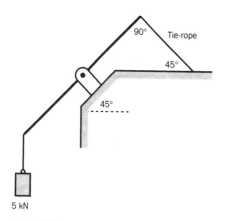

5 kN

Fig. 4.Q13

14 Determine the resultant of the two forces of 2.5 kN and 1.5 kN (Fig. 4.Q14), then determine the tension in the cable and the reaction at the bottom of the pole. Neglect the weight of the pole.

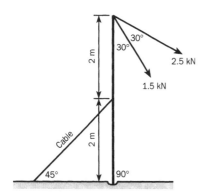

Fig. 4.Q14

15 A framework consisting of three rods hinged at the joints is supported by a smooth wall at B and is hinged to a wall at A (Fig. 4.Q15). Determine the reactions at A and B and the forces in the three members, neglecting their weight.

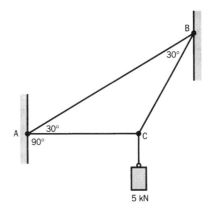

5 kN

Fig. 4.Q15

16 A bent rigid rod has two equal arms AC and BC as shown in Fig. 4.Q16. The wall at A is smooth. Determine the angle x so that the reaction at B is at right angles to the surface of the wall.

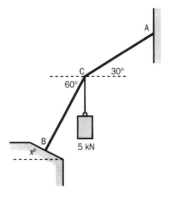

Fig. 4.Q16

17 Determine the tensions in the strings and the reaction at the hook (Fig. 4.Q17).

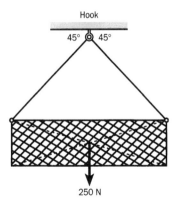

Fig. 4.Q17

18 A horizontal beam, the weight of which may be neglected, is loaded halfway along its length as shown in Fig. 4.Q18. Determine the reactions at A and B.

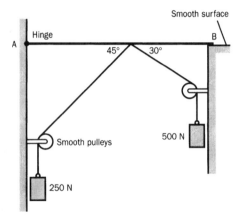

Fig. 4.Q18

19 Neglecting the weight of the beam, determine the reactions at A and B of Fig. 4.Q19.

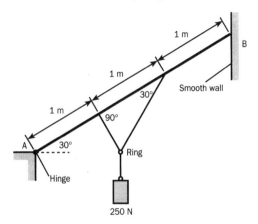

Fig. 4.Q19

20 Neglecting the weight of the rod, determine the reactions at A and B of Fig. 4.Q20.

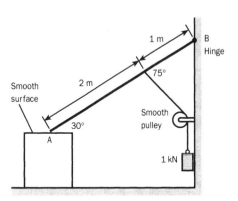

Fig. 4.Q20

21 A horizontal rod, the weight of which may be neglected, is hinged at X as shown in Fig. 4.Q21. Determine by the parallelogram of forces the resultant pull on the smooth pulley, then use the triangle of forces to determine the tension in the tie-rope and the reaction at the hinge.

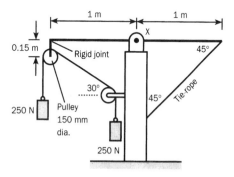

Fig. 4.Q21

22 A rod, the weight of which may be neglected, is hinged at A and supported by a tie-rope connected to B as in Fig. 4.Q22. Determine the resultant pull due to the two weights of 2.5 kN and 4 kN, then determine the forces in the tie and rod.

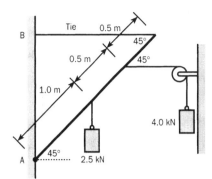

Fig. 4.Q22

23 Determine the reactions A and B to the rod shown in Fig. 4.Q23.

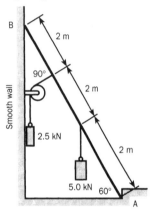

Fig. 4.Q23

24 Determine the position of the resultant for the systems of forces given in Fig. 4.Q24(a), (b) and (c).

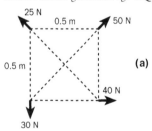

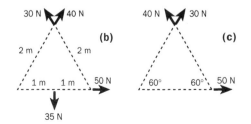

Fig. 4.Q24

25 A mast, fixed at its base, is loaded as shown in Fig. 4.Q25. Determine the magnitude and line of action of the resultant of the given loads.

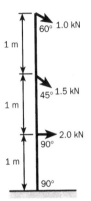

Fig. 4.Q25

26 Determine the magnitude and line of action of the resultant of the four parallel forces acting on the mast of Fig. 4.Q26.

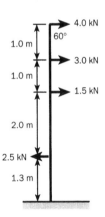

Fig. 4.Q26

27 Determine the magnitude of the vertical reaction at R and the magnitude and line of action of the reaction at L of the system of forces given in Fig. 4.Q27.

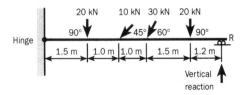

Fig. 4.Q27

28 A beam (Fig. 4.Q28) is hinged at L and supported by means of a cable at R. Determine the reactions at L and R.

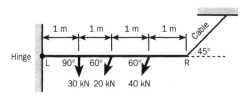

Fig. 4.Q28

29 A vertical mast is supported by means of a smooth ring bolt at X (which means that the reaction at X is horizontal) and by a socket at Y (Fig. 4.Q29). Determine the reactions at X and Y.

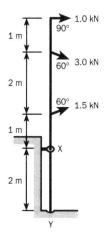

Fig. 4.Q29

30 A tower (Fig. 4.Q30) is supported by roller bearings at R and is hinged at L. Determine the reactions at L and R.

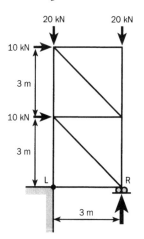

Fig. 4.Q30

31 Determine the reactions at L and R due to the three vertical loads of Fig. 4.Q31.

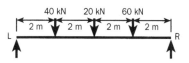

Fig. 4.Q31

32 Assuming the reaction at R of Fig. 4.Q32 to be vertical, determine its magnitude. Also determine the magnitude and direction of the reaction at L.

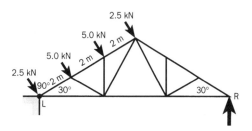

Fig. 4.Q32

33 Repeat Question 32 assuming that the total horizontal component of the applied loads is shared equally by the two supports.

34 The reaction at R to the truss shown in Fig. 4.Q34 is assumed to be vertical. Determine its value due to the wind loading given and also determine the value and line of action of the reaction at L.

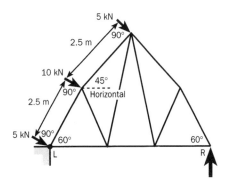

Fig. 4.Q34

35 Repeat Question 34 assuming that the total horizontal component of the applied loads is shared equally by the two supports.

36 Determine the reactions at L and R of Fig. 4.Q36 assuming the reaction at R to be vertical.

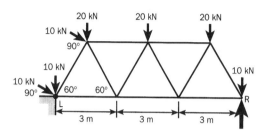

Fig. 4.Q36

37 Repeat Question 36 assuming that the total horizontal component of the applied loads is shared equally by the two supports.

38 The lengths of rafters represented by X and Y of Fig. 4.Q38 are equal and the wind loads are at right angles to the slope of the roof. Determine the reactions at L and R assuming the reactions at L to be vertical.

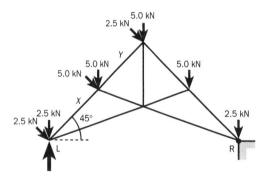

Fig. 4.Q38

39 Figure 3.Q39 shows a truss which is supported at L and R. Determine the reactions, assuming that the total horizontal component of the applied loads is shared equally by the two supports.

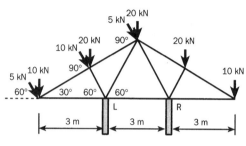

Fig. 4.Q39

Chapter five Moments of forces

The concept of the turning effect or moment of forces is explained in this chapter. The principle of the moment is then applied to the solution of problems connected with the equilibrium of a system of non-concurrent forces, including parallel forces (e.g. beam reactions, couples, etc.).

It is common knowledge that a small force can have a big turning effect or leverage. In mechanics, *moment* is commonly used instead of *turning effect* or *rotational effect*. The word moment is usually thought of as having some connection with time. Moment, however, derives from a Latin word meaning movement and moments of time refer, of course, to movement of time. Similarly, the **moment of a force** can be thought of as the movement of a force, although a more exact definition is 'turning effect of a force'. In many types of problem, there is an obvious **turning point** (or hinge or pivot or fulcrum) about which the body turns or tends to turn as a result of the effect of the force (see Fig. 5.1(a)–(e)). In other cases, where the principle of moments may be used to obtain solutions to problems, there may not be any obvious turning points.

Fig. 5.1 Different forms of moment: (a) anticlockwise; (b) clockwise; (c) and (d) anticlockwise; (e) clockwise at A, anticlockwise at B

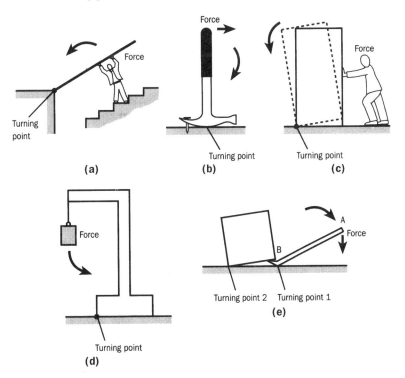

To solve problems by the use of the principle of moments, the effect of the forces must be considered in relation to the direction in which the body turns or tends to turn around the given point. In Fig. 5.1(a) the turning effect or moment of the force is anticlockwise; in (b) the moment is clockwise; and in (c) and (d) the moments are anticlockwise. There are two turning points in Fig. 5.1(e). Considering the point where the crowbar touches the ground, (point 1), the moment of force at A is clockwise. At B, the bar applies a force to the stone and causes it to turn about point 2. The moment of force applied to the stone at B is therefore anticlockwise.

If the line of action of a force passes through a point it can have no turning effect (moment) about that point. For example, a door cannot be opened by pushing at the hinge. Further examples are given in Fig. 5.2. The given forces have no moments about points O.

Fig. 5.2 These forces have no moment about point O

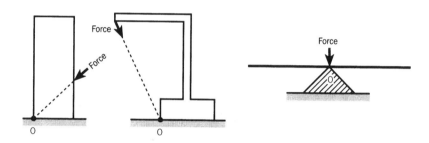

Measurement of moments

Figure 5.3(a) represents a uniform rod which balances on a fulcrum placed halfway along the length of the rod. If a force of 25 N is suspended at 0.3 m from the fulcrum it will be found that a force of 15 N is required at 0.5 m from the fulcrum on the other side in order to maintain equilibrium. Instead of 15 N at 0.5 m from the fulcrum we could place 12.5 N at 0.6 m or 7.5 N at 1.0 m. The results of the experiments would show that the force causing a clockwise moment multiplied by its distance from the turning point equals the force causing an anticlockwise moment multiplied by its distance from the turning point. For example, referring to Fig. 5.3(a), the clockwise moment of the 15 N force is 15 N × 0.5 m or 7.5 N m.

Moments are always expressed in force-length units such as N m, N mm, kN m, etc. A little consideration should show that this must be so, because the turning effect of a force depends on distance as well as on the magnitude of the force.

The distance from the force to the turning point is often called the **lever arm** of the force. In Fig. 5.3(a) the lever arm of the 15 N force is 0.5 m and the lever arm of the 25 N force is 0.3 m. It should be noted that the lever arm should be measured from the turning point to the point where it intersects the line of action of the force at right angles. In Fig. 5.3(a) the lines of action of the weights are vertical and the lever arms are horizontal. In Fig. 5.3(b) the lever arms of the forces are AO and BO respectively and for equilibrium $T \times AO$ must equal $P \times BO$. The reason for measuring the lever arms in this way will be made clear in Example 5.3 on page 73.

Further demonstrations of the principle of moments can be arranged quite easily with the aid of a fulcrum, a rod and a few weights. In Fig. 5.3(c)

Fig. 5.3 Measurement of the 'lever arm' of a force

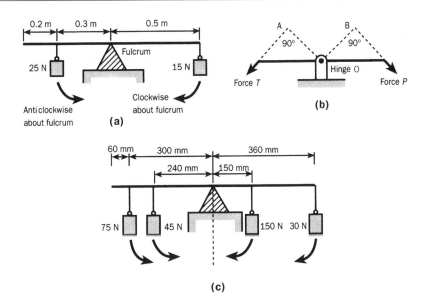

(a)

(b)

(c)

the rod itself, since it is supported at its centre of gravity, has no resultant turning effect, clockwise or anticlockwise:

anticlockwise moments = clockwise moments

$$(75 \times 300) + (45 \times 240) = (150 \times 150) + (30 \times 360)$$

$$33\,330 \text{ N mm} = 33\,300 \text{ N mm}$$

Example 5.1

A plank of uniform cross-section 4 m long and weighing 300 N has a support placed under it at 1.2 m from one end as shown in Fig. 5.4(a). Calculate the magnitude of the weight W required to cause the plank to balance.

Solution

It was stated on page 69 that, when considering the equilibrium of forces, the whole weight of a body can be assumed to act at its centre of gravity. In the case of a uniform rod or beam, this is halfway along the rod. Taking moments about the fulcrum or turning point,

$$1.2W \text{ N m} = 300 \times 0.8 \text{ N m}$$

$$W = \frac{240 \text{ N m}}{1.2 \text{ m}} = 200 \text{ N}$$

The whole weight of the plank is not, of course, concentrated at its centre of gravity. Each little particle of the plank is being attracted downwards by the earth and the 1.2 m length of the plank to the left of the fulcrum is exerting an anticlockwise moment, whilst the 2.8 m length of the plank to the right of the fulcrum is exerting a clockwise moment. The total net effect is, however, the same as if the whole of the weight were acting at 0.8 m to the right of the fulcrum.

For example, assume the plank is divided into 0.4 m lengths, each length weighing 30 N, and assume that each 30 N weight acts halfway along its 0.4 m length (Fig. 5.4(b)).

Fig. 5.4 Example 5.1: (a) moment due to a uniform plank; (b) proof that the whole weight of a body acts at its centre

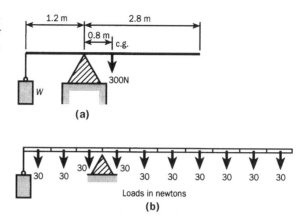

Taking moments about the fulcrum

$$30 \times (0.2 + 0.6 + 1.0) + 1.2W = 30 \times (0.2 + 0.6 + 1.0 + 1.4 \\ + 1.8 + 2.2 + 2.6)$$

$$(54 + 1.2W) \text{ N m} = 294 \text{ N m}$$

$$1.2W \text{ N m} = 240 \text{ N m}$$

$$W = 200 \text{ N}$$

If the plank is divided into a thousand lengths, each length weighing 0.3 N, and moments are taken about the fulcrum, the answer will be the same, so there need be no hesitation in assuming the whole weight of a body to act at its centre of gravity.

Example 5.2

A uniform rod weighing 50 N and carrying weights, as shown in Fig. 5.5(a), is hinged at A. The rod is supported in a horizontal position by a vertical string at B. Calculate the tension in the string and the reaction at the hinge.

Fig. 5.5 Example 5.2: (a) load diagram; (b) free-body diagram

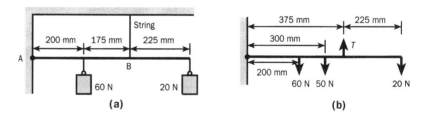

Solution

The free-body diagram is given in Fig. 5.5(b).

If the string were to be cut, all the weights would pull the rod downwards in a clockwise direction about the hinge. The string exerts a counterbalancing anticlockwise moment.

Taking moments about the hinge, assuming T newtons to be the tension in the string:

$$375T = (60 \times 200) + (50 \times 300) + (20 \times 600)$$

$$375T = 39\,000$$

$$T = 39\,000/375 = 104 \text{ N}$$

The total downward force is $60 + 50 + 20$, i.e. 130 N. Since the string is only holding up with a vertical force of 104 N, the reaction at the hinge must be $130 - 104$, i.e. 26 N acting upwards.

Example 5.3

A horizontal rod has a weight of 100 N suspended from it as shown in Fig. 5.6(a). Calculate the tension in the string and the reaction at the hinge, ignoring the weight of the rod.

Fig. 5.6 Example 5.3, Solution 1: (a) load diagram; (b) free-body diagram; (c) total reaction at hinge

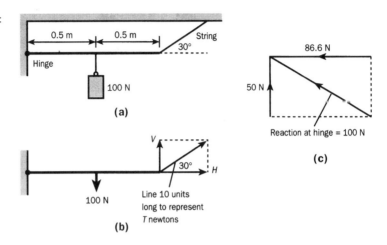

Solution 1

Let the tension in the string be represented by T newtons and resolve this force into its vertical and horizontal components. This may be done by calculation or by drawing a line 10 units long to represent T (Fig. 5.6(b)).

It is found that the vertical component V is 5 units long, i.e. $0.5T$, and that the horizontal component H is 8.66 units long, i.e. $0.866T$. The line of action of the horizontal component of the tension in the string passes through the hinge and therefore has no moment about this point. Taking moments about the hinge:

$$\text{anticlockwise moment} = \text{clockwise moment}$$

$$0.5\ T \times 1\ \text{m} = 100 \times 0.5\ \text{m}$$

$$0.5\ T = 50$$

$$T = 50/0.5 = 100\ \text{N} = \text{tension in string}$$

Reaction at hinge: the total downward force is 100 N. The vertical component of the tension in the string is $0.5T$, i.e. 50 N. (The horizontal component of the tension in the string is ineffective in holding up the suspended weight.) Since there is 100 N pulling down, there must be 100 N pushing up, therefore the hinge must supply a vertical reaction in the upward direction of 50 N. But the string has also a horizontal component of $0.866T$, i.e. 86.6 N, tending to pull the rod away horizontally from the hinge. The hinge must resist this pull. Therefore there must be a horizontal force acting at the hinge in addition to a vertical force.

The total reaction at the hinge is given in Fig. 5.6(c) and may be found from a scale drawing, or, if R is the reaction at the hinge,

$$R^2 = 50^2 + 86.6^2 = 10\,000$$

$$R = \sqrt{10\,000} = 100\ \text{N}$$

Solution 2

In Fig. 5.7, which may be drawn to scale, produce the line of action of the string and draw a line AO at right angles, then taking moments about O,

$$T \times OA \times 100 \times 0.5$$

$$T \times 0.5 = 50$$

$$T = 100 \text{ N}$$

This method is not as convenient as Solution 1 for determining the reaction at the hinge.

Fig. 5.7 Example 5.3, Solution 2

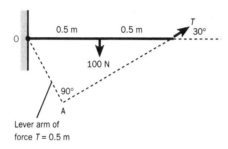

Solution

Since there are only three forces in this problem, the triangle of forces law can be applied (see page 37), and this method may be quicker for problems in which the forces are inclined.

Example 5.4

A ladder rests against a smooth vertical wall and a man weighing 750 N stands on it as indicated in Fig. 5.8. Neglecting the weight of the ladder, determine the reactions at the wall and at the ground.

Solution

This problem is identical with that on page 52 but it is now being solved by using the principle of moments. Since the wall is smooth, the reaction at A is horizontal. There may not appear to be any obvious turning point but B can be considered to be a turning point. (If the wall were to collapse, the ladder would swing round B in an anticlockwise direction due to the force of 750 N. If one could imagine the force at A to be increased above that necessary to maintain equilibrium, it would turn the ladder round B in a clockwise direction.) By taking moments about B, the reaction at B can be ignored since it passes through this point and therefore has no moment about it.

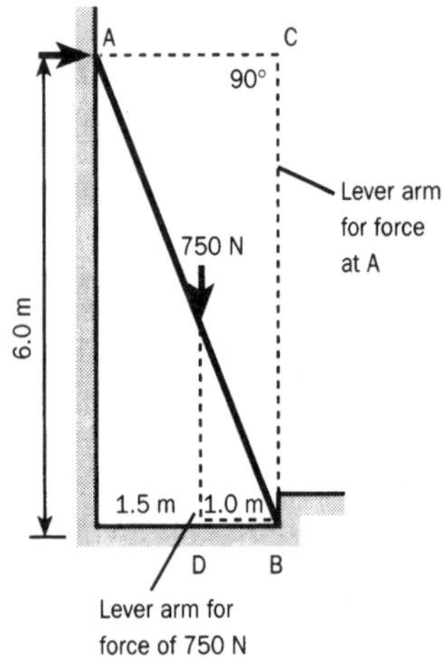

reaction at A × distance CB = 750 N × distance DB

$$R_A \times 6 = 750 \times 1$$

$$R_A = 750/6 = 125 \text{ N}$$

Reaction at B (R_B): since the wall is smooth, it cannot help at all in holding *up* the ladder, therefore the whole weight of 750 N is taken at B. There is therefore an upward reaction at B of 750 N. There must also be a horizontal reaction to prevent the foot of the ladder from moving outwards. This horizontal reaction equals the force at A, i.e. 125 N, therefore the total reaction at B, which can be obtained graphically or by calculation, is 760 N.

Fig. 5.8 Example 5.4

Conditions of equilibrium

The above example is an illustration of the application of the three **laws of equilibrium** which apply to a system of forces acting in one plane. These laws are:

1 The algebraic sum of the vertical forces must equal zero, i.e. if upward forces are called positive and downward forces are called negative, then a force of $+750$ N must be balanced by a force of -750 N and the algebraic sum of the two forces is equal to zero. This can also be expressed by

$$\Sigma V = 0$$

where Σ is the Greek letter 'sigma' (letter S) and in this connection means 'the sum of', whilst V represents 'the vertical forces', upward forces being plus and downward forces being minus.

2 The algebraic sum of the horizontal forces must equal zero, i.e. the sum of the horizontal forces acting towards the left (plus forces) must equal the sum of the horizontal forces acting towards the right (minus forces) or

$$\Sigma H = 0$$

3 The algebraic sum of the moments of the forces must equal zero, i.e. if clockwise moments are called plus and anticlockwise moments are called minus then

$$\Sigma M = 0$$

Example 5.5

A vertical pole (Fig. 5.9(a)) is supported by being hinged in a socket at B and by a smooth ring which can give no vertical reaction at A. Calculate the reactions at A and B neglecting the weight of the pole.

Solution

First, resolve the pull of 5 kN into its vertical and horizontal components as shown in Fig. 5.9(b) and take moments about B.

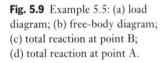

Fig. 5.9 Example 5.5: (a) load diagram; (b) free-body diagram; (c) total reaction at point B; (d) total reaction at point A.

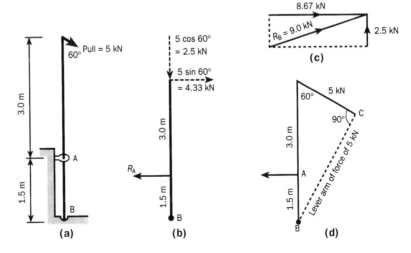

$\sum M = 0$ or clockwise moments equal anticlockwise moments:

$$4.33 \times 4.5 = R_A \times 1.5$$

$$R_A = 13.0 \, \text{kN}$$

(Note that the vertical component of 2.5 kN has no moment about B since its line of action passes through B.)

$$\sum V = 0$$

The vertical component of the reaction at B must equal the vertical component of the pull at the top of the pole since the support of A can only *hold back* and cannot *hold up*. The vertical component of the reaction at B therefore equals 2.5 kN and acts upwards.

$$\sum H = 0$$

The horizontal force at the top of the mast is 4.33 kN acting $\rightarrow$ and the reaction at A is acting $\leftarrow$. There must be, therefore, a horizontal force at B acting $\rightarrow$ equal to $13.0 - 4.33$, i.e. 8.67 kN. The complete reaction at B is given by Fig. 4.9(c). Note that, if the pole itself weighs, say, 0.5 kN, this weight would be taken entirely at B (since the pole is vertical) and the vertical component of the reaction at B would be 3 kN instead of 2.5 kN as shown in Fig. 5.9(c). The complete reaction at B would then be

$$R_B = \sqrt{(8.67^2 + 3^2)} = 9.2 \, \text{kN}$$

Another method of finding the reaction at A (neglecting the weight of the pole) is shown in Fig. 5.9(d).

$$5 \, \text{kN} \times \text{distance BC} = R_A \times 1.5 \, \text{m}$$

Note also that the triangle of forces principle could be applied to determine both reactions.

Example 5.6

A cantilever truss (Fig. 5.10(a)) is supported at A and B in such a manner that the reaction at A is horizontal. Neglecting the weight of the truss, calculate the reactions at A and B.

Fig. 5.10 Example 5.6: (a) load diagram; (b) total reaction at point B

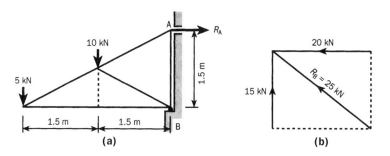

Solution

Take moments about B to determine the reaction at A:

$$\sum M = 0$$

$$1.5 \, R_A = 10 \times 1.5 + 5 \times 3$$

$$R_A = 20 \text{ kN acting towards the right}$$

$\sum H = 0$, therefore horizontal component of reaction at B is 20 kN, acting towards the left.

$\sum V = 0$, therefore vertical component of reaction at B is 15 kN acting upwards.

Referring to Fig. 5.10(b)

$$R_B = \sqrt{(20^2 + 15^2)} = \sqrt{625} = 25 \text{ kN}$$

Resultant of parallel forces

It was stated in Chapter 3 in connection with concurrent forces that the resultant of a given number of forces is the single force which, when substituted for the given forces, has the same effect on the *state of equilibrium* of the body. This definition applies also to parallel forces.

Consider Fig. 5.11(a) which represents a beam of negligible weight supporting three vertical loads of 13, 18 and 27 kN respectively.

The force in the vertical cable can be obtained by taking moments about the hinge, i.e.

$$F \times 3 \text{ m} = 13 \times 1.4 + 18 \times 2.4 + 27 \times 3.3$$

$$= 150.5 \text{ kN m}$$

$$F = 150.5/3 = 50.2 \text{ kN}$$

Now, if one force (the resultant) is substituted for the three downward loads without altering the state of equilibrium, its magnitude must be 13 + 18 + 27 = 58 kN, and if the reaction in the cable is to remain as before, i.e. 50 kN, the moment of the single resultant force about the hinge must equal the sum of the moments of the forces taken separately.

Referring to Fig. 5.11(b), the moment of the resultant force about the hinge is 58x kN m, and this must equal 150.5 kN m, therefore

$$58x = 150.5$$

$$x = 150.5/58 = 2.59$$

The resultant of the three forces is, therefore, a single force of 58 kN acting at 2.59 m from the hinge. The reactions in the cable and at the hinge will be the same for this one force as for the three forces.

Fig. 5.11 Resultant of parallel forces: (a) load diagram; (b) free-body diagram.

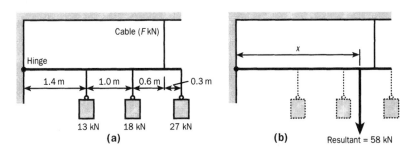

Fig. 5.12 Example 5.7:
(a) load diagram; (b) free-
body diagram

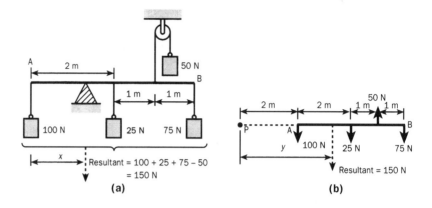

(a)

(b)

Example 5.7

Calculate the magnitude and position of the resultant of the system of forces shown in Fig. 5.12(a) and determine where a fulcrum must be placed to maintain equilibrium. (Neglect the weight of the rod.)

Solution

This is an example of a system of *unlike parallel forces* since all the forces are not acting in the same sense. The moment of this resultant about *any point* must equal the algebraic sum of the moments of the separate forces. Taking moments about end A,

$$150x = 25 \times 2 - 50 \times 3 + 75 \times 4$$
$$= 50 - 150 + 300$$
$$x = 200/150 = 1.33 \text{ m}$$

The fulcrum must therefore be placed at this point in order to maintain the rod in balance (and the reaction at the fulcrum = 150 N).

The same answer would be obtained by taking moments about any point, as for example, the point P, 2 m from A (see Fig. 5.12(b)).

$$150y = 100 \times 2 + 25 \times 4 - 50 \times 5 + 75 \times 6$$
$$= 200 + 100 - 250 + 450$$
$$y = 500/150 = 3.33 \text{ m}$$

The position of the resultant is therefore 3.33 m from P or 1.33 m from A. Problems can often be simplified by working with a resultant instead of with a number of separate forces. This is demonstrated in the following example.

Example 5.8

Calculate the resultant of the three downward forces of Fig. 5.13(a), then use the triangle of forces principle to determine the reaction in the cable and the reaction at the hinge.

Solution

The resultant of the three downward forces is 6 kN and, taking moments about the hinge,

$$6x = 3 \times 0.6 + 1 \times 1.2 + 2 \times 2.5$$
$$x = 8/6 = 1.33 \text{ m}$$

Now, using the method of Chapter 4, page 52, the reactions can be found as shown in Fig. 5.13(b) and (c).

Fig. 5.13 Example 5.8:
(a) load diagram;
(b) free-body diagram;
(c) force diagram at point O

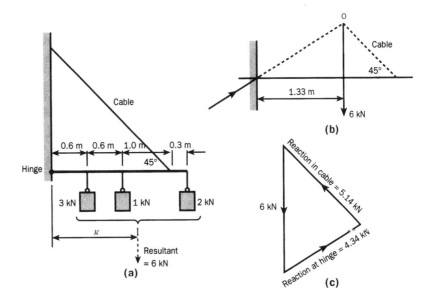

<table>
</table>

Couples

Two equal, unlike (i.e. acting in opposite directions) parallel forces are said to form a **couple**.

Imagine a rod resting on a smooth table, as shown in Fig. 5.14(a), and assume that two equal unlike parallel forces of 20 N are applied horizontally as shown. The rod will rotate in a clockwise direction, and the moment causing this rotation is obtained by multiplying *one* of the forces by the distance AB between the two forces. The perpendicular distance AB is called the arm of the couple or in general terms, the **lever arm**, l_a. Therefore

$$\text{moment of couple} = 20\,\text{N} \times 150\,\text{mm}$$

$$= 3000\,\text{N mm}$$

No matter about what point the moment is considered, the answer is equal to $W \times \text{AB}$. For example, taking moments about any point O between the two forces (Fig. 5.14(b)), the moment causing rotation is

$$(W \times \text{AO}) + (W \times \text{BO})$$

Fig. 5.14 Couples: (a) load diagrams; (b) and (c) considering moments about different points

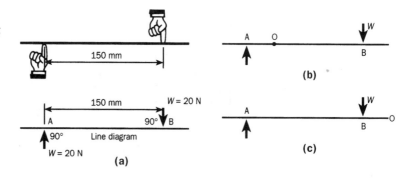

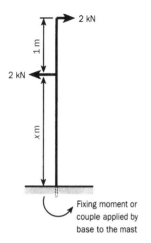

Fig. 5.15 Fixing moment

This equals

$$W(AO + BO) = W \times AB$$

In Fig. 5.14(c) the clockwise moment about O is $W \times AO$, and the anti-clockwise moment about O is $W \times BO$.

The net moment causing rotation is

$$W \times AO - W \times BO = W(AO - BO)$$

$$= W \times AB$$

Note that a couple acting on a body produces rotation and the forces cannot be balanced by a single force. To produce equilibrium another couple of equal and opposite moment is required.

Figure 5.15 represents a mast fixed at its base. The moment tending to break the mast at its base is

$$[2(x + 1) - 2x] \text{ kN m} = 2x + 2 - 2x = 2 \text{ kN m}$$

No matter what the distance x may be, the moment at the base will be 2 kN m. This moment or couple is resisted by the fixing moment or couple applied by the base to the mast.

A knowledge of the principles of couples is useful in understanding beam problems (see Chapter 11).

Beam reactions

The application of the principle of moments of forces to the calculation of beam reactions can be verified by means of the following experiment.

A beam of wood, a little longer than 1.2 m and weighing 15 N, has weights suspended from it as shown in Fig. 5.16(a). It is required to determine the reactions at supports A and B.

Fig. 5.16 Beam reactions:
(a) balance attached at point B;
(b) balance attached at point A

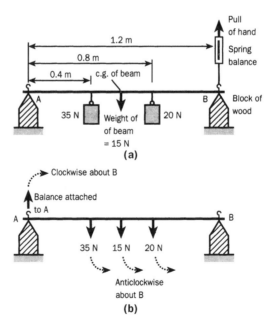

If a spring balance is attached to the beam at B, the beam can be lifted just clear of the support and it will be observed that the reading on the balance is 32.5 N. When the balance is released so that the beam rests again on block B, it follows that the beam must be pushing down on the support with a force of 32.5 N. Similarly, the upward reaction of the support on the beam is 32.5 N.

If the hand holding the balance is moved up and down, the beam will turn round the point where it is supported at A. Taking moments about this point, representing the reaction at B by R_B,

$$R_B \times 1.2 \text{ m (anticlockwise about A)}$$
$$= 35 \times 0.4 + 15 \times 0.6 + 20 \times 0.8$$

therefore

$$1.2R_B = 39 \text{ and } R_B = \frac{39}{1.2} = 32.5 \text{ N}$$

If the balance is transferred to A and the beam lifted off its support, the balance will read 37.5 N (Fig. 5.16(b)).

By calculation, taking B as the turning point,

$$R_A \times 1.2 = 20 \times 0.4 + 15 \times 0.6 + 35 \times 0.8$$

clockwise = anticlockwise

$$R_A = \frac{45}{1.2} = 37.5 \text{ N}$$

Note that the sum of the reactions is 70 N and that this is also the sum of the downward forces ($\sum V = 0$).

Note also that to determine the reaction at A, moments are taken about B, i.e. B is considered as the turning point and the moment of each force or load is the force multiplied by its distance from B.

Similarly, to determine the reaction at B, moments are taken about A, i.e. A is considered as the turning point.

The two weights of 35 N and 20 N in the discussion above are **point loads** or **concentrated loads**, since they act at definite points on the beam. The beam itself is a **uniformly distributed load (UDL)** since its weight is spread uniformly over its whole length. Figure 5.17(a) represents the test beam supporting a block of lead 0.4 m long of uniform cross-section and weighing 30 N. For calculating the reactions at the supports, the whole of this uniformly distributed weight can be assumed to act at its centre of gravity.

Taking moments about A:

$$1.2R_B = 30 \times 0.4 + 15 \times 0.6$$

$$R_A = \frac{21}{1.2} = 17.5 \text{ N}$$

Similarly, taking moments about B: 15

$$1.2R_A = \cancel{15} \times 0.6 + 30 \times 0.8$$

$$R_A = \frac{33}{1.2} = 27.5 \text{ N}$$

$$R_A + R_B = 17.5 + 27.5 = 45 \text{ N} = \text{total load}$$

Fig. 5.17 Uniformly distributed loads: (a) load diagram; (b) and (c) representations of a UDL

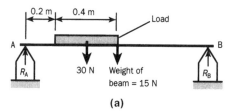

(a)

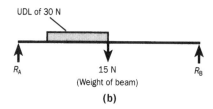

(b)

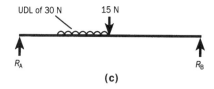

(c)

These calculated values of the reactions can be verified by means of the spring balances.

Instead of drawing the beam every time, a free-body diagram can be drawn as in Fig. 5.17(b) and (c) where two methods of indicating uniformly distributed loads are shown.

Further experiments can be performed with the simple apparatus shown in Fig. 5.17 by placing the supports at different points. Figure 5.18 shows the supports placed 0.8 m apart. In this case, to determine experimentally the reaction at A, the spring balance must be connected to a point on the beam directly above A.

Fig. 5.18 Combination of point and uniformly distributed loads

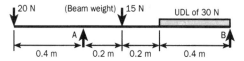

Taking moments about B, the reaction at A produces a clockwise moment and all the loads anticlockwise moments.

$$0.8R_A = 30 \times 0.2 + 15 \times 0.6 + 20 \times 1.2$$

$$R_A = 39/0.8 = 48.75 \text{ N}$$

Taking moments about A, both the reactions at B and the load of 20 N produce anticlockwise moments:

$$0.8R_B + 20 \times 0.4 = 15 \times 0.2 + 30 \times 0.6$$

$$0.8R_B = 3.0 + 18.0 - 8.0$$

$$R_B = \frac{13}{0.8} = 16.25 \text{ N}$$

$$R_A + R_B = 48.75 + 16.25 = 65 \text{ N}$$

$$= \text{sum of the vertical loads}$$

Practical cases

In buildings, point loads on beams are usually due to loads from other beams or from columns, and uniformly distributed loads are due to floors, walls and partitions and the weights of the beams themselves.

In Fig. 5.19(a), assume that the beam AB weighs 10 kN and that it supports two beams which transmit loads of 60 kN and 90 kN respectively to beam AB. In addition, a column transmits a load of 240 kN. The reactions are usually taken as acting at the middle of the bearing area supplied by the supporting brickwork or column. Figure 5.19(b) shows the free-body diagram.

Taking moments about B,

$$6R_A = 90 \times 2.0 + 10 \times 3.0 + 240 \times 3.5 + 60 \times 4.0$$

$$= 180 + 30 + 840 + 240$$

$$R_A = \frac{1290}{6} = 215 \text{ kN}$$

Since the total load on the beam is 400 kN, the reaction at B is 400−215, i.e. 185 kN. It is advisable, however, to determine the reaction

Fig. 5.19 Practical example: (a) loading of beam; (b) free-body diagram

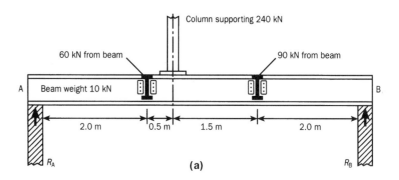

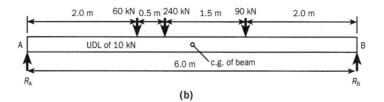

at B by taking moments about A and thereby to obtain a check on the calculations.

A beam weighing 9.6 kN supports brickwork 225 mm thick weighing $20\,\text{kN/m}^3$ as indicated in Fig. 5.20(a). Calculate the reactions at the supports.

Both the brickwork and the beam are uniformly distributed loads.

$$\text{volume of brickwork} = 5.0 \times 2.5 \times 0.225 = 2.8125\,\text{m}^3$$

$$\text{weight of brickwork} = 2.8125 \times 20 = 56.25\,\text{kN}$$

$$\text{weight of beam} = 9.60\,\text{kN}$$

$$\text{total weight} = 65.85\,\text{kN}$$

The free-body diagram is shown in Fig. 5.20(b).

Fig. 5.20 Example 5.9: (a) load diagram; (b) free-body diagram

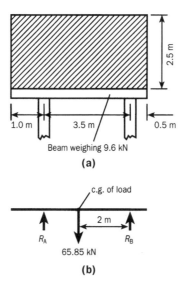

Taking moments about B,

$$3.5R_{\text{A}} = 65.85 \times 2$$

$$R_{\text{A}} = \frac{131.70}{3.5} = 37.63\,\text{kN}$$

$$R_{\text{B}} = 65.85 - 37.63 = 28.22\,\text{kN}$$

or R_{B} can be obtained by taking moments about A, thus providing a useful check.

Summary

Moment   The turning effect of a force.

- Moments are considered in relation to a turning point, real or imaginary.
- A moment is always obtained by multiplying a force by a distance and this distance must be measured from the turning point to where it cuts the line of action of the force at right angles.
- Moments are expressed in force-length units such as newton metre (N m), newton millimetre (N mm), etc.
- If the line of action of a force passes through a given point, it can have no moment about that point.

Conditions of equilibrium   For a system of forces acting in one plane, conditions of equilibrium are:

$$\sum V = 0 \quad \sum H = 0 \quad \sum M = 0$$

Resultant   For a number of parallel forces the resultant is their algebraic sum. Its position can be found by taking moments about *any point* and moment of resultant equals the algebraic sum of moments of the individual forces.

Couple   Two equal unlike parallel forces. The distance between the forces is called the arm of the couple, and the moment of a couple equals one of the forces multiplied by the arm of the couple.

Exercises

1 A uniform rod is in equilibrium under the action of weights as shown in Fig. 5.Q1. Calculate the value of W and the reaction at the fulcrum, ignoring the weight of the rod.

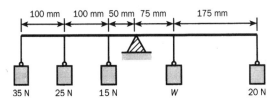

Fig. 5.Q1

2 Calculate the value of x metres so that the uniform rod of Fig. 5.Q2 will balance.

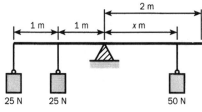

Fig. 5.Q2

3 A uniform rod weighing 100 N and supporting 500 N is hinged to a wall and kept horizontal by a vertical rope (Fig. 5.Q3). Calculate the tension in the rope and the reaction at the hinge.

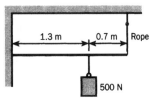

Fig. 5.Q3

4 A uniform rod weighing 40 N is maintained in equilibrium as shown in Fig. 5.Q4. Calculate the distance x and the reaction at the fulcrum.

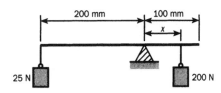

Fig. 5.Q4

5 A uniform rod 0.4 m long weighing 10 N supports loads as shown in Fig. 5.Q5. Calculate the distance x if the rod is in equilibrium.

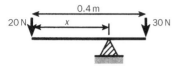

Fig. 5.Q5

6 A uniform horizontal beam 1.3 m long and weighing 200 N is hinged to a wall and supported by a vertical prop as shown in Fig. 5.Q6. Calculate the reactions at the prop and hinge.

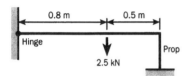

Fig. 5.Q6

7 Calculate the reaction at the prop and hinge in Fig. 5.Q7. The uniform beam weighs 200 N.

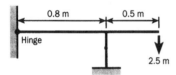

Fig. 5.Q7

8 A compound lever is shown in Fig. 5.Q8. Each of the horizontal rods weighs 50 N and the vertical rod weighs 40 N. Calculate the value of the force W which is required for equilibrium, and calculate the reactions at the fulcrums. (First take moments about A to find the force in the vertical rod, then take moments about B.)

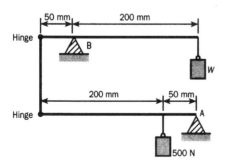

Fig. 5.Q8

9 Solve Exercises 1 to 13 of Chapter 4, with the aid of the principle of moments.

10 Determine the tension in the chain and the reaction at the hinge (Fig. 5.Q10). The uniform rod weighs 500 N.

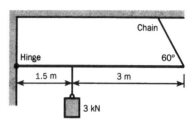

Fig. 5.Q10

11 The horizontal cable in Fig. 5.Q11 is attached to the rod halfway along its length. The uniform rod weighs 250 N. Calculate the tension in the cable and the reaction at the hinge.

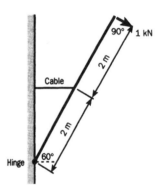

Fig. 5.Q11

12 Calculate the tension in the guy rope of Fig. 5.Q12, then use the triangle of forces law to determine the forces in the mast, tie and jib. Neglect the weight of the members.

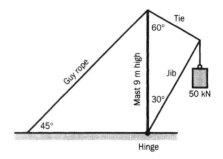

Fig. 5.Q12

13 A uniform rod weighing 50 N is loaded as shown in Fig. 5.Q13. Calculate the tension in the cable and the reaction at the hinge.

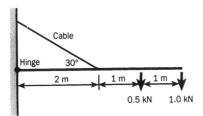

Fig. 5.Q13

14 Two uniform rods each 3 m long and weighing 250 N are connected by a link bar (Fig. 5.Q14). Calculate the tensions in the link bar and chain neglecting their weights.

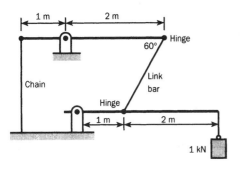

Fig. 5.Q14

15 For each of the structures shown in Fig. 5.Q15(a), (b) and (c) determine the position of the resultant of the vertical loads; then, by using the triangle of forces, determine the tension in the cable and the reaction at the hinge. Neglect the weight of the rod.

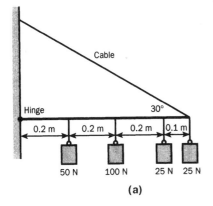

(a)

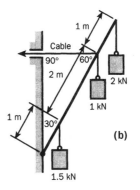

(b)

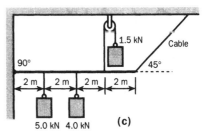

(c)

Fig. 5.Q15

16 The weights of the structures shown in Fig. 5.Q16(a), (b), (c) and (d) may be neglected. In each case the reaction at A is horizontal, and B is a hinge. Determine the position of the resultant of the vertical loads; then, by applying the principle of the triangle of forces, determine the reactions at A and B.

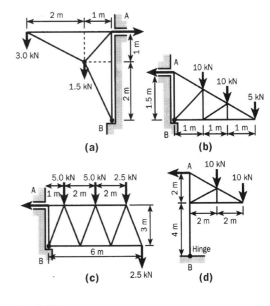

(a)

(b)

(c)

(d)

Fig. 5.Q16

17 The roof truss of Fig. 5.Q17 is supported on rollers at B and the reaction at B is therefore vertical. Determine the resultant of the four wind loads, then use the triangle of forces to determine the reactions at A and B.

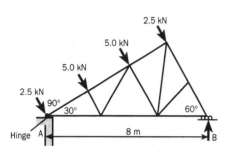

Fig. 5.Q17

18 Reduce the system of forces shown in Fig. 5.Q18 to a couple. Determine the arm and moment of the couple.

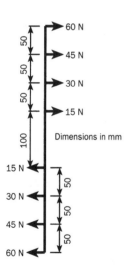

Fig. 5.Q18

19 Calculate the reactions at A and B due to the given point loads in Fig. 5.Q19(a) to (e). The loads are in kilonewtons (kN).

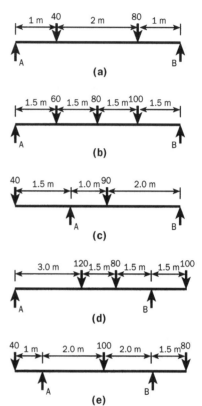

Fig. 5.Q19

20 Calculate the reactions at A and B due to the given system of loads in Fig. 5.Q20(a) to (g).

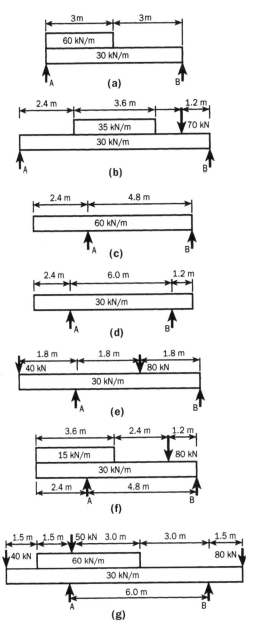

Fig. 5.Q20

21 A beam AB carries two point loads and is supported by two beams CD and EF as shown in plan in Fig. 5.Q21. Neglecting the weights of the beams, calculate the reactions at C, D, E and F.

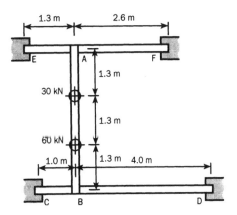

Fig. 5.Q21

22 A reinforced concrete bridge between two buildings spans 8 m. The cross-section of the bridge is shown in Fig. 5.Q22. The brickwork weighs 20 kN/m³, the reinforced concrete weighs 23 kN/m³, and the stone coping weighs 0.5 kN/m. The floor of the bridge has to carry a uniformly distributed load of 1.5 kN/m² in addition to its own weight. Calculate the force on the supporting buildings from one end of each beam.

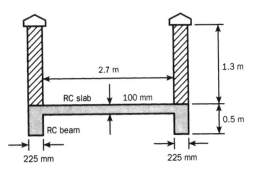

Fig. 5.Q22

Chapter six Framed structures

The framed structures dealt with in this chapter are pin-jointed, determinate, and plane and three-dimensional (or space) frameworks.

The graphical solutions of plane frames by means of force diagrams are explained in some detail. The three analytical or calculation methods presented here are the method of sections, resolution at joints, and tension coefficients. Whilst the first two methods are applied to plane frames only, the usefulness of the application of the third one to space frames is also demonstrated.

A **frame** is a structure built up of three or more members which are normally considered as being pinned or hinged at the various joints. Any loads which are applied to the frame are usually transmitted to it at the joints, so that the individual members are in pure tension or compression.

A very simple frame is shown in Fig. 6.1. It consists of three individual members hinged at the ends to form a triangle, and the only applied loading consists of a vertical load of W at the apex. There are also, of course, reactions at the lower corners.

Under the action of the loads the frame tends to take the form shown in broken lines, i.e. the bottom joints move outward putting the member C in tension, and the members A and B in compression. Members A and B are termed **struts** and member C is termed a **tie**.

Fig. 6.1 A simple frame

Perfect, imperfect and redundant pin-jointed frames

A perfect frame is one which has just sufficient members to prevent the frame from being unstable. An imperfect frame is one which contains too few members to prevent collapse. A redundant frame is one which contains more than the number of members which would constitute a perfect frame.

Figure 6.2(a) shows examples of perfect frames. Figure 6.2(b) shows two examples of imperfect frames. In each of the two cases in Fig. 6.2(b) the frames would collapse as suggested by the broken lines; the addition of one member in each case would restore stability and produce perfect frames as in Fig. 6.2(a).

Figure 6.2(c), for example, shows a perfect frame, and Fig. 6.2(d) (the same frame with the addition of member AB) a redundant frame.

Note: A redundant member is not necessarily a member having no load. The member AB (Fig. 6.2(d)), for example, will be stressed, and will serve to create a stronger frame than that in Fig. 6.2(c), but the **redundant frame** cannot be solved by the ordinary methods of statics. It is, therefore, also called a **statically indeterminate** or hyperstatic frame.

Fig. 6.2 Perfect, imperfect and redundant frames: (a) and (c) perfect; (b) imperfect; (d) redundant

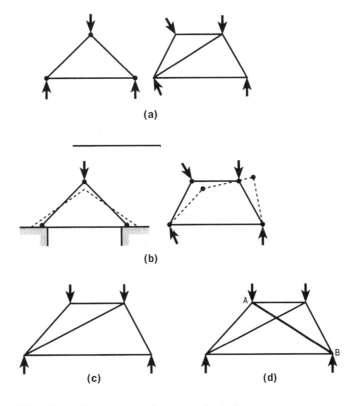

Number of members in a perfect frame

The simplest perfect frame consists of three members in the form of a triangle as in Fig. 6.3(a). The frame also has three joints or node points, A, B and C.

If *two* more members are added to form another triangle as in Fig. 6.3(b), then *one* more joint has been added.

So long as triangles only are added, then the frame will remain a perfect frame.

If J = number of joints, and m = number of members, then

$$2(J - 3) = m - 3$$

$$2J - 6 = m - 3$$

$$2J = m - 3 + 6$$

$$2J = m + 3 \quad \text{or} \quad m = 2J - 3$$

Fig. 6.3 Number of members in a perfect frame: (a) simplest form – 3 members and 3 joints; (b) adding a triangle to a perfect frame

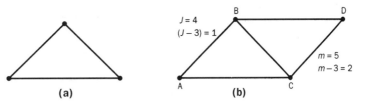

Fig. 6.4 Examples of perfect frames showing the relationship between numbers of members and joints

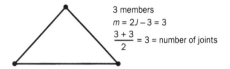

3 members
$m = 2J - 3 = 3$
$\dfrac{3 + 3}{2} = 3$ = number of joints

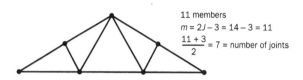

11 members
$m = 2J - 3 = 14 - 3 = 11$
$\dfrac{11 + 3}{2} = 7$ = number of joints

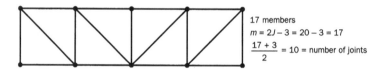

17 members
$m = 2J - 3 = 20 - 3 = 17$
$\dfrac{17 + 3}{2} = 10$ = number of joints

Thus we have the following:

$$\text{Number of members to form a perfect frame}$$
$$= \text{(twice the number of joints)} - 3$$

or

$$\text{Number of joints} = \frac{\text{number of members} + 3}{2}$$

The frames shown in Fig. 6.4 all comply with this requirement.

The frames shown in Fig. 6.5 do not comply with this requirement, i.e. there are more than $2J - 3$ members, and the frames contain one or more redundant members.

In Fig. 6.5(a) the number of members is 8, which is more than $(2 \times 5) - 3$. Similarly, in Fig. 6.5(b) the number of members is 6, which is greater than $(2 \times 4) - 3$.

Fig. 6.5 These frames contain one or more redundant members

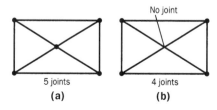

No joint

5 joints
(a)

4 joints
(b)

Fig. 6.6 Frame force diagrams: (a) load diagram for a simple frame; (b) reading clockwise the forces are AB, B1 and 1A; (c) force diagram (note that the arrows follow each other round the triangle)

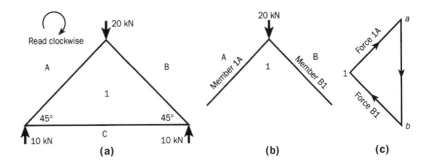

(a)   (b)   (c)

Graphical solutions for frames – force diagrams

Consider the simple perfect frame shown in Fig. 6.6(a). The known forces are the load at the apex and the two equal reactions of 10 kN.

In describing the loads, it is convenient to use Bow's notation (see Chapter 3). Reading clockwise, the 20 kN load at the apex is 'load AB', the reaction at the right support is 'load BC', and the reaction at the left support is 'load CA'. Similarly, the spaces *inside* the frame are usually denoted by numbers so that the force in the horizontal member is 'force C1' or 'force 1C'.

As explained in Chapter 3, the forces in the two inclined members can be obtained by considering the equilibrium of the joint at the apex (Fig. 6.6(b) and (c)).

The force in the horizontal tie member can be obtained by considering either the joint at the left reaction (Fig. 6.7) or the joint at the right reaction (Fig. 6.8).

The three triangles of forces drawn separately for the three joints may, for convenience, be superimposed upon each other to form in one diagram the means of determining all unknown forces. This resultant diagram, which is called a **force diagram**, is shown in Fig. 6.9(a).

Note that no arrows are shown on the combined force diagram. The directions of the arrows in the frame diagram are obtained by considering each joint as in Figs. 6.6–6.8. One must consider each joint to be the centre of a clock, and the letters are read clockwise round this centre. Thus, having drawn the combined diagram, consider, say, the joint at the left reaction (Fig. 6.7). Reading clockwise round this joint, the inclined member is A1. On the force diagram, the direction from *a* to 1 is downwards to the left, therefore the arrow is placed in the frame diagram as shown, near the joint. The tie is member 1C and 1 to *c* on the force diagram is a direction left to right, so the arrow is placed in this direction on the frame (or free-body or space) diagram.

Fig. 6.7 Reaction at left–hand joint of frame in Fig. 6.6: (a) load diagram; (b) force diagram

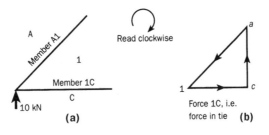

(a)   (b)

Fig. 6.8 Reaction at right-hand joint of frame in Fig. 6.6: (a) load diagram; (b) force diagram

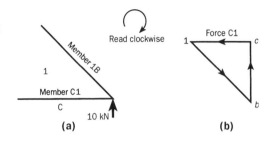

Considering the joint at the apex (Fig. 6.6(b) and (c)), the left-hand inclined member is member 1A (not A1 as was the case when the joint at the left-hand reaction was being considered). From the force diagram 1 to *a* is upwards to the right and this fixes the direction of the arrow on the frame diagram. Similarly, for the same joint, the other inclined member is B1. From the force diagram *b* to 1 is upwards towards the left. (Study the force diagrams and the directions of the arrows in Figs. 6.6–6.8.)

Table 6.1 Forces in frame in Fig. 6.6

Note the directions of the arrows indicating compression (strut) and tension (tie) respectively (Fig. 6.9(b)). These arrows represent the directions of the internal resistances in the members. Member A1 is a strut, which means that it has shortened as a result of the force in it. If the force were removed, the member would revert to its original length (i.e. it would lengthen) and the arrows indicate the attempt of the member to revert to its original length.

Member	Force (kN) in:	
	Strut	Tie
A1	14	
B1	14	
C1		10

Member C1 is a tie and has therefore been stretched. The arrows indicate the attempt of the member to revert to its original (shorter) length.

The forces in the various members may be entered as shown in Table 6.1.

Fig. 6.9 For frame given in Fig. 6.6: (a) force diagram; (b) frame diagram

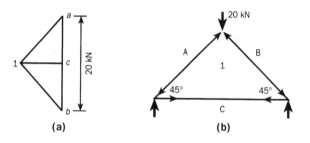

Example 6.1

Symmetrical frame and loading

The principles described above may be applied to frames having any number of members. The work of determining the values of all the unknown forces may be enormously simplified by drawing *one* combined force diagram as shown in Fig. 6.10(b); for this example the operations are explained in detail.

1 Starting from force AB (Fig. 6.10(a)) the *known* forces AB, BC, CD, DE, EF, FG and GA, working clockwise round the frame, are set down in order and to scale as *ab*, *bc*, *cd*, *de*, *ef*, *fg* and *ga* in Fig. 6.10(b). (Since the loading is symmetrical each of the reactions FG and GA is 20 kN.)

2 Point 1 on the force diagram is found by drawing *b*–1 parallel to B1, and *g*–1 parallel to G1. Point 1 lies at the intersection of these lines.

Fig. 6.10 Example 6.1:
(a) free-body, space or frame
diagram; (b) force diagram

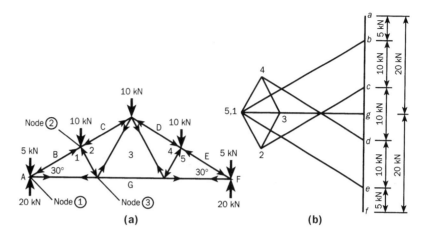

(a)   (b)

3 Point 2 on the force diagram is found by drawing *c*–2 parallel to C2 and
1–2 parallel to 1–2 from the frame (node 2). The point 2 is the
intersection of these two lines.
4 Point 3 on the force diagram is found in the same manner by drawing
2–3 parallel to 2–3, and *g*–3 parallel to G3 from the frame, point 3 being
the intersection.
5 The remaining points 4 and 5 are found in the same way as above, though
in this case, as the frame is symmetrical and forces E5 and 4–5 are the
same as B1 and 1–2, etc., only half the diagram need have been drawn.
6 The directions of the arrows may now be transferred to the frame
diagram (Fig. 6.10(a)) by working clockwise around each node point.
   Taking node 1 for example: moving clockwise, force AB is known to
be downward, and *ab* (5 kN) is read downward on the force diagram.
The next force in order (still moving clockwise round the joint) is force
B1, which as *b*–1 on the force diagram acts to the left and downwards.
Thus force B1 must act towards node 1 as indicated in Fig. 6.10(a). The
next force in order (still moving clockwise) is 1G. 1–*g* on the force
diagram acts from left to right, and thus the arrow is put in on the frame
diagram acting to the right from node 1.
7 Node 2: BC is known to act downward. The next force (moving
clockwise) is C2, which, as *c*–2 in the force diagram acts downward to
the left, thus must act towards node 2. The arrow is shown accordingly.
The next force (still moving clockwise) is 2–1, and as 2–1 on the force
diagram acts upward to the left, this must act toward the node 2, as
shown by the arrow.
8 When all nodes have been considered in this way, it will be seen that
each member of the frame in Fig. 6.10(a) has two arrows, one at each
end of the member. As explained earlier, these arrows indicate not what
is being done to the member, but what the member is doing to the
nodes at each end of it.
   Hence a member with arrows thus ↔ is in compression, i.e. it is
being compressed and is reacting by pressing outward. This member is
called a **strut**.
   Similarly, two arrows thus →← indicate tension in the member.
This member is called a **tie**.

Table 6.2 Example 6.1 – forces

Member		Forces (kN) in:	
		Strut	Tie
B1	E5	30.0	
C2	D4	25.0	
G1	G5		26.0
–	G3		17.5
1–2	4–5	9.0	
2–3	3–4		9.0

9 The *amounts* of all the forces B1, C2, 1–2, etc., are now scaled from the force diagram, and the amounts and types of force can be tabulated as in Table 6.2.

It should be noted that constructing a force diagram involves considering each joint of the frame in turn and fixing the position of one figure on the force diagram by drawing two lines which intersect. No joint can be dealt with unless *all except two* of the forces meeting at the joint are known. This is why node 1 was taken as the first joint since, of the four forces meeting at this joint, only two, B1 and 1G, are unknown. Solving this joint gives the forces in the members B1 and 1G. Node 2 can now be dealt with since only two forces, C2 and 2–1, are unknown. Node 3 cannot be dealt with before node 2 because, although the force in the member G1 is known, there are still three unknown forces, 1–2, 2–3, and 3G.

Example 6.2

Frame containing a member having no force (Fig. 6.11)
The external loads and the two equal reactions are set off to scale as shown in Fig. 6.11(b).

Starting with node 1 the two unknown forces are A1 and 1E.
From *a* on the force diagram a vertical line must be drawn (parallel to A1) and point 1 must lie on this vertical line where it meets a horizontal line drawn parallel to member 1E. The point 1 must, therefore, be on the same point as *e* and force 1E = 0.
Considering node 2, point 2 is found by drawing 1–2 to intersect *b*–2.
Point 3 is found by drawing *c*–3 and 3–2 on the force diagram.

Fig. 6.11 Example 6.2: (a) frame diagram; (b) force diagram

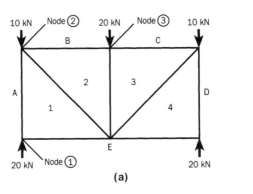

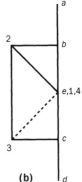

Since the frame and loading are symmetrical there is no need to draw further lines, but in this simple example, one more line, shown dotted, completes the force diagram.

Example 6.3

Frame with unsymmetrical loading

When the load and/or the frame is unsymmetrical, *every* joint must be considered when drawing the force diagram. Figure 6.12 is an example. The reactions may be determined graphically using the link polygon (Chapter 4) or by calculation (Chapter 5).

Fig. 6.12 Example 6.3: (a) frame diagram; (b) force diagram

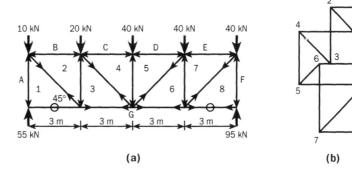

(a)   (b)

Example 6.4

Frame with loads suspended from the bottom chord of the frame in addition to loads on the top chord

The reactions in Fig. 6.13 are determined in the usual manner and are 40 kN at the left support and 50 kN at the right support. The vertical load line is then drawn as follows: *a–b* is 20 kN down; *b–c* is 40 kN down. The next force is the reaction CD of 50 kN, therefore from *c* to *d* is measured upwards a distance equal to 50 kN. The next load is DE and is 20 kN down, therefore *d–e* represents this load on the force diagram. Similarly, *e–f* represents the load of 10 kN (EF) and finally *f–a* represents the vertical reaction FA of 40 kN.

The remaining force diagram can now be drawn in the usual manner, e.g. starting with node 1, two lines *a*–1 and *f*–1 are drawn to intersect at point 1, and so on, as explained in Example 6.1.

Fig. 6.13 Example 6.4: (a) frame diagram; (b) force diagram

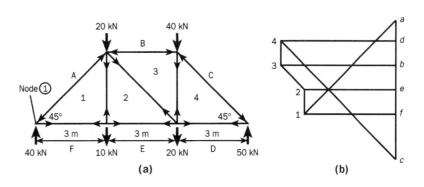

(a)   (b)

Example 6.5

Frame cantilevering over one support
Figure 6.14 shows a frame cantilevering 2 m over the left support. By calculation, the reactions are found to be 7.5 kN and 4.0 kN respectively. The load line is drawn as explained in the previous example, *a–b*, *b–c*, *c–d*, *d–e* (all downwards), *e–f* upwards (4.0 kN), *f–g* downwards (1.5 kN), *g–a* upwards (7.5 kN).

Example 6.6

Cantilever frames
In Fig. 6.15, there is no need to determine the reactions before starting on the force diagram. The loads are first plotted in order, *a–b*, *b–c*, *c–d*, *d–e*.

Fig. 6.14 Example 6.5: (a) frame diagram; (b) force diagram

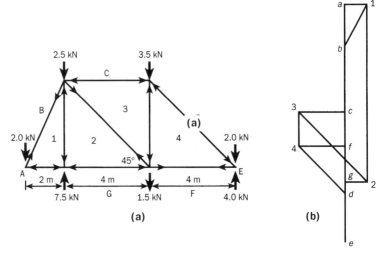

Fig. 6.15 Example 6.6: (a) frame diagram; (b) force diagram

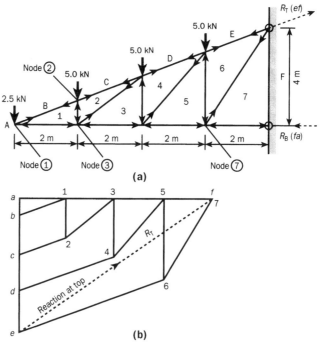

Starting with node 1, *b*–1 and 1–*a* fix the position of point 1 on the force diagram, and so on, finishing with node 7, i.e. 6–7 and 7–*a*. Since member 7A is in compression, the reaction at the bottom support will be equal in value to the force in the member, therefore *f* is at the same point as 7. The reaction at the top is given by the line *e–f* on the force diagram, whilst line *a–f* represents the reaction at the bottom.

Frames subjected to inclined loads

The frames dealt with in the preceding examples have all carried purely vertical loads, and so the end reactions have also been vertical.

Sometimes roof trusses have to be designed to withstand the effect of wind, and this produces loads which are assumed to be applied to the truss (via the purlins) at the panel points, in such a way that the panel loads are acting normal to the rafter line. As the inclined loads have a horizontal thrust effect on the truss as well as a vertical thrust effect, it follows that at least one of the reactions will have to resist the horizontal thrust effect.

It was stated in Chapter 4 that, when a truss is subjected to inclined loads, the usual assumptions are either that one of the reactions is vertical, or the horizontal components of the two reactions are equal. The method of determining such reactions by the use of the link polygon has been described in Chapter 4.

When only symmetrical inclined loads have to be considered, and one of the reactions is assumed to be vertical as in Fig. 6.16(a), the reactions can be found more easily by the method described below than by the link polygon.

If the four loads are considered to be replaced temporarily by the resultant *R*, then only three loads act on the truss, i.e. *R*, R_L and R_R. The three loads must act through one point for equilibrium, and the loads *R* and R_R obviously act through point *x*, the point of intersection of the vertical reaction R_R and the resultant inclined load *R*. As R_L must act through point *x*, the direction of R_L is found by joining point A to point *x*.

The direction of all three loads is thus known, together with the magnitude of the resultant *R*, so the amounts of R_L and R_R may be found graphically as in Fig. 6.16(b).

The resultant *R* is set down to scale (*b* to *c*). The point *d* is found by drawing in the direction of R_R from *c* and the direction of R_L from *b* to their intersection point *d*.

This vector diagram *bcd* is, of course, the triangle of forces for the loads, and *cd* = reaction R_R to scale, and *db* = reaction R_L to scale. From these three points, *b*, *c*, and *d*, the force diagram for the frame can be drawn in the usual way.

The vertical reaction can also be calculated by taking moments about the other end of the truss. For example, from Fig. 6.16(c), taking moments about A:

$$R_R \times L = R \times y$$

$$R_R = \frac{R \times y}{L}$$

Fig. 6.16 Inclined loads: (a) load diagram; (b) force diagram; (c) taking moments about A

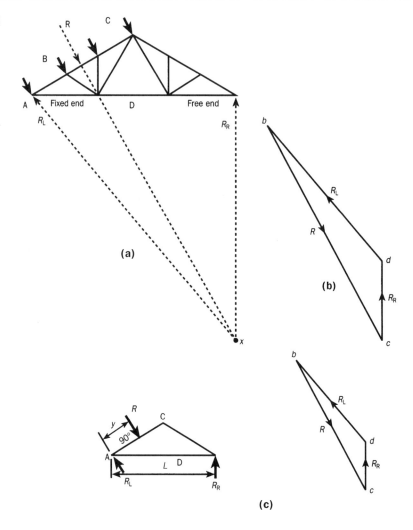

(a)

(b)

(c)

Point *d* can now be found by setting down the resultant load *R* (line *bc*) followed by the vertical reaction R_R as calculated to give the point *d*. Reaction R_L will then be found by scaling force *db*.

Example 6.7

Frame with inclined loads

When the two reactions have been found, either graphically or by calculation, the forces in the individual bars may be found by drawing a force diagram, as in Fig. 6.17(b).

The resultant load *R* (inclined) is produced to intersect the vertical reaction R_R at the point *x*. The direction of the left-hand reaction R_L is then found by joining *x* to A.

The two reactions and the three inclined loads are thus now all known, and may be set down as in Fig. 6.17(b) in the order *ab*, *bc*, *cd* (inclined), *de* (vertical), and *ea* parallel to R_L.

Fig. 6.17 Example 6.7, first frame: (a) load diagram; (b) force diagram

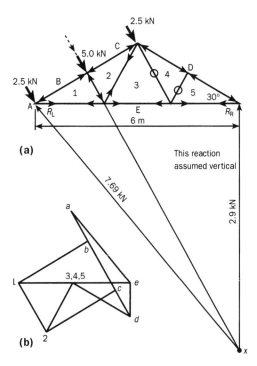

The points 1, 2, 3, 4, 5 on the force diagram are then found in the usual way, as shown in Fig. 6.17(b).

Note that there is no force in members 3–4 and 4–5 from this system of loading (see Table 6.3, first example).

Another example is shown in Fig. 6.18 and Table 6.3, second example.

Table 6.3 Example 6.7 – forces

First example (Fig. 6.17)			Second example (Fig. 6.18)		
Member	Force (kN) in:		Member	Force (kN) in:	
	Strut	Tie		Strut	Tie
B1	7.0		B1	12.7	
C2	7.0		C2	10.0	
D4	5.8		D3	13.1	
D5	5.8		E5	8.7	
E1		9.8	E6	8.7	
E3		5.0	E7	8.7	
E5		5.0	F1		9.7
1–2	5.0		1–2	5.7	
2–3		5.0	2–3	6.0	
			3–4		10.0
No force in 3–4 and 4–5			No force in 4–5, 5–6, 6–7, F4 and F7		

Fig. 6.18 Example 6.7, second
frame: (a) load diagram; (b) force
diagram

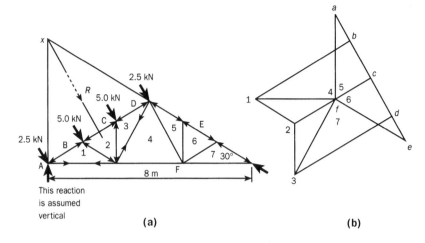

This reaction
is assumed
vertical

(a)

(b)

Calculation methods for frames

When the forces in *all* the members have to be found, the force diagram is usually the most convenient method, but on occasions it is necessary to determine the force in a single individual member. In cases of this type it is normally more convenient to calculate the amount of the force.

Method of sections

Consider the N girder shown in Fig. 6.19(a). It is required to calculate the force in the member 1–2 of the bottom boom.

1 Imagine the girder to be cut completely through, along the section S–S, a section passing through the member 1–2 concerned and two other members, 3–4 and 3–2.

2 Consider the portion of truss to the right of line S–S to be removed. The portion to the left of line S–S would then, of course, collapse, because three forces (the forces in members 1–2, 3–2 and 3–4) which were necessary to retain equilibrium had been removed.

3 If now, three forces x, y and z, equal respectively to force 1–2, force 3–4 and force 3–2, are now applied to the portion of frame concerned, as shown in Fig. 6.19(b), then the portion of frame will remain in equilibrium under the action of the 50 kN reaction, the two 20 kN loads, and the forces x, y and z.

4 These three forces (x, y and z) are as yet unknown in amount and direction, but the force in member 1–2, i.e. force x, is to be determined at this stage, and it will be seen that the other two unknown forces (y and z) intersect at and pass through the point 3 so that they have no moment about that point.

5 Thus, taking moments about the point 3, the portion of frame to the left of S–S is in equilibrium under the moments of the two applied loads, the moment of the reaction, and the moment of the force x.

The moment of the reaction is a clockwise one of (50 × 3), i.e. 150 kN m.

The moment of the applied load is anticlockwise about point 3 and equals (20×3), i.e. 60 kN m.

For equilibrium the moment of force x about point 3 (i.e. $3x$) must equal $(150 - 60)$, i.e. 90 kN m anticlockwise.

Therefore $3x = 90$, whence $x = 30$ kN and the arrow must act in the direction shown in Fig. 6.19(b) (in order that the moment of force x about point 3 is anticlockwise).

The member 1–2 is therefore in tension (pulling away from joint 1).

Note: It will be seen that, in general, the rule must be to take a cut through three members (which include the one whose force is to be determined) and to take moments about the point through which the lines of action of the other two forces intersect.

To determine the force in the member 3–4:

Take moments about the point 2 where forces x and z intersect:

$$\text{Moment of reaction} = (50 \times 6)$$

$$= 300 \text{ kN m clockwise}$$

$$\text{Moments of applied loads} = (20 \times 6) + (20 \times 3)$$

$$= 180 \text{ kN m anticlockwise}$$

Therefore

$$\text{Moment of force } y = 3y = 300 - 180$$

$$= 120 \text{ kN m anticlockwise}$$

$$y = 40 \text{ kN}$$

and, for the moment to be anticlockwise about point 2, the arrow must act towards joint 3, therefore member 3–4 is a strut.

It should be noted that this particular method has its limitations. It would not appear to apply, for example, to member 3–2 (force z), for no point exists through which forces x and y intersect, as x and y are, in fact, parallel.

However, when force y has been calculated, the force z may be found by taking moments about point 1, including the known value of force y (member 3–4).

Fig. 6.19 Section method: (a) load diagram; (b) load diagram of left-hand section

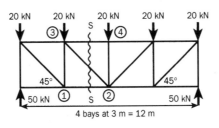

(a)

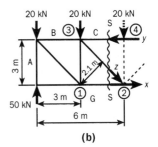

(b)

Note: force y is 40 kN and is anticlockwise about point 1.

Moment of the reaction about point 1

$$= (50 \times 3) = 150 \text{ kN m}$$

Moments of the applied loads and the force in member 3–4

$$= (20 \times 3) + (40 \times 3) = 180 \text{ kN m}$$

Therefore

Moment of force z about point 1

$$= 180 - 150 = 30 \text{ kN m clockwise}$$

Hence

$$2.1z = 30$$

$$z = 14.2 \text{ kN}$$

The arrow must act as shown in Fig. 6.19(b), therefore member 3–2 is a tie.

It should be noted that these arrows must be considered in relation to the nearest joints of that portion of the frame which remains after the imaginary cut through the three members has been made. These results should be checked by drawing a force diagram for the entire frame.

Example 6.8

Roof truss with vertical loads

Calculate the forces in the members marked 2–3, 1–6 and 2–6 of the frame shown in Fig. 6.20.

Fig. 6.20 Example 6.8

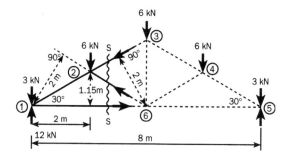

Solution

Member 2–3: take moments about point 6 (section at S–S):

Moment of reaction $= 12 \times 4$

$$= 48 \text{ kN m clockwise}$$

Moment of applied loads $= (3 \times 4) + (6 \times 2)$

$$= 24 \text{ kN m anticlockwise}$$

Moment of force in member $2-3 = 48 - 24$

$$= 24 \text{ kN m anticlockwise}$$

Therefore

$$\text{Force } 2\text{--}3 \times 2 \text{ m} = 24 \text{ kN m}$$

$$\text{Force } 2\text{--}3 = 12 \text{ kN (struct)}$$

Member 1–6: take moments about point 2:

$$\text{Moment of reaction} = 12 \times 2$$

$$= 24 \text{ kN m clockwise}$$

$$\text{Moment of loads} = 3 \times 2$$

$$= 6 \text{ kN m anticlockwise}$$

$$\text{Force } 1\text{--}6 \times 1.15 \text{ m (anticlockwise)} = 24 - 6$$

$$\text{Force } 1\text{--}6 = 18/1.15$$

$$= 15.7 \text{ kN (tie)}$$

Member 2–6: take moments about point 1:

$$\text{Moment of reaction} = 0$$

$$\text{Moment of loads} = 6 \times 2$$

$$= 12 \text{ kN m clockwise}$$

$$\text{Force } 2\text{--}6 \times 2 \text{ m (anticlockwise)} = 12 \text{ kN m}$$

$$\text{Force } 2\text{--}6 = 6 \text{ kN (struct)}$$

Remember that, for determining whether members are struts or ties, the arrows as shown in Fig. 6.20 must be considered as acting towards or away from the nearest joint in that portion of the frame which remains after the *cut* has been made.

Example 6.9 Girder with parallel flanges

The application of this method may be simplified considerably when the frame concerned has parallel flanges. Consider, for example, the Warren girder shown in Fig. 6.21(a).

In using the method of sections to find the force in member EF, the cut would be made through the section X–X, and moments taken about the point C.

Taking moments about C to the right,

$$(\text{Force in EF}) \times h = \text{moments of forces } R_R \text{ and } W_3$$

$$= (R_R \times x_2) - (W_3 \times x_1)$$

$$\text{Force in EF} = \frac{(R_R \times x_2) - (W_3 \times x_1)}{h}$$

Fig. 6.21 Example 6.9: (a) load diagram; (b) girder considered as a simple beam

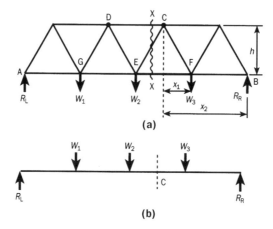

(a)

(b)

But the value $(R_R \times x_2) - (W_3 \times x_1)$ is obviously the bending moment at C, considering the girder as an ordinary beam as shown in Fig. 6.21(b).

So, to calculate the force in any top or bottom boom member of a girder of this type, calculate the bending moment at the node point opposite to the member and divide by the vertical height of the girder, e.g.

$$\text{Force in EF} = \text{BM at point C} \div h$$

$$\text{Force in DC} = \text{BM at point E} \div h$$

$$\text{Force in GE} = \text{BM at point D} \div h$$

Note: This simplification of the method is sometimes known as the **method of moments**.

Example 6.10

In Fig. 6.22, by calculation, the reactions are

$$R_l = 45\,\text{kN} \quad \text{and} \quad R_R = 55\,\text{kN}$$

$$\text{BM at B and G} = (45 \times 3) = 135\,\text{kN m}$$

$$\text{BM at C and H} = (45 \times 6) - (20 \times 3) = 210\,\text{kN m}$$

$$\text{BM at D and J} = (55 \times 3) = 165\,\text{kN m}$$

Fig. 6.22 Example 6.10

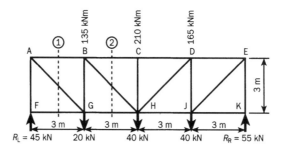

Therefore, considering members cut by section 1,

$$\text{Force in AB} = \frac{\text{BM at G}}{h} = 135/3 = 45 \text{ kN}$$

because the two other members meet at G.

$$\text{Force in FG} = \frac{\text{BM at A}}{h} = 0/3 = 0$$

because the two other members meet at A.

Next, consider members cut by section 2:

$$\text{Force in BC} = \frac{\text{BM at H}}{h} = 210/3 = 70 \text{ kN}$$

because the other two members meet at H.

$$\text{Force in GH} = \frac{\text{BM at B}}{h} = 135/3 = 45 \text{ kN}$$

because the other two members meet at B.

Hence

$$\text{Force in CD} = \frac{\text{BM at H}}{h} = 210/3 = 70 \text{ kN}$$

$$\text{Force in HJ} = \frac{\text{BM at D}}{h} = 165/3 = 55 \text{ kN}$$

$$\text{Force in DE} = \frac{\text{BM at J}}{h} = 165/3 = 55 \text{ kN}$$

$$\text{Force in JK} = \frac{\text{BM at E}}{h} = 0/3 = 0$$

Method of resolution of forces at joints

The forces in the individual members of loaded frames may also be determined by considering the various forces acting at each node point. This method is particularly useful for dealing with those members not so easily tackled by the method of sections, e.g. the vertical members of the frame dealt with in the previous example.

Example 6.11

Consider the node F in Fig. 6.23(a).

There is a load of 45 kN *upward* at F. Obviously, member AF must itself counteract this (as FG, a horizontal member, cannot resist vertical force). Hence the force in member AF must equal 45 kN, and it must, at the node point F, be acting downwards in opposition to the reaction R_L, as shown by the arrowhead 1. This shows member AF to be a strut, and the other arrowhead 2 may be put in, acting towards A.

Consider the node A.

It is now known that a force of 45 kN acts upwards at this point (arrowhead 2). Again, member AB, being horizontal, cannot resist a vertical load, so member AG must resist the upward 45 kN at A. Thus the force in AG must act downwards at A, and the arrow 3 may be placed in position. The amount of force AG must be such that its vertical component equals 45 kN as in Fig. 6.23(b). Thus

$$\frac{45}{\text{force AG}} = \sin 45° = 0.7071$$

Therefore

$$\text{Force in AG} = 45/0.7071 = 63.3 \text{ kN}$$

and AG is in tension, so the arrow 4 may be drawn in.

The load of 63.6 kN acting down and to the right at A tends to pull point A to the right, horizontally. Thus the force in member AB at A must *push* point A to the left for equilibrium (arrow 5, showing member AB to be a strut); thus arrow 6 may be placed as shown in Fig. 6.23(c). The amount of force in AB is such that it counteracts the horizontal component of force 63.6 kN (arrow 3).

Thus, as in Fig. 6.23(d),

$$\frac{\text{Force in AB}}{63.6} = \cos 45° = 0.7071$$

Thus force AB = 63.6 × 0.7071 = 45 kN compression, as was seen by the method of moments.

Consider the node G.

It has already been seen that there is no force in member FG, since the bending moment is zero at the opposite point A. Thus, as there is a force of 63.6 kN at G, acting upwards and to the left, the force in BG must act in such a way that it counteracts the resultant of the vertical forces at G. These vertical forces are:

• the vertical resultant upward of the 63.6 kN (i.e. 45 kN ↑)
• the downward applied load of 20 kN ↓

Fig. 6.23 Example 6.11: (a) load diagram; (b) force AG; (c) force in AB; (d) force AB; (e) force in BG

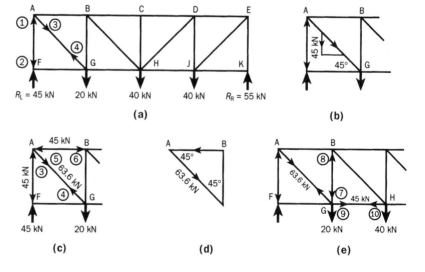

Their resultant is obviously 25 kN upward, so member BG must act downwards to the extent of 25 kN (arrows 7 and 8 may thus be placed) and BG is thus a strut with a force of 25 kN (Fig. 6.23(e)). Also, as AG pulls to the *left* at G to the extent of 45 kN (the horizontal component of the force AG) then member GH must pull to the *right* to the extent of 45 kN (arrow 9), and member GH is seen to be a tie with a force of 45 kN. This again agrees with the result obtained by the method of moments.

This process may be continued from each node to the next, and although its application is admittedly somewhat more laborious than the drawing of a force diagram, this method is strongly recommended. Its use most definitely results in a better and more complete understanding of the forces acting at each node, and by its use struts and ties are more easily distinguished.

Method of tension coefficients

This is a modified version of the previous method and it is particularly useful in the case of pin-jointed, three-dimensional or space frames.

The basis of this method is the resolution into components of the external and internal forces acting on each joint of the frame, using the lengths of members and the coordinates of joints. All the members of the frame are initially assumed to be in tension.

Figure 6.24 shows bar AB, a member of a pin-jointed frame, assumed to be subject to a tensile force T_{AB}.

Let the length of bar AB be L_{AB} and the coordinates of joint A be x_A, y_A and those of joint B, x_B, y_B. Then, resolving the force T_{AB} in the x direction:

$$T_{AB} \times \frac{x_B - x_A}{L_{AB}} = \frac{T_{AB}}{L_{AB}} \times (x_B - x_A)$$

In this expression the factor T_{AB}/L_{AB}, i.e. tension in bar AB divided by the length of the bar, is called the **tension coefficient** for the bar and is denoted by the symbol t_{AB}.

Hence the component in the x direction is

$$t_{AB}(x_B - x_A)$$

and, similarly, the component in the y direction is

$$t_{AB}(y_B - y_A)$$

Where there are a number of members and applied loads acting at a joint, which is in equilibrium under the action of the internal and external

Fig. 6.24 Tension coefficients method

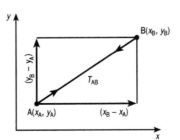

forces, two sets of equations can be formed, one for each of the directions
x and y.

Example 6.12

A load of 7.2 kN is suspended from a soffit by two ropes PQ and QR
(Fig. 6.25). Determine the forces in the ropes.

Fig. 6.25 Example 6.12

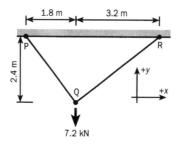

7.2 kN

Solution

The equations for joint Q (i.e. assuming the origin of coordinates is at Q)
in terms of tension coefficients are

$$\text{In direction } x \quad -1.8t_{PQ} + 3.2t_{QR} = 0 \tag{1}$$

$$\text{In direction } y \quad +2.4t_{PQ} + 2.4t_{QR} - 7.2 = 0 \tag{2}$$

Then from (2)

$$t_{PQ} = 3.0 - t_{QR}$$

and substituting into (1)

$$-1.8(3.0 - t_{QR}) + 3.2t_{QR} = 0$$

$$t_{QR} = 5.4/5.0 = 1.08$$

and

$$t_{PQ} = 3.0 - 1.08 = 1.92$$

But $t_{PQ} = T_{PQ}/L_{PQ}$ and thus $T_{PQ} = t_{PQ} \times L_{PQ}$. Therefore

$$T_{PQ} = 1.92 \times \sqrt{(2.4^2 + 1.8^2)}$$

$$= 1.92 \times 3.0 = 5.76 \text{ kN}$$

and

$$T_{QR} = 1.08 \times \sqrt{(2.4^2 + 3.2^2)}$$

$$= 1.08 \times 4.0 = 4.32 \text{ kN}$$

It should be noted that both T_{PQ} and T_{QR} have positive values and, therefore, they are ties. A negative solution denotes a compression member, i.e.
a strut.

From the above examples it follows that, for a frame having J joints, $2J$
equations can be formed. This, incidentally, is the same number of equations
as that used in the method of resolution at joints, which may explain the lack
of popularity of the method of tension coefficients for plane frames. It is,

however, very useful for computer application, particularly in the solution of space frames.

Space frames

A space frame has to be considered in three dimensions, hence the components of the forces at each joint are resolved in three directions x, y and z. Consequently, there will be $3\mathcal{J}$ equations for the frame and the number of members in a perfect pin-jointed space frame is $m = 3\mathcal{J} - 6$. Most space frames, however, will have pinned supports, so the number of members in such a frame is $m = 3\mathcal{J}_f$, where $\mathcal{J}_f$ denotes a free joint (i.e. a joint other than at a support).

Example 6.13

Space frame with vertical load
The three equations for joint A of the shear legs shown in Fig. 6.26 are formed as follows:

In direction x $+2t_{AB} - 2t_{AD} = 0$ (4)

In direction y $+3t_{AB} + 3t_{AC} + 3t_{AD} + 21 = 0$ (5)

In direction z $+2t_{AB} + 4t_{AC} + 2t_{AD} = 0$ (6)

Fig. 6.26 Example 6.13: (a) plan; (b) elevation

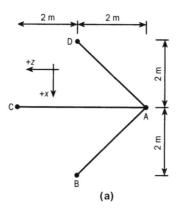

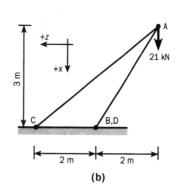

(a)

(b)

Then from (1), $t_{AB} = t_{AD}$, and adding (1) and (3) gives

$t_{AB} = -t_{AC}$

and substituting both into (2),

$-3t_{AC} + 3t_{AC} - 3t_{AC} = -21$

$t_{AC} = 7$ and $t_{AB} = -7 = t_{AD}$

Therefore

$$T_{AB} = -7 \times \sqrt{(2^2 + 2^2 + 3^2)}$$

$$= 28.86 \text{ kN (strut)}$$

$$T_{AC} = +7 \times \sqrt{(4^2 + 3^2)}$$

$$= 35.00 \text{ kN (tie)}$$

$$T_{AD} = 28.86 \text{ kN (strut)}$$

Example 6.14

Space frame with inclined load

In the case of Fig. 6.27 the simple space frame is supporting an inclined load. It is, therefore, necessary to resolve the load into components in directions x, y and z, before writing down the usual equations.

Fig. 6.27 Example 6.14: (a) plan; (b) elevation

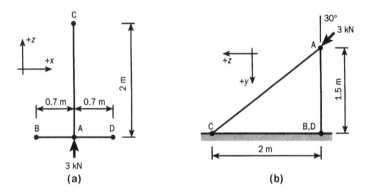

Let the load be F and its components F_x, F_y and F_z, respectively. Therefore

$$F_x = 0$$

$$F_y = 3 \times \cos 30° = 2.6 \text{ kN}$$

$$F_z = 3 \times \sin 30° = 1.5 \text{ kN}$$

It is usually helpful, particularly for a many-jointed frame, to present the equations and the resulting tension coefficients in tabular form as shown in Table 6.4.

Hence

$$T_{AB} = T_{AD} = -0.49 \times \sqrt{(0.7^2 + 1.5^2)}$$

$$= -0.811 \text{ kN (negative, i.e. strut)}$$

and

$$T_{AC} = -0.75 \times \sqrt{(2.0^2 + 1.5^2)}$$

$$= -1.875 \text{ kN (also a strut)}$$

Table 6.4

Joint	Direction	Equation	Tension coefficient
A	x	$-0.7t_{AB} + 0.7t_{AD} = 0$	$t_{AB} = t_{AD}$
	y	$+ 2.6 + 1.5t_{AB} + 1.5t_{AC} + 1.5t_{AD} = 0$	$t_{AB} = -0.49$
	z	$1.5 + 2.0t_{AC} = 0$	$t_{AC} = -0.75$

Exercises

Note: All loads are in kN.

1–15 Determine the reactions and the type and magnitude of the forces in the members of the frames shown in Figs. 6.Q1–6.Q15. Show the results in tabular form.

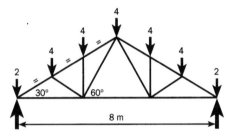

Fig. 6.Q1

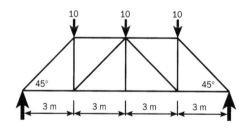

Fig. 6.Q2

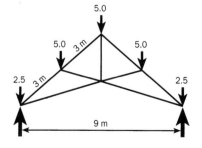

Fig. 6.Q3

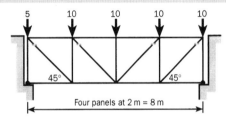

Fig. 6.Q4

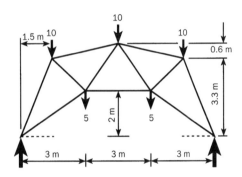

Fig. 6.Q5

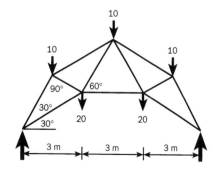

Fig. 6.Q6

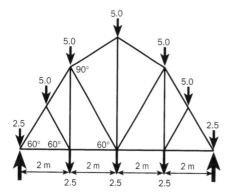

Fig. 6.Q7

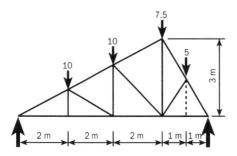

Fig. 6.Q8

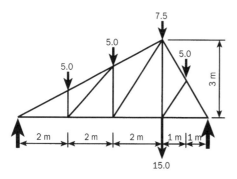

Fig. 6.Q9

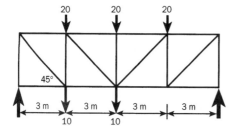

Fig. 6.Q10

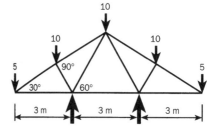

Fig. 6.Q11

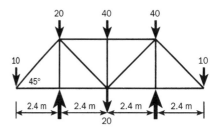

Fig. 6.Q12

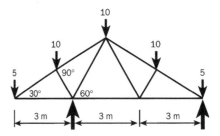

Fig. 6.Q13

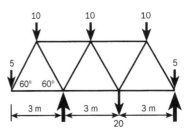

Fig. 6.Q14

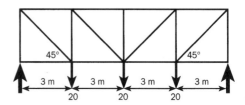

Fig. 6.Q15

16–18 For the frames shown in Figs. 6.Q16–6.Q18 determine:

- the directions and values of the reactions
- the type and magnitude of the forces in all the members.

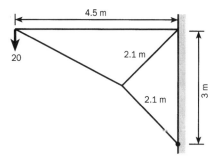

Fig. 6.Q16

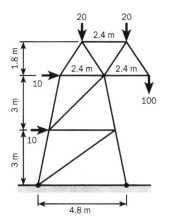

Fig. 6.Q17

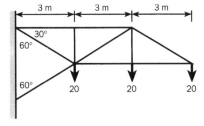

Fig. 6.Q18

19 A framed structure is supported at A and B in such a manner that the reaction at A is horizontal. Calculate the values of the two reactions and determine the type and magnitude of the forces in the members of that structure shown in Fig. 6.Q19.

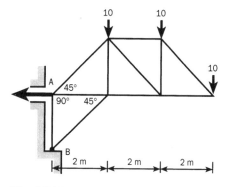

Fig. 6.Q19

20–23 The right-hand ends of the frames shown in Figs. 6.Q20–6.Q23 are supported on rollers, i.e. the reaction there is vertical as indicated. Determine the reactions and the type and magnitude of the forces in the members of the frames.

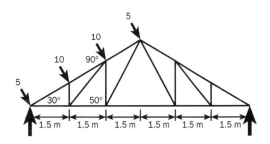

Fig. 6.Q20

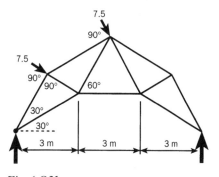

Fig. 6.Q21

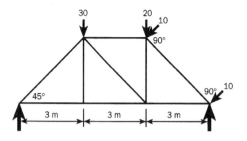

Fig. 6.Q22

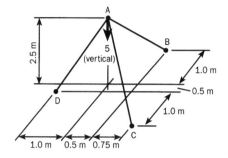

Fig. 6.Q32

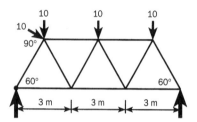

Fig. 6.Q23

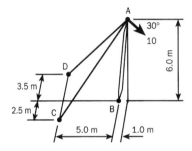

Fig. 6.Q33

24–27 Repeat Questions 20–23 assuming that the roller supports are now at the left-hand ends of the frames.

28–31 Determine the reactions and the type and magnitude of the forces in members in the frames shown in Figs. 6.Q20–6.Q23, assuming that the two supports resist equally the horizontal components of the inclined loads.

32–34 Using the method of tension coefficients determine the forces in the space frames shown in Figs. 6.Q32– 6.Q34.

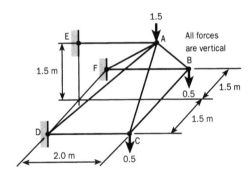

Fig. 6.Q34

Chapter seven Construction materials

Construction materials are dealt with in the context of their suitability for use, properties and cost. Stress, strain and elasticity are dealt with in the context of direct tensile and compressive stresses and their relationship is explained in terms of the modulus of elasticity. The ultimate and yield stresses are considered, and the need for a reduced working or permissible stress, material strength and the idea of a factor of safety are identified. The modular ratio for composite elements is introduced.

General

Building materials that can be formed into many shapes and sizes are classified as either natural or manufactured construction materials.

Natural construction materials such as timber, natural rock and stones are very variable, having a wide range of strengths and durability. They often contain natural defects. For example, timber in its natural form may contain defects such as knots, fissures (resin pockets or similar), blue stain, wormholes and decay. This inevitably causes a wide variation in its properties.

Manufactured construction materials such as steel, aluminium, etc., are less variable in their strength and durability. Their properties depend greatly on the degree of quality control placed on manufacture and construction. The quality control factors and scheme must demonstrate that the requirements of the relevant code of practice are consistently satisfied. This quality control scheme is commonly available to the designer.

Understanding the quality control factors placed on manufacture and construction is important in understanding both the subsequent performance of the final products in real structural environments and what can be done to improve their quality for use in different loading and environmental conditions; conditions such as temperature variation and/or when the material comes into contact with severe exposure conditions, for example, when a concrete surface is exposed to severe rain or alternate wetting and drying (weathering).

The structural elements or units supplied are required to have specified strength and quality limits termed as the 'acceptance limits or the material characteristic strength'. For example, if the specified compressive strength limit for brick masonry units is set by the designer as 10 N/mm^2, then, in any sample of these masonry units, the average compressive strength must not have more than a 2.5 per cent chance of falling below 10 N/mm^2.

Both natural and man-made construction materials used in building construction should comply with the requirements set out in the code of practice and the current building regulations.

Suitability for use

The structural engineer or practitioner has to consider the criterion of *suitability for use* in selecting which materials to use for building a particular structure. Their *suitability for use* is an essential criterion as it is a matter of ensuring that the materials will perform adequately both during construction and in service after the completion of the structure. Meeting this criterion is likely to satisfy all the main strength and durability properties, aesthetic appearance, cost, and effects on the environment. The materials must be strong enough and *not fail, deform excessively, or degrade significantly under the combined action of the applied loads and environmental effects during the design life of the structure.* In addition, water tightness, cost of manufacturing, fabrication and transportation, and speed of construction might become important issues in making decisions on which materials to use.

Structural engineers or practitioners base their selection of a particular material on either their experience or, commonly, on the characteristic material properties, such as strength, durability, deformation, etc., which are known from tests on samples or specimens of the materials that have been prepared and tested under loading and environmental conditions to simulate the conditions that would be undergone by the material in a real structure. As an example of the variation of construction material strengths, Fig. 7.1 shows the range of stiffness and limiting stress of six construction materials.

$$\text{Elastic modulus } (E) = \frac{\text{stress}}{\text{strain}}$$

The elastic modulus is a measure of the stiffness of a material, i.e. the material's resistance to deformation under the applied load. There may be different values for E for the same material depending on the testing environments, such as the testing temperature and the state of the applied load, i.e. static or dynamic.

It can be seen from Fig. 7.1 that the strength properties of carbon fibre, different species of timber, and different mixtures of normal weight concrete are more variable than in steel and glass fibre, as expected. For example, the compressive strength of normal weight concrete is affected by the strength, amount and quality of the aggregate, cement, sand and water used to prepare the mixture, in addition to the quality of mixing and curing techniques. The higher the quality control, the lower the difference in the compressive strength of the individual tested samples. The materials that showed high variation in their elastic stiffness in Fig. 7.1 also have large coefficients of variation, C_v, where C_v is non-dimensional, and is commonly used by materials engineers to allow comparisons to be made of different materials or different kinds of the same material.

$$C_v = S/f_m$$

where S (standard deviation) $= \sqrt{\sum (X - f_m)^2/n - 1}$ = the variability of the variant and x = individual sample tested values, f_m = mean value of tested samples and n = number of tested samples.

Fig. 7.1 Stiffness and limiting stress of six construction materials. *$G=10^9$, **$M=10^6$. (Source: Owens and Knowles, *Steel Designers' Manual*, 5th edn. Blackwell Scientific Publications, London, 1994.)

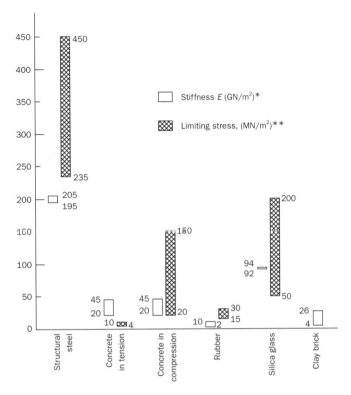

Example 7.1

Calculation of C_v, the coefficient of strength variation

Table 7.1 shows the results of 14 compressive strength tests of steel bars made under the same conditions.

1 Determine the mean strength of the material tested.

2 Determine the coefficient of strength variation C_v.

Table 7.1 Example 7.1 – test values

Test values, x (N/mm²)						
278.0	280.0	274.0	274.5	281.0	283.0	284.0
290.0	282.5	278.5	277.0	282.0	276.0	279.5

$$\sum x = 3920 \, \text{N/mm}^2$$

$$\text{mean value}, f_m = \frac{\sum x}{n} = 280 \, \text{N/mm}^2$$

$$\text{therefore } S \text{ (standard deviation)} = \sqrt{\left(\frac{\sum(x - f_m)^2}{n - 1}\right)}$$

$$= \sqrt{\left(\frac{229}{13}\right)} = 4.197 \, \text{N/mm}^2$$

$$C_v = \frac{S}{f_m} = \frac{4.197}{280} = 0.015$$

The coefficient of variation of the tested samples is very small, which indicates that a high-quality control scheme was in place when the steel was manufactured. This is quite obvious, as the table shows only small differences in the compressive strength of the individual tested samples.

The most common construction materials are steel, reinforced concrete, masonry, timber, aluminium, glass, reinforced polyester, glass fibre, fibre reinforced cement and composite materials. Steel, reinforced concrete and timber are the construction materials that will be considered in more detail in this textbook. For further information on construction materials see the reference list at the end of the book.

Steel, concrete and timber as construction materials

Steel

Steel has been produced for many years from iron ore by removing the ore's naturally occurring impurities. The material has high strength in tension and compression and is able to undergo large deformation without fracture. It is a ductile material.

Elasticity

A typical curve for steel in tension is shown in Fig. 7.2. As the load is applied the relationship between stress and strain is linear up to the yield points, and the steel behaves almost as an elastic material with virtually full recovery if the load is removed. This is called its elastic range or permissible stress range. Beyond this range the steel behaves more and more as a plastic material. If the load is removed within the plastic range, recovery is no longer full and the tested element will be permanently strained (damaged) and permanent deformation will remain. The slope of the stress/strain curve in the elastic range defines the modulus of elasticity of the material. The modulus of elasticity for most structural steel tends to be a constant value of approximately 205 kN/mm².

Basic grades

The three basic grades of steel are S460, S355 and S275. The numbers represent the minimum tensile strength of each grade in N/mm². Grade S275 is the most commonly used in steel structures at the start of the twenty-first century (see Fig. 7.2).

Fig. 7.2 Short term design stress–strain curves for steel aluminum and concrete

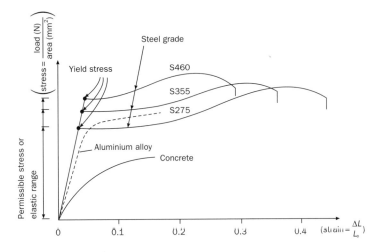

Steel sections

Figure 7.3 shows the most commonly used steel sections. These sections may be rolled into their particular shape while the material is still very hot (at the mills), or may be fabricated using welded or bolted connections. The number of rolled or fabricated sections is very large and the geometric properties of these sections formed from any one of the three grades of steel have been tabulated in the British Steel publication entitled *Structural sections to BS 4: Part 1 and BS 4848: Part 4.*

Fig. 7.3 Sections of steel elements

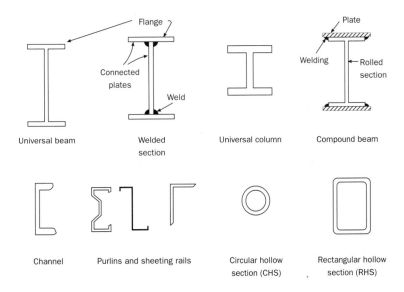

Steel durability

Steel structures should be designed to be long lasting and require little maintenance. The durability of steel is influenced by exposure conditions, steel quality and fire protection. Steel elements can be exposed to a wide range of conditions, such as the atmosphere, soil, seawater, or stored chemicals. Rusting and surface degradation caused by atmospheric conditions should be prevented by anti-corrosion treatment, such as surface preparation or painting. Metal coating, such as galvanizing or zinc spraying, can provide very good protection giving a rust and surface degradation-free life of about 20 years.

Fire protection

The effect of fire on the strength of steel is shown in Fig. 7.4. It can be seen that the rate of loss of strength is very high at temperatures more than 300 °C. Therefore an early fire protection system to protect the steel structure from fire is very important. Concrete, brickwork and light encasement (such as recommended paint, or wet plaster spread on metal lathing or directly onto the steelwork) are commonly used.

Form of steel structures

A steel frame can take many shapes to suit the requirements of a client (see Fig. 7.5) and can be constructed or fabricated out of hot rolled structural steel shapes or cold formed steel sections. It provides adaptability, speed, lower monitoring/control costs and lower preliminaries. Economic benefits, through earlier rents or uses, can be guaranteed by fast track construction and faster completion. The effects of environmental and other conditions on the final shape of a steel building should be discussed by the design/construction team in the initial stages of planning the layout and preparing the construction plan.

The architect should consider aspects such as the integration of facilities for all types of usage; environmental aspects; internal non-commercial values and spaces; creation of a building that would add some excitement, colour,

Fig. 7.4 Relationship between steel strength and temperature

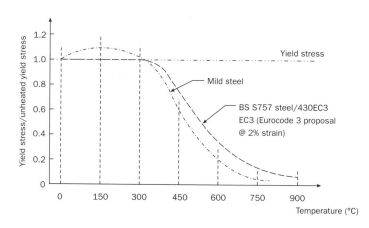

Fig. 7.5 Three common types of steel buildings: (a) single-storey rigid portal frame building; (b) simple stanchion and truss frame; (c) multi-storey buildings

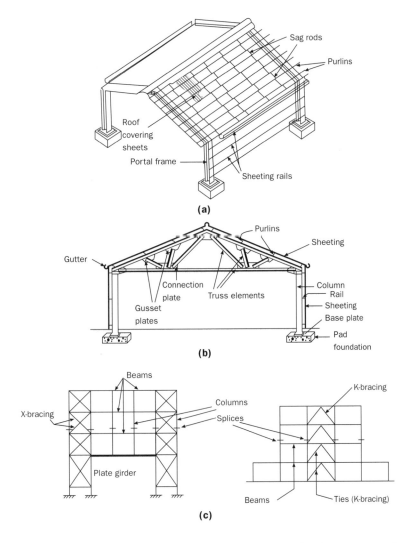

light and size to provide visual interest; and provision for the easy expansion of the area of the building.

All steel buildings must comply with the current Building Regulations. These are statutory instruments approved by parliament covering all aspects of building construction.

Figure 7.5 shows three common types of steel buildings: simple stanchion and truss frame buildings, single-storey rigid portal frame buildings, and multi-storey buildings. The three types cover a wide range of uses for steel frame buildings, such as factories, warehouses, offices, flats, schools, hospitals and car parks.

Reinforced concrete

BS 8110: Part 1: 1997 gives recommendations on concrete structural design and construction.

Concrete

Concrete is a mixture of aggregates, sand and cement which when mixed with water and left to set can form a very hard mass. Concrete strength and durability depend on many factors, among them:

- the quantities and quality of the aggregates, sand, cement and water
- the age of the concrete when used, since the time the water was added to the mixture
- the qualities and quantities of any added substances used to improve the concrete properties, such as plastisizers, microsilica, pulverized fuel ash (PFA)
- the quality of mixing
- quality control during construction.

Figure 7.6 shows the increase in concrete compressive strength with age. The precise relationship between a concrete compressive strength and its age depends upon the type of cementitious material used. The rate of increase in concrete strength is most marked over the first three weeks and slows down thereafter. Laboratory tests conducted on concrete cubes show that at 28 days, concrete has already reached about three-quarters of the maximum compressive strength it will attain even if the concrete is left to cure for many years. The 28 days concrete compressive strength value is therefore important, as both researchers and designers commonly use this value in their calculations.

Reinforced concrete elements

Concrete is a hard and brittle material. It is a material that is strong in its resistance to compression, but very weak in tension. A good concrete will safely take a stress of up to $70 \, \text{N/mm}^2$ in compression, but the safe stress in tension is usually limited to no more than one-tenth of its compressive stress, i.e. this concrete will fail under tension stress of more than $7 \, \text{N/mm}^2$. For example, in a homogeneous concrete beam the stress distribution is as shown in Fig. 7.7(b) and in the case of a section symmetrical

Fig. 7.6 Concrete strength with age, using ordinary portland cement (OPC)

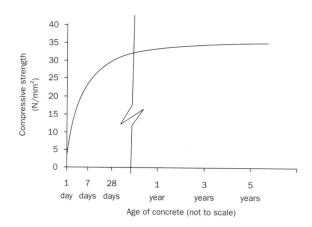

Fig. 7.7 Reinforced concrete beam: (a) beam; (b) stress distribution; (c) stress in steel; (d) cross-section

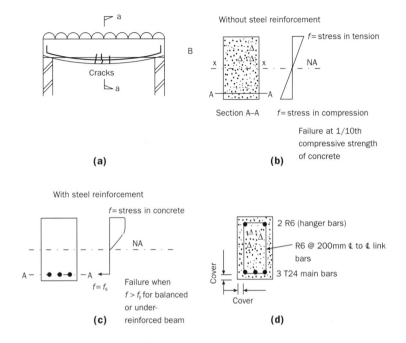

about the X–X axis, the actual stress in tension equals the actual stress in compression. If such a beam were of ordinary unreinforced concrete, then the stresses in tension and in compression would out of necessity have to be limited to avoid overstressing in the tension face, whilst the compressive fibres would be taking only one-tenth of their safe allowable stress. The result would be very uneconomical since the compression concrete would be under-stressed.

In order to increase the safe load-carrying capacity of the beam, and allow the compression concrete to use its compressive resistance to the full, steel bars are introduced into the tension zone of the beam to carry the whole of the tensile forces.

This may, at first, seem to be a little unreasonable – for the concrete is capable of carrying some tension, even though its safe tensile stress is very low. In fact it will be seen from Fig. 7.7(b) that the strains in the steel and in the concrete at A–A are the same. Therefore, if the stress in steel is 140 N/mm², the concrete would be apparently stressed to *1/m* times this and, since *m*, the ratio of the elastic moduli of steel and concrete, is usually taken to be 15, the tensile stress in concrete would be in excess of 9 N/mm². The concrete would crack through failure in tension at a stress very much lower than this and thus its resistance to tension is disregarded in the design process.

Most concrete elements, such as concrete beams, columns, foundations, retaining walls, slabs and floors, tend to bend under the action of the applied loads, and therefore experience tension in one side and compression in the other side during their design life. Therefore, they are commonly reinforced with steel bars inserted in their tension side to bring into action the full strength of that member both in tension and in compression, thus an economical section can be achieved.

Design strength of reinforced concrete

As mentioned in Chapter 1 design strength is given by the following:

$$\text{design strength} = \frac{\text{characteristic strength}}{\gamma_m}$$

Reinforced concrete consists of two materials: concrete and steel bars. The characteristic strength of concrete, f_{cu}, otherwise known as the grade of the concrete, is the strength of a large specified number of concrete cubes tested to destruction (see the relevant code of practice). The characteristic strength, f_{cu}, is calculated as detailed in Chapter 1. The value of the partial factor of safety, γ_m, for concrete in flexure, shear and bond equals 1.5 and for steel reinforcement equals 1.05.

Table 7.2 shows the most common grades of concrete, f_{cu}, as well as the characteristic strength of steel, f_y, commonly known as the steel grade.

The code of practice shows how:

*   Engineers believe the stresses in beams, slabs, and column sections, etc., are distributed.
*   Designers can calculate the sizes required for each concrete element and the quantities of steel bars needed to meet the strength and durability requirements.
*   Designers can obtain an adequate nominal cover for the reinforcement, so as to meet the specified periods of fire resistance and the durability requirements, so it will not deteriorate unduly under the action of the environment over the intended design life.

Table 7.2 Grade of concrete and steel reinforcement

Grade of concrete, f_{cu} (N/mm²)	Grade of steel reinforcement, f_y (N/mm²) (specific characteristic strength)	
C20	Hot rolled (mild steel)	250
C25	High yield steel	
	(hot rolled or cold worked)	460
C30		
C35		
C40		
C501		

Durability of concrete

Serviceable and durable concrete structures require suitable materials, well-designed, well-detailed and properly constructed/erected structures. Quality control supervision is equally important. The designer should take into account the following:

*   Use a high quality mix design with an appropriate water/cement ratio. Make sure high quality control is in place when the concrete is produced for any particular use. The concrete should be dense and possess high

resistance to deleterious substances such as carbon dioxide, oxygen, chloride ions and other potentially deleterious materials.
- The environmental conditions to which the concrete structure or the concrete element will be exposed as well as the shape and the bulk of the structure.
- The concrete's nominal cover to the steel reinforcement (including link bars; see Fig. 7.7). Nominal cover is normally taken to be the greatest values obtained from Tables 3.3 or 3.4 of BS 8110: Part 1: 1997. These tables give the minimum concrete cover to the steel bars for protection against corrosion and fire respectively. For durability of concrete in general, see BS 5328: Part 1: 1997 and Tables 3 and 6 of BS 5328: Part 2: 1997.

Timber

BS 5268: Part 2: 1996 *Code of practice for permissible stress design, materials and workmanship* gives recommendations on the design and use of timber.

There are varieties of softwood and hardwood species listed in the code of practice that are suitable for use in buildings. BS 5268: Part 2 gives a series of strength classes which for design use can be considered as being independent of species and grade. Hence, structural designers commonly specify the timber strength, or strength class, as being softwood or hardwood, rather than specifying a species and grade. A builder uses timber, the strength of which can easily be identified by a mark on each piece of timber. If the choice of timber material is limited by factors other than strength, such as natural durability, suitability to preservation, glues and fasteners, then the structural designer will specify the particular species required, or exclude them from within the specified strength class.

Grade of timber

According to BS 5268, timber can be stress graded by either:

- **Visual inspection** – carried out by a well-trained inspector or observer. According to the code of practice, the inspector considers the general quality of the timber, the number and size of knots, the slope of the grain, the size and number of fissures, and the influence of other natural defects on the timber strength before specifying the timber as grade GS or SS (see Fig. 7.8) where GS = general structural grade, SS = special structural grade.
- **Machine strength grading** – a grading machine approved by the UK Timber Grading Committee, and which meets the requirements of EN 519, is used to measure the strength and stiffness of a species of timber using non-destructive testing techniques. The four machine grades are:

MGS – general structural timber
MSS – special structural timber
M75 – 75 per cent free of defects
M50 – 50 per cent free of defects

Fig. 7.8 Defects in timber
species

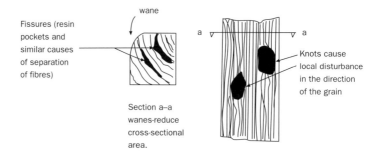

Dimension of stress graded timber

Basic sawn thicknesses and widths for stress graded timber that is available
on the market are shown in Table 7.3.

Table 7.3 Sawn thicknesses and widths for stress graded
timber commonly available in the market

Width (mm)	Thickness (mm)					
	38	44	47	50	63	75
75	✔	✔	✔	✔		
100	✔	✔	✔	✔		
125	✔	✔	✔	✔		
150	✔	✔	✔	✔	✔	
175	✔	✔	✔	✔	✔	✔
200	✔	✔	✔	✔	✔	✔
225	✔		✔	✔	✔	✔

Note: The actual size of *planed* timber is smaller than its
specified basic size.

Softwoods graded according to BS 4978 have strength classes ranging from
C14 to C30, whereas tropical hardwoods graded according to BS 5756
have strength classes ranging from D40 to D70. Table 7.4 shows some samples
of standard names of softwoods and hardwoods and their strength classes.

Stress

When a member is subjected to a load of any type, the many fibres or particles of which the member is made up transmit the load throughout the length
and section of the member, and the fibres doing this work are said to be in a
state of stress.

There are different types of stress, but the principal kinds are tensile and
compressive stresses.

Tensile stress

Consider the steel bar shown in Fig. 7.9(a) having a cross–sectional area of
A mm^2 and pulled out at each end by forces W. Note that the total force in

Table 7.4 Some samples of standard names of softwood for service classes 1 and 2 (dry stress) and their strength classes[e]

(*Sources*: Adapted from Tables 2 and 7, BS 5268-2: 1996 structural use of timber)

| Standard name/species | Grade | Strength class | Grade stress parallel to the grain[a] (N/mm²) | | | | Grade stress perpendicular to the grain[a] (N/mm²) | Modulus of elasticity, E (N/mm²) | |
			Bending	Tension	Compression	Shear	Compression	Mean	Minimum
Imported timber									
Douglas fir–larch	GS	C16	5.3	3.2	6.8	0.67	2.2[b]/1.7[c]	8800[d]	5800
Redwood	GS	C16	5.3	3.2	6.8	0.67	2.2/1.7	8800	5800
Whitewood	GS	C16	5.3	3.2	6.8	0.67	2.2/1.7	8800	5800
Douglas fir–larch	SS	C24	7.5	4.5	7.9	0.71	2.4/1.9	10800	7200
Redwood	SS	C24	7.5	4.5	7.9	0.71	2.4/1.9	10800	7200
Whitewood	SS	C24	7.5	4.5	7.9	0.71	2.4/1.9	10800	7200
British grown									
Douglas fir	GS	C14	4.1	2.5	5.2	0.6	2.1/1.6	6800	4600
British pine	GS	C14	4.1	2.5	5.2	0.6	2.1/1.6	6800	4600
Douglas fir	SS	C18	5.8	3.5	7.1	0.67	2.2/1.7	9100	6000
British pine	SS	C22	6.8	4.1	7.5	0.71	2.3/1.7	9700	6500

[a]The timber microstructure consists of the grains, which act like long strong fibres that are separated by a softer material (see Fig. 7.8). If the stress is applied parallel to the grains, the fibres are compressed at their ends and therefore they can carry higher stress. When they are compressed together, as is the case when stress is applied perpendicular to the grains, the softer material between the grains will tend to fail at lower applied stress.

[b,c]Use the higher value when the specification specifically prohibits wane at the bearing area. Otherwise the lower values apply.

[d]Use the mean elastic modulus when the load sharing factor applies. Otherwise use *E* (minimum).

[e]Grade stress values for service class 3 (wet exposure) should be obtained by multiplying the tabulated stresses and moduli given in Table 7 of the code by the modification factor K2 from Table 13 of BS 5268.

Fig. 7.9 Stress in a steel bar:
(a) tensile; (b) compressive

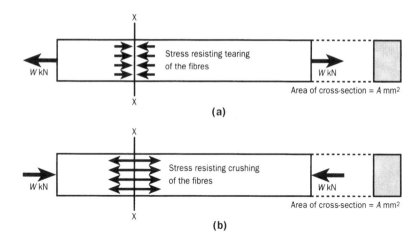

the member is W kN (not $2W$ kN). There would be no force at all in the member if only one load W was present as the member would not be in equilibrium.

At *any* plane such as X–X taken across the section, there exists a state of stress between the fibres on one side of the plane and those on the other. Here the stress is tensile by nature – the fibres on one side exerting stress to resist the tendency of being pulled away from the fibres on the other side of the plane. This type of resistance to the external loads is set up along the whole length of the bar and not at one plane only, just as resisting forces are set up by every link in a chain.

In this particular case (Fig. 7.9(a)), if the loads act through the centroid of the shape, the stress is provided equally by the many fibres and each square millimetre of cross-section provides the same resistance to the 'pulling apart' tendency. This is known as **direct** or **axial stress**.

Figure 7.10 represents a bar of cast steel which is thinner at the middle of its length than elsewhere, and which is subjected to an axial pull of 45 kN. If the bar were to fail in tension, it would be due to the bar snapping where the amount of material is a minimum. The total force tending to cause the bar to fracture is 45 kN at all cross-sections, but whereas 45 kN is being resisted by a cross-sectional area of 1200 mm^2 for part of its length, it is being resisted by only 300 mm^2 at the middle portion of the bar. The intensity of load is greatest at this middle section and is at the rate of 150 N/mm^2 of cross-section. At other points along the bar, 45 kN is resisted by 1200 mm^2 and the stress is equal to 37.5 N/mm^2.

In cases of direct tension, therefore

$$\text{stress} = \frac{\text{applied load}}{\text{area of cross–section of member}} = \frac{W}{A}$$

The SI unit for stress is the pascal, Pa = 1 N/m^2. However, since the square metre is a rather inconvenient unit of area when measuring stress, British Standards have adopted the multiple N/mm^2 as a stress unit for materials.

It should be noted that 1 N/mm^2 = 1 MN/m^2 = 1 MPa.

Fig. 7.10 Tensile stress in a non-uniform steel bar

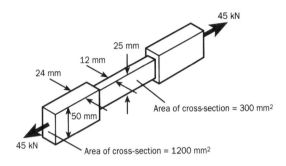

Example 7.2

A bar of steel 2000 mm² in cross-sectional area is being pulled with an axial force of 180 kN. Find the stress in the steel.

Solution

Since 2000 mm² of cross-section is resisting 180 kN, or 180 000 N, it means that each mm² is resisting 90 N. In other words,

$$\text{stress} = \frac{W}{A} = \frac{180\,000\text{ N}}{2000\text{ mm}^2} = 90\text{ N/mm}^2$$

Compressive stress

Figure 7.9(b) shows a similar member to that of Fig. 7.9(a) but with two axial forces of W kN each acting inwards towards the other and thus putting the bar into a state of compression.

Again, at any plane section such as X–X there is a state of stress between the fibres, but this time the stress which is generated is resisting the tendency of the fibres to be crushed. Once more the stress is shared equally beween the fibres, and the stress is

$$\frac{\text{load}}{\text{area}} \quad \text{i.e.} \quad \frac{W}{A} \quad \text{MN/m}^2 \text{ or N/mm}^2$$

Example 7.3

A brick pier is 0.7 m square and 3 m high and weighs 19 kN/m³. It is supporting an axial load from a column of 490 kN (Fig. 7.11). The load is

Fig. 7.11 Example 7.3

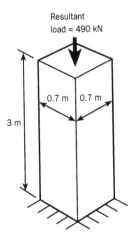

uniformly spread over the top of the pier so the arrow shown merely represents the resultant of the load. Calculate t he stress in the brickwork immediately under the column and the stress at the bottom of the pier.

Solution   Area of cross-section = 0.49 m². Thus 490 kN on 0.49 m² is equivalent to 1000 kN on each square metre. Immediately under the column, therefore

Stress = 490 kN/0.49 m² = 1 MN/m² or 1 N/mm²

Weight of pier = 0.7 × 0.7 × 3.0 × 19 = 28 kN, hence

Total load = 490 + 28 = 518 kN

and at the bottom of the pier

Stress = 518 kN/0.49 m² = 1057 kN/m² or 1.06 N/mm²

Strain

All materials slightly alter their shape when they are stressed. A member which is subjected to tensile stress increases in length and its cross-section becomes slightly smaller. Similarly, a compression member becomes shorter and slightly larger in cross-section. The very slight alteration in cross-section of building members is not as a rule important and only the alterations in length will be considered in this book.

When a tension force is applied to a rubber band it becomes longer, and when the force is removed, the material reverts to its original length. This property is common to some building materials such as steel, although under its normal working loads the amount of elongation is usually too small to be detected by the naked eye. Another property of elastic materials is that the alteration in length is directly proportional to the load. For example, if in a given member

10 kN produces an extension of 2 mm

20 kN will produce an extension of 4 mm

30 kN will produce an extension of 6 mm

and so on, provided the *elastic limit* of the material is not exceeded.
This law:

Change in length is proportional to the force

was first stated by Robert Hooke (1635–1703) and is therefore known as Hooke's law. Hooke was a mathematician and scientist and a contemporary of Sir Christopher Wren.

Although all materials alter in length when stressed, materials which are not elastic do not obey Hooke's law. That is, the changes in length are not directly proportional to the load. Structural steel and timber are almost perfect elastic materials.

Just as it is convenient to represent the stress in a member as force per unit area, so it is convenient to represent the change in length of a member in terms of change per unit length, i.e.

$$\text{strain} = \frac{\text{change in length}}{\text{original length}} = \frac{\delta l}{l}$$

where δ is the Greek letter delta, which is frequently used to denote small changes.

Since strain is the direct outcome of stress, it, too, is classified as tensile and compressive.

Elasticity

Suppose that a general formula is required for determining the amount of elongation (or shortening) in any member composed of an elastic material. Let that material be rubber. By carrying out experiments on pieces of rubber of different lengths, different areas of cross-section and different qualities, the following facts would be demonstrated:

- Increase in the load W produces proportionate increase of elongation (Hooke's law), therefore W is one term of the required formula.
- Increase in length l of the member means increase in elongation. It will be found that the elongation is directly proportional to the length of the member, i.e. for a given load and a given area of cross-section a member 500 mm long will stretch twice as much as a member of the same material 250 mm long. l is therefore a term in the formula.
- It is harder to stretch a member of large cross-sectional area A than a member of small cross-sectional area. Experiments will show that the elongation is inversely proportional to the area, i.e. for a given load and a given length, a member 10 mm^2 in cross-sectional area will stretch only half as much as a member 5 mm^2 in area. A is therefore a term in the formula.
- Rubber can be obtained of different qualities, i.e. of different stiffnesses. Other conditions being constant (i.e. length, area, etc.) it is more difficult to stretch a 'stiff' rubber than a more flexible rubber. Let this quality of stiffness be represented by the symbol E, then the greater the value of E, i.e. the greater the stiffness of the material, the smaller will be the elongation, other conditions (length, load and area) being constant.

The formula for elongation (or shortening) of an elastic material can now be expressed as

$$\delta l = \frac{Wl}{AE}$$

W and l are in the numerator of the fraction, because the greater their values, the greater the elongation. A and E are in the denominator of the fraction because the greater their values (i.e. the greater the area and the greater the stiffness) the smaller will be the elongation.

The symbol E in the above formula represents the stiffness of a given material, i.e. it is a measure of its elasticity and is called the **modulus of elasticity**. If it can be expressed in certain definite units, the formula can be used to determine elongations of members of structures. Now if $\delta l = Wl/AE$, by transposing the formula,

$$E = Wl/A\,\delta l$$

and the value of E for different materials can be obtained experimentally by compression or tension tests.

The foregoing explanation is lengthy and the usual more concise treatment of the subject is as follows. In an elastic material stress is proportional to strain (Hooke's law), i.e.

$$\frac{\text{stress}}{\text{strain}} = \text{a constant value}$$

This value is called the modulus (i.e. measure) of elasticity of the material and is denoted by E.

Therefore

$$\frac{\text{stress}}{\text{strain}} = E$$

but

$$\text{stress} = \frac{\text{load}}{\text{area}} = \frac{W}{A}$$

and

$$\text{strain} = \frac{\text{change in length}}{\text{original length}} = \frac{\delta l}{l}$$

Thus

$$\frac{W}{A} \left| \frac{\delta l}{l} \right. = E \quad \text{or} \quad E = \frac{Wl}{A\,\delta l}$$

The modulus of elasticity is often called Young's modulus, after the scientist Thomas Young (1773–1829). Note that modulus of elasticity is expressed in the same units as stress. This is because stress is load per unit area and strain has no dimensions, it is just a number and therefore

$$\frac{\text{stress}}{\text{strain}} \left(\frac{\text{MN/m}^2 \text{ or N/mm}^2}{\text{a number}} \right) = E \text{ N/mm}^2$$

Figure 7.12(a) indicates in essentials the manner of testing a block of timber 75 mm × 100 mm in cross-section and 300 mm high. In the actual

Fig. 7.12 Testing a block under compression: (a) experimental method; (b) graph of results given in Table 7.5

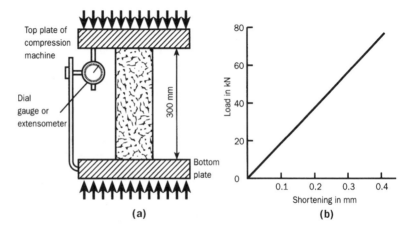

Table 7.5 Timber block – compression (Fig. 7.12)

Reading	Load on specimen W (kN)	Shortening of specimen δl (mm)
1	0	0.0
2	20	0.1
3	40	0.2
4	60	0.3
5	80	0.4

experiment four gauges (extensometers) were used, one at each corner, and the average of the four readings was taken for each increment of load. A gauge was used at each corner because possible small inequalities in the top and bottom surfaces of the timber might result in unequal shortening of the timber block. The readings in Table 7.5, which have been slightly adjusted for purposes of clearer explanation, were recorded as the compressive load was gradually increased.

Now

$$E = \frac{Wl}{A\,\delta l}, \text{ where } l = 300\,\text{mm} \quad \text{and} \quad A = 7500\,\text{mm}^2$$

Hence

$$\text{From reading 2} \quad E = \frac{20\,000 \times 300}{7500 \times 0.1} = 8000\,\text{N/mm}^2$$

$$\text{From reading 3} \quad E = \frac{40\,000 \times 300}{7500 \times 0.2} = 8000\,\text{N/mm}^2$$

$$\text{From reading 4} \quad E = \frac{60\,000 \times 300}{7500 \times 0.3} = 8000\,\text{N/mm}^2$$

$$\text{From reading 5} \quad E = \frac{80\,000 \times 300}{7500 \times 0.4} = 8000\,\text{N/mm}^2$$

The results of the experiment can be plotted as shown in Fig. 7.12(b). In an actual experiment results might not be as uniform as given in Table 7.5, but the resultant graph would approximate very closely to a straight line.

This indicates that E is a constant value, and it is this value which is taken as the modulus of elasticity for the timber.

It has to be noted, however, that materials such as timber and concrete are subject to the phenomenon of creep. This is a time-dependent strain deformation caused not by an increase in stress but by the duration of the applied load. Under these conditions the value of E is not constant since the stress/strain diagram is not a straight line. These strain deformations have to be taken into account in the design of structural elements but the topic is outside the scope of this book.

Example 7.4

A post of timber similar to that used in the above test is 150 mm square and 4 m high. How much will the post shorten when an axial load of 108 kN is applied?

Solution

The modulus of elasticity E of this type of timber is known to be 8000 N/mm².

$$A = 22\ 500\ mm^2 \qquad l = 4000\ mm$$

Therefore,

$$\delta l = \frac{Wl}{AE} = \frac{108\ 000 \times 4000}{22\ 500 \times 8000} = 2.4\ mm$$

This problem demonstrates that if the modulus of elasticity of a material is known, the amount a given member will shorten (or lengthen) under a given load can be calculated.

Behaviour of steel in tension – yield point

A steel bar, 12 mm in diameter, was gripped in the jaws of a testing machine and subjected to a gradually increasing pull (Fig. 7.13). A gauge was attached to two points on the bar 250 mm apart (gauge length = 250 mm) and the readings recorded in Table 7.6. (The elongations given in the table have been slightly adjusted.)

The area A of a 12 mm diameter circle is 113.1 mm².

$$l = 250\ mm$$

From reading 2,

$$E = \frac{Wl}{A\ \delta l} = \frac{3800 \times 250}{113.1 \times 0.04} = 210\ 000\ N/mm^2$$

From reading 3,

$$E = \frac{Wl}{A\ \delta l} = \frac{7600 \times 250}{113.1 \times 0.08} = 210\ 000\ N/mm^2$$

and so on.

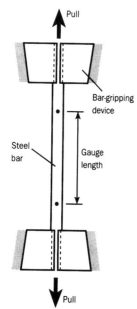

Fig. 7.13 Testing a steel bar in tension

Table 7.6 Steel – elongation (Fig. 7.13)

Reading	Load (kN)	Elongation (mm)
1	0.0	0.00
2	3.8	0.04
3	7.6	0.08
4	11.4	0.12
5	15.2	0.16
6	19.0	0.20

The constant value of E in this case is 210 000 N/mm².

Note that the modulus of elasticity of steel is much greater than that of timber. Steel, of course, is much more difficult to stretch than is timber.

Within the range of the above experiment, where the highest recorded load was 19 kN, the steel behaved as an elastic material. This means that, when the load is taken off, the bar will revert to its original length; on re-loading, the bar will again elongate; and on unloading the elongation will disappear, and so on.

If the load is gradually increased beyond 19 kN, the bar will continue to stretch proportionately to the applied load until a loading of between 30 and 32 kN is reached. (These loads on a bar 12 mm in diameter are equivalent to stresses of 265–280 N/mm².) At about this point the steel reaches what is called its **elastic limit** (i.e. it ceases to behave as an elastic material) and begins to stretch a great amount compared with the previous small elongations. This stretching takes place without the application of any further load and the steel is said to have reached its **yield point**.

At this point the **plastic** or **ductile behaviour** of the steel begins. If the load is removed from the steel after it has been loaded beyond its yield point, the steel will not revert to its original length. The elongation which remains is called *permanent set*.

After a short time the steel recovers a little and ceases to stretch. Additional load can now be applied, but the steel has been considerably weakened and stretches a great deal for each small increment of load. Finally, at a stress of about 450 N/mm² the bar breaks, but just before it breaks, it *waists* at the point of failure as indicated in Fig. 7.14(a). If the two fractured ends are placed together and the distance between the original gauge points is measured, it will be found that the total elongation is 20 per cent or more of the original length.

Figure 7.14(b) shows a graph which has been constructed from the results of a test on a 12 mm diameter bar, the gauge length being 250 mm.

Actually, the limit of proportionality (stress proportional to strain) is reached a little before the yield point, but for most practical purposes it is quite justifiable to consider the limit of proportionality, the elastic limit and the yield point to be identical. In addition, referring to Fig. 7.14(a), the bar reduces considerably in cross-section just before it fails, and some load can be taken off

Fig. 7.14 Results of a tensile test of steel: (a) waisting of bar at failure point; (b) typical graph of results; (c) 0.2% strain proof stress

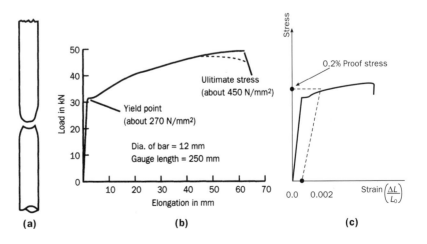

the bar so that the actual load which causes failure is less than the maximum recorded load. This is indicated on the graph by the dotted line. The ultimate or failing stress, i.e. the tensile strength of the steel, is calculated on the original cross-sectional area of the bar and not on the final reduced cross-section.

As you have seen from the above information, stress and strain are very important to both designers and material engineers in selecting construction materials for a particular building or structure. From them they can calculate the strength, stiffness and plastic properties of the selected materials. In general the stress should always be kept below the yield value throughout the useful life of the structure. Where the construction material does not have a distinct yield stress or yield point, a specified proof stress should be used instead for strength calculations. A 0.2 per cent strain proof stress is defined in Fig. 7.14(c) for high yield steel. To obtain the proof stress you simply draw a line parallel to the linear elastic portion of the stress strain curve.

Stiffness is a measure of the deflection of a material under the action of an applied load. It is also very useful in calculating the buckling and deflection of flexural and compression members. The strength of a material is very important for calculating the size of the structural element as well as determining the maximum load the structure can carry.

The strength of construction materials upon which design or selection of the size of an individual structural element is based is given by:

$$\text{Design strength} = \frac{\text{characteristic strength } (f_k)}{\text{partial factor of safety } (\gamma_m)}$$

The *characteristic strength*, f_k, *material strength* and *partial factor of safety* (γ_m) are explained in Chapter 1.

Table 7.7 gives some values of γ_m adopted in practice for use with ultimate limit states for steelwork, concrete, timberwork and masonry work. For serviceability limit states the reader is referred to the relevant code of practice.

The importance of using the design strength of materials in design calculation is explained as follows.

Table 7.7 Values of characteristic strength (γ_m) for five construction materials

Concrete	Steel reinforcement	Steelwork	Masonry work	Timberwork
1.5	1.05	1	2.5 to 3.5[a]	_[b]

[a] depends on the category of quality controls, see the relevant code of practice.
[b] During structural calculations, permissible stress = the grade stress × appropriate modification factors, see BS 5268: Part 2, Tables 13, 14, 15, 19 and 20.

Strength of materials and factor of safety

In Examples 7.2 and 7.3 the stress in the steel was calculated as 90 N/mm² and the stress in the brickwork as 1.0 N/mm² (or 1.06 N/mm²) respectively. Two questions now arise:

- Is it safe to allow these materials to be stressed to this extent?
- Can higher stresses be allowed so that smaller members may be used, thus economizing in material?

From the above discussion of the behaviour of steel in tension it can be deduced that the bar of 2000 mm^2 (Example 7.2) would require about 900 kN to cause it to fail in tension. The bar, however, is only supporting 180 kN, therefore it is amply strong and a smaller bar would suffice. The question now is: How small can the bar be? It would be, surely, unwise to make it of such dimensions that the steel in the bar would be stressed beyond its yield point because the steel would be in an unstable state. Various other factors also argue against using too high a stress for design purposes. The actual stress in the bar might be more (or even less) than the calculated value, due to assumptions made during the calculation stage. For example, the calculations may be based on the presence of a perfect hinge at a certain point in the structure. If the construction is such that there is a certain amount of fixity at the so-called hinge, the actual stresses in the members might be somewhat different from the calculated stresses.

Structural design is not an exact science, and calculated values of reactions, stresses, etc., whilst they may be mathematically correct for the theoretical structure (i.e. the model), may be only approximate as far as the actual behaviour of the structure is concerned.

For these and still other reasons it is necessary to make the design or working stress (or the allowable or permissible stress) less than the ultimate stress or (as intimated in the case of steel) the yield stress to allow for a safety margin against failure. This margin is provided by the introduction of the **factor of safety** γ_m.

The value of the factor of safety γ_m varies between 1.0 and 3.5 (see Table 7.7) at present and depends on many circumstances. It has been progressively reduced as the knowledge of structural behaviour of materials has increased and the quality of supervision of construction has improved. Its application to structural calculations is explained in later chapters.

Stresses in composite members – modular ratio

The knowledge of the modulus of elasticity E is particularly useful in determining stresses in composite members. These are structural elements made up of two (or more) materials (e.g. steel and timber in flitch beams or steel and concrete in reinforced concrete beams), in which the materials are rigidly fixed together so that any changes in the length of the element are the same in each of the constituent materials.

Consider an element consisting of materials A and B. Then

$$E_A = \frac{\text{stress}_A}{\text{strain}_A} \quad \text{and} \quad E_B = \frac{\text{stress}_B}{\text{strain}_B}$$

but strain$_A$ = strain$_B$, from the above definition of composite member, so taking the ratio of the moduli of elasticity

$$\frac{E_A}{E_B} = \frac{\text{stress}_A}{\text{strain}_B}$$

The ratio E_A/E_B, being the ratio of two moduli, is called the **modular ratio** and is denoted by the letter m.

$$\text{Modular ratio } m = \frac{E_A}{E_B}$$

Example 7.5

Two 150 mm × 75 mm × 4 m long timber members are reinforced with a steel plate 150 mm × 6 mm × 4 m long (Fig. 7.15), the three members being adequately bolted together.

The permissible stresses for the timber and the steel are 6 N/mm² and 130 N/mm², respectively, and E for timber is 8200 N/mm² and for steel is 205 000 N/m².

Calculate the permissible tensile load for this composite member and the amount of elongation due to this load.

Solution

Area of timber = 2 × 150 × 75 = 22 500 mm²

Area of steel = 150 × 6 = 900 mm²

The stresses in the composite parts will be in the ratio

$$\frac{E_s}{E_t} = \frac{205\,000}{8200} = 25$$

so that if the stress in timber is 6 N/mm², the stress in steel would have to be 6 × 25 = 150 N/mm² which exceeds the permissible stress for steel of 130 N/mm².

It follows, therefore, that the timber may not be fully stressed and the steel stress is the critical one.

Thus the stress in timber will be 130/25 = 5.2 N/mm².

Hence

$$\text{Safe load for timber} = 5.2 \times 22\,500 = 117 \text{ kN}$$

$$\text{Safe load for steel} = 130 \times 900 = \frac{117 \text{ kN}}{234 \text{ kN}}$$

and the elongation

$$\delta l = \frac{Wl}{AE} = \frac{117\,000 \times 4000}{900 \times 205\,000} = 2.6 \text{ mm or}$$

$$\delta l = \frac{117\,000 \times 4000}{22\,500 \times 8200} = 2.6 \text{ mm}$$

Summary

Stress   Obtained by dividing the applied load by the area of cross-section of the member, i.e.

$$\text{Tensile or compressive stress} = \frac{W}{A}$$

The design, permissible, working or allowable stress for a material depends on the nature of the material, the type of stress and the use of the material in the building, for example, whether it is used in a long column or in a short column.

Factor of safety   This is the failing or ultimate stress of the material divided by the design, permissible, working or allowable stress.

Strain   In tension or compression, strain is given by

$$\frac{\text{change in length}}{\text{original length}} = \frac{\delta l}{l}$$

In elastic materials obeying Hooke's law stress is proportional to strain provided the elastic limit of the material is not exceeded.

Modulus of elasticity   Also called Young's modulus, E, it is a measure of the resistance of an elastic material to being stretched or shortened. The greater the value of E, the more difficult it is to cause shortening or lengthening of the material.

$$E = \frac{\text{stress}}{\text{strain}} = \frac{Wl}{A\,\delta l}$$

and is measured in MN/m^2, N/mm^2, etc.

Modular ratio   When two materials A and B are combined,

$$\frac{\text{stress}_A}{\text{stress}_B} = \frac{E_A}{E_B} = m$$

where m denotes the modular ratio.

Exercises

1   A steel tie-bar 100 mm × 10 mm in cross-section is transmitting a pull of 135 kN. Calculate the stress in the bar.

2   Calculate the safe tension load for a steel bar 75 mm × 6 mm in cross-section, the working stress being 155 N/mm².

3   A tie-bar is 75 mm wide and it has to sustain a pull of 100 kN. Calculate the required thickness of the bar if the permissible stress is 150 N/mm².

4   A bar of steel circular in cross-section is 25 mm in diameter. It sustains a pull of 60 kN. Calculate the stress in the bar.

5   Calculate the safe load for a bar of steel 36 mm in diameter if the working stress is 155 N/mm².

6   A bar of steel, circular in cross-section, is required to transmit a pull of 40 kN. If the permissible stress is 150 N/mm² calculate the required diameter of the bar.

7   A timber tension member is 100 mm square in cross-section. Calculate the safe load for the timber if the permissible stress is 8 N/mm². Calculate the diameter of a steel bar which would be of equal strength to the timber member. Permissible stress for the steel is 150 N/mm².

8   A tie-bar of the shape shown in Fig. 7. Q8 has a uniform thickness of 12 mm and has two holes of 20 mm

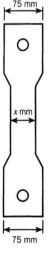

75 mm

x mm

75 mm

Fig. 7.Q8

diameter each. Calculate the width x so that the bar is equally strong throughout its length. Calculate the safe pull for the bar if the permissible stress is 150 N/mm².

9 A tie-bar of steel 150 mm wide is connected to a gusset plate by six rivets as shown in Fig. 7.Q9. The diameter of each rivet hole is 22 mm. Calculate the required thickness of the bar if the working stress is 155 N/mm².

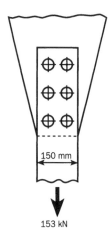

Fig. 7.Q9

10 Calculate the cross-sectional dimensions of a square brick pier to support an axial load of 360 kN, if the permissible stress for the brickwork is 1.7 N/mm².

11 A short specimen of deal timber 50 mm square in cross-section failed in a compression machine at a load of 70 kN. The permissible stress for such timber is 5.6 N/mm². Calculate the factor of safety.

12 A steel stanchion carrying a load of 877.5 kN is to be provided with a square steel base plate to spread the load on to a concrete foundation block. Calculate the minimum length of side (in mm) of the base plate if the stress on the concrete must not exceed 4.5 N/mm².

13 A steel column circular in cross-section is 150 mm in diameter and carries a load of 1.2 MN. Calculate (a) the compressive stress in the column; (b) the length of side of a square steel plate to transmit the column load to a concrete foundation block; the permissible stress on the concrete is 4.5 N/mm²; (c) assuming the

concrete base to weigh 150 kN, calculate the plan dimensions of the concrete foundation so that the stress on the soil does not exceed 200 kN/m².

14 The following data were recorded during a tensile steel test:

Diameter of bar = 20 mm

Distance between gauge points = 200 mm

Elongation due to load of 50 kN = 0.18 mm

Load at yield point = 79 kN

Failing or ultimate load = 127 kN

Calculate, in N/mm², (a) the stress at yield point; (b) the ultimate stress; (c) the modulus of elasticity of the steel.

15 A steel bar 100 mm × 12 mm in cross-section and 3 m long is subjected to an axial pull of 130 kN. How much will it increase in length if the modulus of elasticity of the steel is 210 000 N/mm²?

16 A hollow steel tube of 100 mm external diameter and 80 mm internal diameter and 3 m long is subjected to a tensile load of 400 kN. Calculate the stress in the material and the amount the tube stretches if Young's modulus is 200 000 N/mm².

17 During an experiment on a timber specimen 75 mm × 75 mm in cross-section, a shortening of 0.22 mm was recorded on a gauge length of 300 mm when a load of 36 kN was applied. Calculate the modulus of elasticity of the timber. Using this value of E, determine the amount of shortening of a timber post 150 mm square and 2.4 m high due to an axial load of 130 kN.

18 Assuming the permissible stress for a timber post 150 mm square and 2.7 m high is 6.5 N/mm², calculate the safe axial load for the post. How much will the post shorten under this load, assuming E to be 11 200 N/mm²?

19 Three separate members of steel, copper and brass are of identical dimensions and are equally loaded. Young's moduli for the materials are steel, 205 000 N/mm². copper, 100 000 N/mm²; brass, 95 000 N/mm². If the steel member stretches 0.13 mm calculate the amount of elongation in the copper and brass members.

20 During a compression test, a block of concrete 100 mm square and 200 mm long (guage length = 200 mm) shortened 0.2 mm when a load of 155 kN was applied. Calculate the stress and strain and Young's modulus for the concrete.

21 A tension member is made of timber and steel firmly fixed together side by side. The cross-sectional area of the steel is 1300 mm^2 and that of the timber is 4000 mm^2 and the length of the member is 3 m. If the maximum permissible stresses for the steel and timber when used separately are 140 N/mm^2 and 8 N/mm^2 respectively, calculate the safe load which the member can carry and the increase in length due to the load. Young's modulus for steel is 205 000 N/mm^2 and for timber is 8200 N/mm^2.

22 A structural member made of timber is 125 mm $\times$ 100 mm in cross-section. It is required to carry a tensile force of 300 kN and is to be strengthened by two steel plates 125 mm wide bolted to the 125 mm sides of the timber. Calculate the thickness of steel required if the permissible stresses for the steel and timber are 140 N/mm^2 and 7 N/mm^2 respectively. Assume that Young's modulus for the steel is 25 times that for the timber.

23 A timber post 150 mm square has two steel plates 150 mm $\times$ 6 mm bolted to it on opposite sides along the entire length of the post. Calculate the stresses in the timber and steel due to a vertical axial load of 350 kN. If the post is 3 m high calculate the amount of shortening under the load. E for steel is 205 000 N/mm^2 and E for timber is 8200 N/mm^2.

24 A metal bar consists of a flat strip of steel rigidly fixed alongside a flat strip of brass. The brass has a cross-sectional area of 900 mm^2 and the steel 300 mm^2. The compound bar was placed in a tensile testing machine and the extension measured by means of an extensometer fixed over a 250 mm gauge length. The extension was recorded as 0.13 mm. Calculate the load applied to the bar and the stress in each material.

E for brass = 80 000 N/mm^2

E for steel = 205 000 N/mm^2

In this chapter the twin effects of the action of applied loads on beams are investigated in terms of the shear force and bending moment. The plotting of shear force diagrams (SFDs) and bending moment diagrams (BMDs) is demonstrated, and the significance of the position of the points of zero shear (maximum bending moment M_{max}) and zero bending (contraflexure) is explained.

When a beam is loaded, the applied loads have a tendency to cause failure of the beam, and whether or not the beam actually does fail depends obviously upon the extent or amount of loading and on the size and strength of the beam in question.

It is necessary to provide a beam that will safely carry the estimated loading with a reasonable factor of safety, and which will at the same time be light enough for economy and shallow enough to avoid unnecessary encroachment upon headroom.

In order to calculate the stresses that loading will induce into the fibres of a beam's cross-section, and to compare them with the known safe allowable stress for the material of which the beam is made, it is necessary to study the ways in which loading *punishes* a beam, and to assess the degree of *punishment*.

Loading tends to cause failure in two main ways:

- By 'shearing' the beam across its cross-section, as shown in Fig. 8.1(a).
- By bending the beam to an excessive amount, as shown in Fig. 8.1(b).

These two tendencies to failure or collapse do occur simultaneously, but for a clearer understanding of each they will be examined separately.

Shear force

Consider the portion of beam shown in Fig. 8.2(a). For simplicity of explanation the weight of the beam itself has been ignored.

Figure 8.2(b) shows how the beam would tend to shear at point A. The only load to the left of A is the left-hand (l.h.) reaction (acting upwards) of 55 kN and there is therefore a resultant force of 55 kN tending to shear the portion of beam to the left of A upwards as shown. Note that the loads to the right of A are 70 kN downward and reaction 15 kN upward, so that there is also a resultant of 55 kN tending to shear the portion of beam to the right of A downward.

This 55 kN upward to the left, and 55 kN downward to the right, constitutes the shearing force at the point A.

It follows from the above that **shear force** may be defined as the algebraic sum of the loads to the left or to the right of a point.

Fig. 8.1 Failure caused by loading: (a) shearing; (b) bending

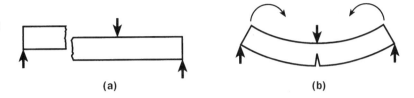

This type of shearing force, where the resultant shear is upwards to the left and downwards to the right, may be called **positive** shear.

Consider now the point B as shown in Fig. 8.2(c). The resultant shear to the left is seen to be $55 - 40 = 15$ kN upward to the left, and $30 - 15 = 15$ kN downward to the right (again positive shear). At point C in Fig. 8.2(d), the shear to the left is $30 + 40 - 55 = 15$ kN down, and to the right equals the r.h. reaction of 15 kN upward. This type of shear, down to the left and up to the right, is called **negative** shear.

The above example of shear force ignored the weight of the beam and the only loading consisted of point loads – that is to say the loading was considered to be applied at a definite point along the span. Loads in actual fact are rarely applied in this fashion to structural members, but many loads applied to beams approximate to point loading and in design are considered as concentrated loads. The main beam shown in Fig. 8.3, for example, carries the reaction from one secondary beam and in addition has sitting on its top flange a short steel post; both these loads are applied to such a short length of the beam that they may be considered as point loads.

On the other hand, there are many cases where the loading is applied at a more or less uniform rate to the span of the beam. Such an example is found where a brick wall is carried on the top flange of a beam, or when a reinforced concrete slab sits upon a steel or reinforced concrete beam. Now, shear force has been described as the algebraic summation of loads to the left (or to the right) of a point. Therefore, the shear alters in such cases only where another

Fig. 8.2 Shear force in a beam

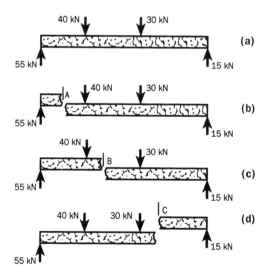

Fig. 8.3 Shear forces considered as point loads

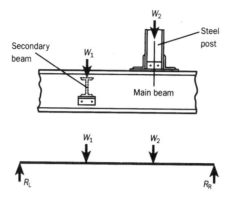

point load occurs. When the loading is uniformly distributed, however, as in Fig. 8.4, then the shear force will vary at a uniform rate also and a sudden jump in the value will only occur at the point of application of the one point load. The shear at point A (Fig. 8.4(a)) upward to the left will be

$$140\text{kN up} - 30\text{ kN down} = +110\text{ kN}$$

Similarly the shear at point B (Fig. 8.4(b)) upward to the left is

$$140\text{kN up} - 60\text{ kN down} = 80\text{ kN (again positive)}$$

The shear just to the left of C (Fig. 8.4(c)) is

$$140 - 90 = 50\text{ kN positive}$$

whilst just to the right of C when the 40 kN point load is included it is

$$140 - (90 + 40) = 10\text{ kN positive}$$

Thus it will be seen that uniform loads cause gradual and uniform change of shear, whilst point loads bring about a sudden change in the value of the shear force.

Fig. 8.4 Shear force due to uniform loading

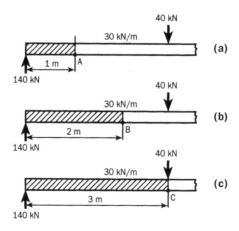

Bending moment

The degree of punishment in bending is measured as bending moment, and the amount of bending tendency is dependent upon the loads and upon the distance between them.

For example, the beam shown in Fig. 8.5(a) tends to split in bending under the 30 kN load because the l.h. reaction of 15 kN acting 1.5 m to the left has a clockwise bending tendency of $15 \times 1.5 = 22.5$ kN m, at the point where splitting of this type would most easily occur in this case.

If the beam had been of 10 m span instead of only 3 m the l.h. reaction (ignoring for a while the self-weight of the beam) would still be only 15 kN, but this time the bending tendency at the point of maximum stress would be $15 \times 5 = 75$ kN m.

Reverting to the beam shown in Fig. 8.5(a), which we have seen has a clockwise bending tendency to the left of point C of 22.5 kN m, there is also an anticlockwise bending tendency to the right of C of $15 \times 1.5 = 22.5$ kN m caused by the action of the r.h. reaction. At *any point* along the span of a simple beam of this type supported at its ends, the bending tendency to the left will always be clockwise, and that to the right anticlockwise but of the same amount. Thus at any point of such a beam the bending will be of the *sagging* type as shown in Fig. 8.1(b), and the fibres towards the lower face of the section will be subjected to tension.

There are types of beams that bend in the opposite way. For example, a cantilever as shown in Fig. 8.5(b) has a *hogging* rather than a sagging tendency and obviously the moments in this type must be anticlockwise to the left and clockwise to the right.

To distinguish between these two types of bending it is normal to describe sagging as positive and hogging as negative.

From the foregoing example, which has been of the very simplest nature, it follows that **bending moment** may be described as the factor which measures the bending effect at any point of a beam's span due to a system of loading. The amount of bending moment is found by taking the moments acting to the left *or* to the right of the point concerned. The beam discussed and shown in Fig. 8.5(a) had only one load acting to the left of the point C at which the bending moment was calculated; where there is more than one load on the portion of beam concerned – as for example in the portion of beam shown in Fig. 8.6(a) – then taking moments to the left of point C, the reaction has a clockwise bending effect and the two downward loads have anticlockwise bending tendency. The net bending moment at C will be the

Fig. 8.5 Bending moments:
(a) sagging of beam; (b) hogging of beam

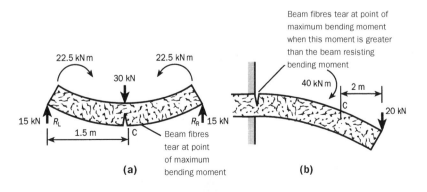

(a)   **(b)**

Fig. 8.6 Example of bending moments

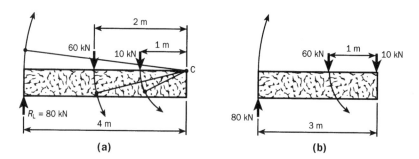

(a)   (b)

difference between these two types of moment, i.e. their algebraic sum. The bending moment at any point is, therefore, defined as the algebraic sum of the moments caused by the forces acting to the left or to the right of that point.

In Fig. 8.6(a), the l.h. reaction exerts a clockwise moment of 80 kN × 4 m = 320 kN m about point C and the downward loads exert moments in an anticlockwise direction equal to 60 kN × 2 m = 120 kN m and 10 kN × 1 m = 10 kN m. Thus the bending moment at C is

$$+80 \times 4 - 60 \times 2 - 10 \times 1 = 320 - 120 - 10 = 190 \text{ kN m}$$

The bending moment at the point of the 10 kN load in the above beam would be

$$80 \times 3 \text{ m} - 60 \text{ kN} \times 1 \text{ m} = 240 - 60 = 180 \text{ kN m}$$

as shown in Fig. 8.6(b). Note that in this case the 10 kN load is ignored, as, passing through the point concerned, it does not exert a moment about that point.

When, on the other hand, a uniformly distributed load is applied to the beam, the punishment caused is less than that exerted by a comparable point load. Therefore, it would be most uneconomical in design to treat such loads as being concentrated at their midpoint.

Where such uniform rate of loading occurs, as in the case of the portion of beam shown in Fig. 8.7 for example, only that portion of loading which lies to the left of the point C (shown shaded) need be considered when the moment at that point to the left is being calculated. The shaded portion of load is 30 kN/m × 4 m = 120 kN, and its resultant lies halfway along its length (2 m from C). Therefore the moment about C of the uniform load is 120 kN × 2 m = 240 kN m and the total bending moment at point C is

$$200 \times 4 - 120 \times 2 = 800 - 240 = 560 \text{ kN m}$$

Fig. 8.7 Bending moments due to uniform loading

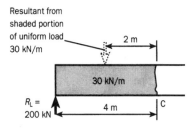

**Shear force and bending
moment diagrams**

Shear force and bending moment have been described in general terms and
it will have been seen that the values of both vary at different points along
the span.

It is often desirable to show this variation by means of diagrams, which
are really graphs, and these diagrams are called shear force diagrams and
bending moment diagrams.

The following examples will serve to show how the diagrams of this type
may be constructed for simple and more complex cases.

Simply supported beam with point load

Figure 8.8(a)(i) shows a simply supported beam of span l carrying one point
load of W kN at the centre of the span. Since the loading is symmetrical the
reactions must be equal to each other, and each reaction will be $\frac{1}{2}W$ kN.
Ignoring the self-weight of the beam, at any point C at x m from the left-
hand reaction, the shear force to the left will be simply the value of the l.h.
reaction, i.e. $\frac{1}{2}W$ kN upward to the left, and wherever point C lies between
the l.h. end and the load, the shear will still be $\frac{1}{2}W$ kN.

Similarly at any point to the right of the load as at point D, the shear to
the left of D is $\frac{1}{2}W$ kN downward and $\frac{1}{2}W$ kN upward, which is equal to
$\frac{1}{2}W$ kN down to the left or $\frac{1}{2}W$ kN up to the right (negative shear). These
variations of the shear values are shown to scale on the **shear force diagram (SFD)** (Fig. 8.8(a)(ii)). The vertical ordinate of the diagram at any
point along the span shows the shear force at that point, and the diagram is
drawn to two scales: a vertical scale of 1 mm = a suitable number of
kilonewtons, and a horizontal scale of 1 mm = a suitable number of metres,
being the same scale as that used in showing the span of the beam.

Fig. 8.8 Shear force and bending
moments: (a)(i) beam with a
single central load point,
(ii) shear force diagram (SFD),
(iii) bending moment diagram
(BMD); (b)(i) beam with a single
non-central load point, (ii) SFD,
(iii) BMD

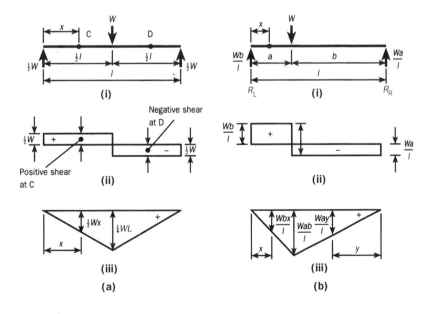

Figure 8.8(a)(iii) shows the variation of bending moment and constitutes a **bending moment diagram (BMD)**.

Obviously the moment at each end is zero, and at any point C at x m from the l.h. end the bending moment (summation of moments to the left) is simply $\frac{1}{2}W \times x = \frac{1}{2}Wx$ kN m. This moment increases as x increases, and will reach a maximum amount of $\frac{1}{2}W \times \frac{1}{2}l = \frac{1}{4}Wl$ at the centre of span as shown.

Again the diagram is drawn horizontally to the same scale as the span of the beam, but this time vertically to a scale of 1 mm = a suitable number of kN m or Nm m, and the vertical ordinate at any point along the span represents the bending moment at that point.

Comparing the two diagrams, it will be seen that the bending moment everywhere on the span is positive, and that the shear force changes its type from positive to negative at the point along the span where the bending moment reaches its maximum amount.

Figure 8.8(b)(i) shows a simply supported beam loaded with one single non-central point load, at distance a from one end and at b from the other.

Taking moments about the l.h. end,

$$\text{r.h. reaction} \times l = W \times a$$

$$\text{r.h. reaction} = Wa/l$$

Similarly, taking moments about the r.h. end,

$$\text{l.h. reaction} \times l = W \times b$$

$$\text{l.h. reaction} = Wb/l$$

At any point between the l.h. end and the point load,

$$\text{shear to the left} = \text{l.h. reaction} = Wb/l \text{ (positive)}$$

Similarly, at any point between the load and the r.h. end,

$$\text{shear to the right} = \text{r.h. reaction of } Wb/l$$
$$\text{(up to the right, thus negative shear)}$$

Thus the shear changes sign from positive to negative at the point load as in the previous example, and the SFD is as shown in Fig. 8.8(b)(ii).

As before, the bending moment is zero at the l.h. end, and at any point between that end and the load is equal to

$$R_L \times x = (Wb/l) \times x = Wbx/l$$

This reaches a maximum value of Wab/l at the point load (where $x = a$).

Also, at any point between the load and the r.h. end, at a distance of y from the r.h. reaction,

$$\text{Bending moment} = \text{r.h. reaction} \times y = Way/l$$

and this also reaches a maximum at the point load (where $y = b$) of Wab/l.

The full bending moment diagram (Fig. 8.8(b)(iii)) is thus a triangle with a maximum vertical height of Wab/l at the load, and the bending moment at any point along the span may thus be scaled from the diagram to the same scale which was used in setting up the maximum ordinate of Wab/l.

Simply supported beam with uniformly distributed load

The next case will deal with a simply supported beam of span l carrying a uniformly distributed load of intensity w kN/metre run (Fig. 8.9(a)). The span consists of l m of load, and thus the total load is wl kN, and, as the beam is symmetrical, each reaction will be half of the total load:

$$R_{\mathrm{L}} = R_{\mathrm{R}} = \tfrac{1}{2}wl \text{ kN}$$

The shear (up to the left) at a point just in the span and very very near to R_{L} is quite obviously simply the l.h. reaction of $\tfrac{1}{2}wl$.

When the point concerned is, say, 1 m from R_{L}, however, the shear to the left (summation of loads to the left of the point) is then

$$\tfrac{1}{2}wl \text{ upwards} + 1 \text{ m of load } w \text{ downwards}$$

$$= \tfrac{1}{2}wl - w$$

Similarly when the point concerned is 2 m from R_{L}, the shear up to the left is

$$\tfrac{1}{2}wl \text{ upwards} - 2w \text{ (2m of load) downwards}$$

$$= \tfrac{1}{2}wl - 2w$$

Fig. 8.9 Simply supported beam carrying a uniform load: (a) load diagram; (b) SFD; (c) BMD

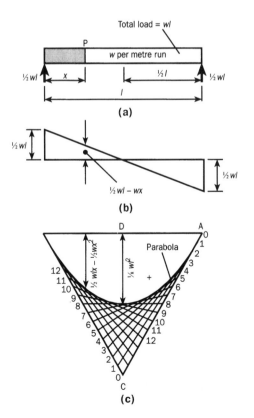

Putting this in general terms, the shear at any point P on the span at distance x from R_L is

$$\tfrac{1}{2}wl - x \text{ metres of uniform load} = \tfrac{1}{2}wl - wx$$

This will give a positive result wherever x is less than $\tfrac{1}{2}l$, and a negative result where x exceeds $\tfrac{1}{2}l$, so the shear will 'change sign' where $x = \tfrac{1}{2}l$ (at the point of midspan), and the SFD will be as shown in Fig. 8.9(b).

Referring again to Fig. 8.9(a), the bending moment at the l.h. end is again zero, and at a point 1 m from R_L, the bending moment (summation of moment to the left) is simply the algebraic sum of:

- the clockwise moment of the l.h. reaction ($\tfrac{1}{2}wl \times l$)
- the anticlockwise moment from 1 m of downward load ($w \times l \times \tfrac{1}{2} = \tfrac{1}{2}w$).

Thus, the bending moment $= \tfrac{1}{2}wl - \tfrac{1}{2}w$.

If the bending moment is required at a point P, at x m from R_L, then the bending moment is the algebraic sum again of:

- the clockwise moment of the l.h. reaction ($\tfrac{1}{2}wl \times x$)
- the anticlockwise moment of x m of load ($wx \times \tfrac{1}{2}x$)

Thus, the bending moment $= \tfrac{1}{2}wlx - \tfrac{1}{2}wx^2$.

This bending moment will be positive for any value of x, and will reach a maximum value of

$$\tfrac{1}{2}(wl \times \tfrac{1}{2}l) - \tfrac{1}{2}w(\tfrac{1}{2}l)^2 = \tfrac{1}{4}wl^2 - \tfrac{1}{8}wl^2 = \tfrac{1}{8}wl^2$$

It should be most carefully noted that the maximum bending moment (M_{max}) is $\tfrac{1}{8}wl^2$, where w is the amount of uniform load per metre.

Sometimes it is more convenient to think in terms of the *total load* W (i.e. capital W), and in this case $W = wl$, and M_{max} in terms of the total load will then be

$$\tfrac{1}{8}wl \times l = \tfrac{1}{8}Wl$$

If the values of this bending moment at points along the span are plotted as a graph, the resulting BMD will be a parabola with a maximum ordinate of $\tfrac{1}{8}wl^2$ or $\tfrac{1}{8}Wl$ as shown in Fig. 8.9(c). Where the diagram has to be drawn, it will be necessary only to draw a parabola having a central height of $\tfrac{1}{8}Wl$, and any other ordinates at points away from the centre may be scaled or calculated as required.

Shear force diagrams

It is advisable, at this stage, to study the shear force diagrams already drawn carefully and to note that these diagrams are drawn on a horizontal base. The upward loads (reactions) are projected upward on this base, and at the point loads, the SFD drops vertically by the amount of the point load.

Where a uniform load occurs, however, the SFD slopes down at a uniform rate, dropping for each metre of span an amount equal to the amount of uniform load per metre as shown in Fig. 8.10. It should be observed that in every case, the value of the bending moment has been a maximum at the point where the shear force changes its sign (generally referred to as the point of zero shear).

Fig. 8.10 Slope of SFD due to uniform loading

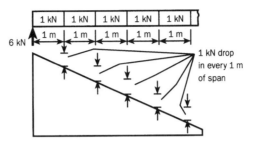

Bearing this in mind, it is common practice to draw the shear force diagram simply to discover where the maximum bending moment occurs. This is quite unnecessary, as will be seen from consideration of the two cases shown in Fig. 8.11.

In Fig. 8.11(a), the l.h. reaction is 130 kN, and the shear force will change sign at the point of M_{max} where the downward loads starting from the l.h. end just equal 130 kN.

From A to C, the downward uniform load is 2 m at 15 kN/m = 30 kN. Then at the point load, there is a further drop of 70 kN, making a drop in the shear diagram from A to C of 100 kN.

Hence, if D is the point of M_{max} (zero shear) then the portion of load shown shaded must equal $130 - 100 = 30$ kN, so that the downward loads up to D equal the l.h. reaction.

Thus, if the distance from C to D is x, then x m at 15 kN/m must equal 30 kN so $15x = 30$, i.e. $x = 2$ m and the maximum bending moment occurs at $2 + 2 = 4$ m from R_L.

Similarly in the case shown in Fig. 8.11(b) the load from the l.h. end up to and including the 70 kN point load is $2 \times 15 + 70 = 100$ kN.

Therefore, a further $145 - 100 = 45$ kN of load is required beyond C to D, the point of maximum bending moment. Therefore

$$15 \times x = 45 \text{ kN}$$

$$x = \frac{45}{15} = 3 \text{ m}$$

and M_{max} occurs at point D which is $2 + 3 = 5$ m from the l.h. end.

Fig. 8.11 SFDs for beams carrying uniform loads

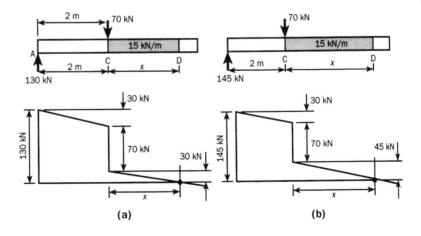

It will be seen that where the loading is uniform, the position of the maximum bending moment (M_{max}) occurs at a point on the span such that the sum of the downward loads from the l.h. end exactly equals the l.h. reaction.

Similarly, the downward loads taken from the r.h. end will equal the r.h. reaction at this point.

Uniformly distributed and point loads

When the loading includes point loading, however, as it does in the example shown in Fig. 8.12, it may well be that the point at which the bending moment is a maximum coincides with the position of a point load – and in the beam illustrated in Fig. 8.12 the maximum value occurs at the 40 kN load.

Summing up the loads (starting from the l.h. end) it will be seen that just to the left of the 40 kN load they are $30 \times 1.7 + 10 = 61$ kN, that is just less than R_L.

Immediately to the right of the 40 kN load they add up to $30 \times 1.7 + 10 + 40 = 101$ kN which is more than R_L.

The rule for finding the position of the **maximum bending moment** may therefore be stated as follows: add the downward loads together, starting from one reaction to the point where they equal (or suddenly become greater than) that reaction. This is the point of zero shear, and the point of maximum bending moment.

Cantilever

A beam which is supported at one end only by being firmly built into a wall, or which is held horizontally at one end only by other means, is called a **cantilever**. Figure 8.13(a)(i) shows such a cantilever AB having a length of l m and loaded with one point load of W kN at the free end B.

The only downward load is W kN, so for equilibrium the l.h. upward reaction will also be W kN, but in addition, to prevent the beam from rotating, the wall or other form of restraint at the fixed end A must exert a moment of Wl in an anticlockwise direction.

Fig. 8.12 SFD for a beam carrying uniform and point loads

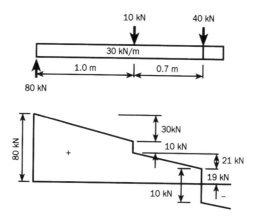

Fig. 8.13 SFDs and BMDs for cantilever beams: (a)(i) point load, (ii) SFD, (iii) BMD; (b)(i) uniformly distributed load, (ii) SFD, (iii) BMD

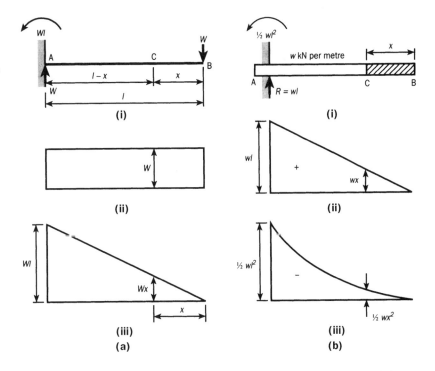

The shear at any point C between A and B (sum of the loads to the left or to the right of C) will be W kN down to the right, or W kN up to the left. This is positive shear, and the value remains the same at all points along the span, so that the SFD is a rectangle as shown in Fig. 8.13(a)(ii).

The bending moment (taken as the sum of moments to the right) at any point C, distance x m from the free end B, will be the downward load of W multiplied by x m.

Thus the bending moment at any point is Wx kN m. This is a negative or hogging bending moment and its value varies directly as the value of x varies, so that the BMD will be a triangle as shown in Fig. 8.13(a)(iii).

The maximum value will occur at the fixed end when $x = l$, and its amount is Wl kN m.

Note: The previous rule (page 153) for finding the position of M_{max} will not apply to cantilevers. In the case of cantilevers M_{max} will always occur at the fixed end.

Cantilever with uniformly distributed load

Figure 8.13(b)(i) shows a cantilever with a uniformly distributed load of w kN/m. The total load will be $w \times l = wl$ kN and the upward reaction at A will also be wl kN.

The moment of the downward load wl about the support A is

$$wl \times \tfrac{1}{2}l = \tfrac{1}{2}wl^2 \text{ kN m}$$

Thus the wall (or other form of restraint) at A must exert an anticlockwise moment of $\tfrac{1}{2}wl^2$ on the beam to prevent its rotating under the couple formed by the upward reaction and the resultant of the downward load.

The shear at any point C at distance x from the free end will be the portion of load to the right of C which is x m at w kN/m = wx kN (positive shear). This varies at the same rate as x varies, and so the shear force diagram will be a triangle having a maximum value at the reaction (where $x = l$) of wl kN, as shown in Fig. 8.13(b)(ii).

Similarly the bending moment at C will be the moment (clockwise) of the portion of load to the right of C, which is $(wx) \times \frac{1}{2}x = \frac{1}{2}wx^2$ kN m (negative or sagging moment).

This again reaches a maximum value of $\frac{1}{2}wl^2$ kN m at the fixed end A where $x = l$ and the rate of increase will be found to form a parabola as shown in the BMD in Fig. 8.13(b)(iii).

Note: If the total load W is used instead of the value of the load per metre run, w, then $W = wl$, and the maximum bending moment will be $\frac{1}{2}Wl$ kN m.

Example 8.1

Draw the SFD and BMD for the cantilever shown in Fig. 8.14(a) indicating all important values.

Solution

Reaction
There is only one reaction, R_L at A:

$$R_L = 30 + 20 + 10 = 60 \text{ kN}$$

Shear force
At any point between A and C, the shear downward to the right is the sum of the three downward loads which is 60 kN positive.

Similarly between C and D the shear is the sum of the two loads to the right which is 30 kN positive. Between D and the free end B the shear is the single load of 10 kN positive.

This is shown in the SFD constructed in Fig. 8.14(b).

Fig. 8.14 Example 8.1: (a) load diagram; (b) SFD; (c) BMD

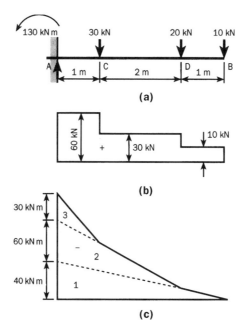

Bending moment
The BMD is most easily drawn by treating the three loads separately: the 10 kN load produces a bending moment of $10 \times 4 = 40$ kN m at A and the BMD for this load alone is the triangle 1 shown in Fig. 8.20(c).

The 20 kN load causes a bending moment at A of $20 \times 3 = 60$ kN m and the BMD for this load alone is the triangle 2.

Finally, the 30 kN load causes a $30 \times 1 = 30$ kN m bending moment at A and the BMD for this load is the triangle 3.

As in fact all three loads are on the beam at the same time, then the final BMD is the sum of the three triangles 1, 2 and 3 as shown in Fig. 8.14(c).

Example 8.2

Draw the SFD and BMD for the beam loaded as shown in Fig. 8.15(a)(i) indicating all important values.

Solution

Reactions
Taking moments about A:

$$R_R \times 4 = 20 \times 4 \times \tfrac{1}{2} \times 4 + 40 \times 0.8 + 30 \times 16 + 20 \times 32$$

$$= 160 + 32 + 48 + 64 = 304$$

$$R_R = 76 \text{ kN}$$

and

$$R_L = 80 + 40 + 30 + 20 - 76 = 94 \text{ kN}$$

Shear force

Note: It should be obvious by now that SFDs represent all point loads as vertical lines – upward in the case of reactions, and downward in the case of downward loads.

Downward uniform loads are shown as sloping lines (gradual change in the value of the shear force) and therefore the SFD may be plotted quickly from a horizontal base by merely plotting the loads in this way as they occur.

The horizontal base is shown as line A1–B1 in Fig. 8.15(a)(ii). The procedure is as follows:

1 Draw up from A1 the vertical reaction R_L of 94 kN (A1–F).
2 Draw the gradual change of (20×0.8) kN between A and C as the sloping line F–G.
3 Draw G–H of 40 kN vertically downward to represent the sudden change of shear at the 40 kN load at point C.
4 Draw the gradual change of 16 kN between C and D as the sloping line H–J.
5 Draw J–K of 30 kN vertically downward to represent the sudden change of shear at the 30 kN load at point D.
6 Draw the gradual change of (20×1.6) kN between D and E as the sloping line K–L.
7 Draw L–M of 20 kN vertically downward to represent the sudden change of shear at the 20 kN point load at E.
8 Draw the gradual change of 16 kN between E and B as the sloping line M–P.
9 Finally draw the vertical right hand reaction of 76 kN upward from P to join the horizontal base at B1.

Figure 8.15(a)(ii) shows the construction of this SFD, and the final diagram with its important values is shown in Fig. 8.15(a)(iii).

Bending moment

The bending moment at any point or the maximum bending moment may be calculated in the usual way by taking moments to the left or to the right of that point. But if the final BMD is required (as it is in this example) then it may be drawn by constructing separately:

- the BMD for the uniformly distributed load (above baseline – Fig. 8.15(b)(i)
- the BMD for the point loads (below baseline – Fig. 8.15(b)(ii)

Fig. 8.15 Example 8.2 (a)(i) shear force load diagram, (ii) construction of SFD, (iii) final SFD; (b)(i) BMD for uniformly distributed loads, (ii) BMD for point loads, (iii) addition of BMDs, (iv) final BMD

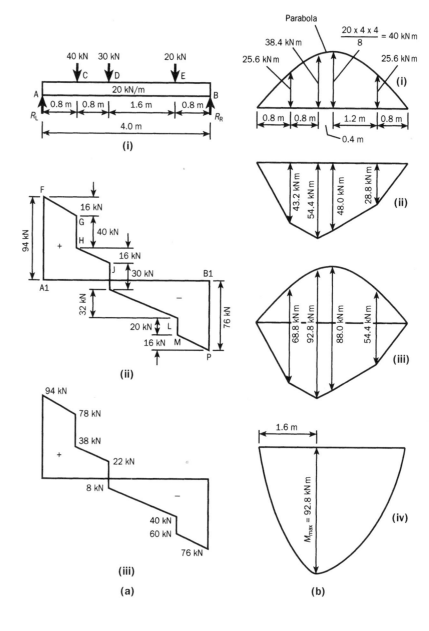

Then, by adding the calculated ordinates (Fig. 8.15(b)(iii)), the final BMD may be drawn, as shown in Fig. 8.15(b)(iv).

The value of the bending moment at any point may also be obtained by computing the area of the SFD to the left or to the right of that point.

Consider the bending moment at the centre of the span. The positive shear force area is

$$\tfrac{1}{2}(94 + 78) \times 0.8 + \tfrac{1}{2}(38 + 22) \times 0.8$$

$$= 68.8 + 24.0 = 92.8 \text{ kN m}$$

To obtain the negative shear force area, it is necessary to determine the value of the shear force at the centre of the span which is 0.4 m to the right of the 30 kN point load.

So, the shear force at the centre of span is

$$8 + 20 \times 0.4 = 16 \text{ kN}$$

Hence the negative shear force area to the left of the centre of the span is

$$\tfrac{1}{2}(8 + 16) \times 0.4 = 4.8 \text{ kN m}$$

Thus the bending moment at the centre of the span is

$$92.8 - 4.8 = 88.0 \text{ kN m}$$

Example 8.3 Draw the SFD for the beam shown in Fig. 8.16, and calculate the value of M_{max}. Determine also the bending moment at a point C at 2 m from the l.h. reaction.

Solution Taking moments about A to determine reactions:

$$R_{\text{R}} \times 9 = 30 \times 9 \times (\tfrac{1}{2} \times 9) + 60 \times 6 \times (\tfrac{1}{2} \times 6)$$

$$= 1215 + 1080 = 2295 \text{ kN}$$

$$R_{\text{R}} = 255 \text{ kN}$$

and

$$R_{\text{L}} = 60 \times 6 + 30 \times 9 - 255 = 375 \text{ kN}$$

Fig. 8.16 Example 8.3: (a) load diagram; (b) SFD

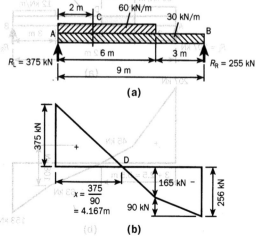

M_{max} will occur at x m from A where

$$x \text{ m} \times (60 \times 30) \text{ kN/m} = R_L = 375 \text{ kN}$$

$$x = \frac{375}{60 + 30} = 4.167 \text{ m}$$

The value of M_{max} will, therefore, be

$$M_{max} = 375 \times 4.167 - (60 + 30) \times 4.167 \times (\tfrac{1}{2} \times 4.167)$$

$$= 1562.50 - 781.25 = 781.25 \text{ kN m}$$

or by computing the areas of the SFD

$$M_{max} = 375 \times (\tfrac{1}{2} \times 4.167) = 781.25 \text{ kN m (as before)}$$

Bending moment at C, 2 m from R_L, is

$$375 \times 2 - (30 + 60) \times 2 \times (\tfrac{1}{2} \times 2) = 570 \text{ kN m}$$

If a BMD is required, bending moment values may be calculated at several points of the span and these values drawn to scale vertically from a horizontal line representing the span of the beam. The BMD is obtained by joining the tops of these lines.

Example 8.4

Draw the SFD for the loaded beam shown in Fig. 8.17, and determine the position and amount of the maximum bending moment (M_{max}).

Solution

Taking moments about A to determine reactions,

$$R_R \times 12 = 36 \times 3 \times 1.5 + 18 \times 12 \times 6 + 12 \times 3 \times 10.5$$

$$= 162 + 1296 + 378 = 1836$$

$$R_R = 153 \text{ kN}$$

Fig 8.17 Example 8.4: (a) load diagram; (b) SFD

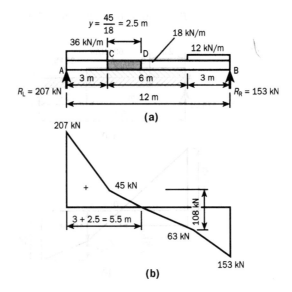

and

$$R_L = 36 \times 3 + 18 \times 12 + 12 \times 3 - 153$$
$$= 108 + 216 + 36 - 153$$
$$= 207 \, kN$$

Position of M_{max}
At point C, 3 m from l.h. end, downward load from l.h. end is

$$(36 + 18) \times 3 = 162 \, kN$$

M_{max} will occur where this has increased to 207 kN. Thus the M_{max} will occur at point D where the amount of load between C and D (shown shaded) is

$$207 - 162 = 45 \, kN$$

Thus

$$\text{Distance } v \times 18 \, kN/m = 45 \, kN$$

$$v = 45/18 = 2.5 \, m$$

Therefore M_{max} occurs at D at $(3.0 + 2.5) = 5.5$ m from the l.h. end

$$M_{max} = 207 \times 55 - 162 \times (2.5 - (\tfrac{1}{2} \times 3.0))$$
$$- 18 \times 2.5 \times \tfrac{1}{2} \times 2.5$$
$$= 1138.5 - 648 - 56.25$$
$$= 434.25 \, kN \, m$$

Check by shear force areas:

$$M_{max} = \tfrac{1}{2}(207 + 45) \times 3 + 45 \times \tfrac{1}{2} \times 2.5$$
$$= 378 + 56.25$$
$$= 434.25 \, kN \, m$$

Example 8.5

Draw the SFD and BMD for the beam shown in Fig. 8.18(a), and determine the position and amount of the maximum bending moment.

Solution

Taking moments about A:

$$R_R \times 1.8 = 45 \times 0.6 + 54 \times 1.2 + 18 \times 2.4$$
$$= 27.0 + 64.8 + 43.2$$
$$= 135$$
$$R_R = 75 \, kN$$

and

$$R_L = 45 + 54 + 18 - 75 = 42 \, kN$$

Fig. 8.18 Example 8.5: (a) load diagram; (b) SFD; (c) BMD; (d) deflected shape of beam

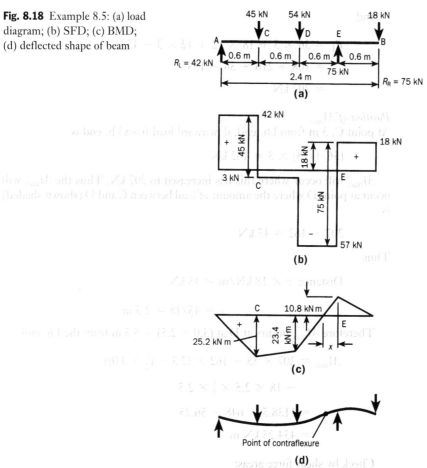

The shear force diagram is as shown in Fig. 8.18(b). The values of bending moments at

$$45 \text{ kN load} = 42 \times 0.6 = 25.2 \text{ kN m} \quad (+)$$

$$54 \text{ kN load} = 42 \times 1.2 - 45 \times 0.6 = 23.4 \text{ kN m} \quad (+)$$

$$\text{r.h. support} = 42 \times 1.8 - 45 \times 1.2 - 59 \times 0.6$$

$$= 10.8 \text{ kN m}$$

The bending moment diagram is as shown in Fig. 8.18(c).

Note that, due to the overhanging of the r.h. end, part of the beam is subjected to negative bending moment.

There are, therefore, two points of zero shear: one at C which marks the point of maximum positive moment, and one at E where the maximum negative moment occurs. The absolute maximum bending moment is thus 25.2 kN m at C, and the shape of the deflected beam is as shown in Fig. 8.18(d).

The point at which the negative bending moment changes to positive (and vice versa) is called the **point of contraflexure**. The value of the bending moment at that point is zero.

Example 8.6

Draw the SFD and BMD for the beam shown in Fig. 8.19(a) and determine the position and amount of the maximum bending moment.

Solution

Since the loading and the beam are symmetrical, the reactions are each equal to $\frac{1}{2}(20 + 70 + 20) = 55$ kN

$$\text{Bending moment at D} = 55 \times 1.3 - 20 \times 2.3$$
$$= 71.5 - 46.0$$
$$= 25.5 \text{ kN m} \quad (+)$$
$$\text{Bending moment at C} = -20 + 1.0$$
$$= -20.0 \text{ kN m} \quad (-)$$

Fig. 8.19 Example 8.6: (a) load diagram; (b) SFD; (c) BMD; (d) deflected shape of beam

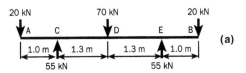

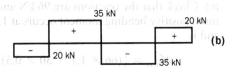

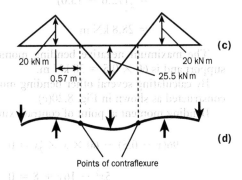

Points of contraflexure

Example 8.7

A beam 4 m long carrying a uniformly distributed load of 60 kN/m cantilevers over both supports as shown in Fig. 8.20(a). Sketch the SFD and BMD and determine the position of the point of contraflexure.

Fig. 8.20 Example 8.7: (a) load
diagram; (b) SFD; (c) BMD

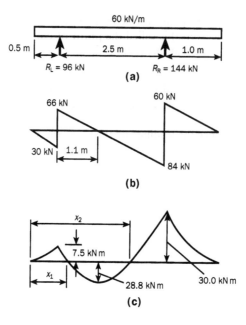

(a)

(b)

(c)

Solution The shear force and bending moment diagrams are given in Fig. 8.20(b) and
(c). Check that the reactions are 96 kN and 144 kN respectively. The maximum positive bending moment occurs at 1.6 m from the left end of the beam
and is

$$M_{\max} = \tfrac{1}{2}(66 \times 1.1 - 30 \times 0.5)$$

$$= \tfrac{1}{2}(72.6 - 15.0)$$

$$= 28.8 \text{ kN m}$$

The maximum negative bending moment occurs over the right-hand
support and is $60 \times 0.5 = 30$ kN m.

By calculating several other bending moment values the diagram can be
constructed as shown in Fig. 8.20(c).

Bending moment at point of contraflexure is

$$96(x - 0.5) - 60 \times x \times \tfrac{1}{2}x = 0$$

$$5x^2 - 16x + 8 = 0$$

$$x = \frac{16 \pm \sqrt{(256 - 160)}}{10}$$

$$x_1 = 0.62 \text{ m}$$

$$x_2 = 2.58 \text{ m}$$

Summary

Shear force   At a point in the span of a beam, shear force is defined as the algebraic sum of the loads to the left or to the right of the point.

Bending moment   At a point in the span of the beam, bending moment is defined as the algebraic sum of all the moments caused by the loads acting to the left or to the right of that point.

Maximum bending moment (M_{max})   The algebraic sum of the moments of the loads and reaction acting to one side of the point of maximum bending moment.

To determine the position of the maximum bending moment in a beam simply supported at its two ends, add together the downward loads starting from one reaction to the point where they equal (or suddenly become greater than) that reaction. This is the point of zero shear and the point of maximum bending moment.

The maximum bending moment in cantilevers will always occur at the fixed end and equals the sum of the moments of all the loads about the fixed end.

Point of contraflexure   This occurs where the bending moments change sign and their value is zero.

Exercises

1–14   Calculate the reactions, determine the position and amount of the maximum bending moment for the beams shown in Figs. 8.Q1 to 8.Q14.

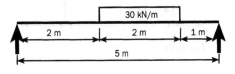

Fig. 8.Q3

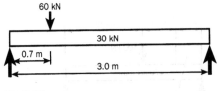

Fig. 8.Q1

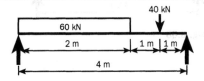

Fig. 8.Q4

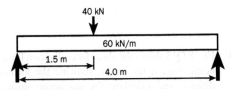

Fig. 8.Q2

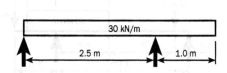

Fig. 8.Q5

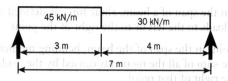

Fig. 8.Q6

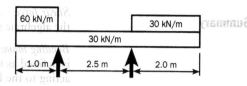

Fig. 8.Q11

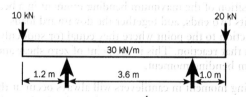

Fig. 8.Q7

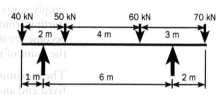

Fig. 8.Q12

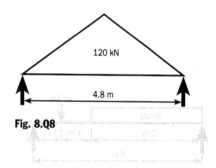

Fig. 8.Q8

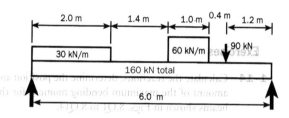

Fig. 8.Q13

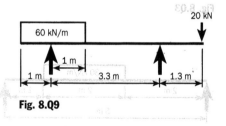

Fig. 8.Q9

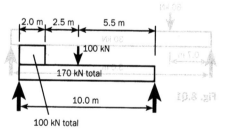

Fig. 8.Q14

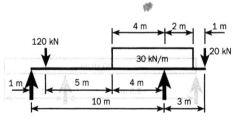

Fig. 8.Q10

15   Fig. 8.Q15 shows the shear force diagram for a loaded
beam. Sketch the beam, showing the loading conditions, and calculate the maximum positive and
negative bending moments.

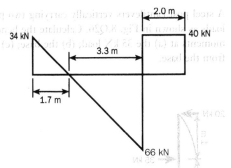

Fig. 8.Q15

16 A steel beam spans 4.8 m and carries on its whole length a brick wall 220 mm thick and 2.7 m high. The brickwork weighs 20 kN/m³ and the self-weight of beam and casing is estimated as being 5.7 kN. What is the maximum bending moment on the beam?

17 A steel beam is as shown in Fig. 8.Q17. The portion between supports A and B carries a uniform load of 30 kN/m and there are point loads at the free ends as shown. What is the length l in metres between A and B if the bending moment at a point C midway between these supports is just zero?

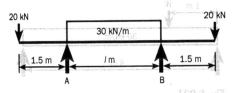

Fig. 8.Q17

18 For the beam loaded as shown in Fig. 8.Q18 (a) calculate the reactions; (b) determine the position and amount of the maximum positive and maximum negative bending moments.

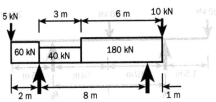

Fig. 8.Q18

19 Referring to Fig. 8.Q19, calculate (a) end reactions; (b) position and amount of M_{max}.

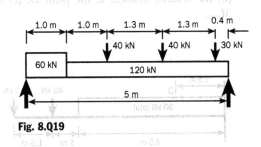

Fig. 8.Q19

20 Calculate the maximum bending moment in Fig. 8.Q20.

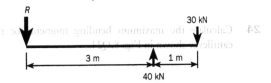

Fig. 8.Q20

21 Determine the position and amount of the maximum shear force and bending moment in Fig. 8.Q21.

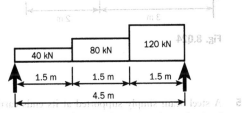

Fig. 8.Q21

22 Referring to Fig. 8.Q22 (a) calculate the end reactions; (b) determine the position and amount of the maximum bending moment, M_{max}.

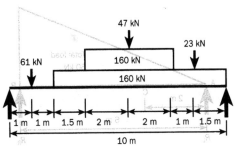

Fig. 8.Q22

23 For the cantilever loaded as shown in Fig. 8.Q23, calculate (a) the bending moment at the 40 kN load; (b) the bending moment at the point D; (c) the M_{max}.

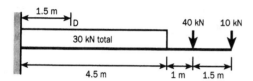

Fig. 8.Q23

24 Calculate the maximum bending moment for the cantilever shown in Fig. 8.Q24.

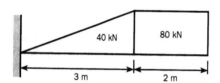

Fig. 8.Q24

25 A steel beam simply supported at its ends carries a load of varying intensity as shown (Fig. 8.Q25). Determine (a) the end reactions R_L and R_R; (b) the M_{max}; (c) the bending moment at point C.

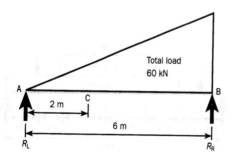

Fig. 8.Q25

26 A steel post cantilevers vertically carrying two point loads as shown in Fig. 8.Q26. Calculate the bending moments at (a) the 35 kN load; (b) the base; (c) 1 m from the base.

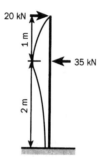

Fig. 8.Q26

27 A steel beam loaded as shown in Fig. 8.Q27 has a maximum bending moment (occurring at the point load) of 135 kN m. What is the value W kN of the point load?

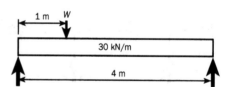

Fig. 8.Q27

28 In the beam shown in Fig. 8.Q28, the maximum bending moment occurs at the supports R_L and R_R and the bending moment at the central 20 kN load is zero. What is the length in metres of span l?

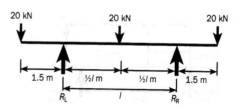

Fig. 8.Q28

29 In the beam shown in Fig. 8.Q29, the maximum negative bending moment is twice the amount of the maximum positive bending moment. What is the value in kN of the central load W

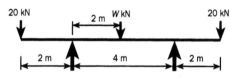

Fig. 8.Q29

30 Derive an expression for the beam as shown in Fig. 8.Q30 for (a) the maximum bending moment; (b) the bending moment at C.

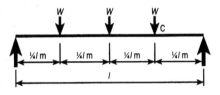

Fig. 8.Q30

This chapter considers the effect of the shape or profile of a beam's section on the beam's resistance to the punishing forces and moments induced by the loading. The beam itself must, of course, be made just strong enough to withstand these punishing effects with a reasonable factor of safety, and the strength of the beam or its degree of resistance to bending moment and shear force is built up in terms of:

- the shape and size of the beam's section
- the strength of the particular material of which the beam is made.

The final degree of 'defence' or resistance will be measured in units which take into account both of these two factors, but for a clearer understanding they will be treated separately to begin with. The present chapter will show how the shape or profile of a beam's section affects its strength.

The properties which various sections have by virtue of their shape alone are:

- cross-sectional area
- position of centre of gravity or area (centroid)
- moment of inertia or the second moment of area
- section modulus or modulus of section
- radius of gyration.

Section modulus and radius of gyration will be considered in Chapters 11 and 14 respectively.

The cross-sectional area should need no description and may be calculated with ease for most structural members.

The centre of gravity or centroid

The **centre of gravity** of a body is a point in or near the body through which the resultant attraction of the earth, i.e. the weight of the body, acts for all positions of the body.

It should be noted, however, that the section of a beam is a plane figure without weight, and therefore the term **centre of area** or **centroid** is more appropriate and is frequently used in this case. The determination of the position of the centre of gravity of a body or centroid of a section is equivalent to determining the resultant of a number of like, parallel forces.

Figure 9.1(a) shows a thin sheet of tinplate with several small holes drilled in it. To one of these holes (hole A) is attached a string. It should be obvious that the position shown for the sheet in Fig. 9.1(a) is an impossible one. The

Fig. 9.1 Determining the centre of gravity of a metal sheet

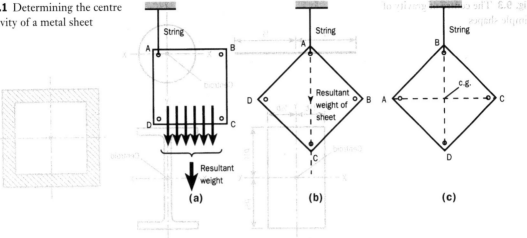

(a)   **(b)**   **(c)**

sheet will swing round in a clockwise direction and come to rest as indicated in Fig. 9.1(b). Each particle of the sheet in Fig. 9.1(a) is attracted vertically downwards by the force of gravity and the parallel lines indicate the direction of the gravity forces. The resultant of these parallel forces is the total weight of the sheet. The sheet comes to rest in such a position that the line of the resultant weight and the line of the vertical reaction in the string form one continuous line. When the string is attached to point B the sheet will hang as in Fig. 9.1(c) and the intersection of the two lines AC and BD is called the centre of gravity of the body. The actual position of the centre of gravity is in the middle of the thickness of the metal immediately behind the intersection of the two lines AC and BD. If the thickness is infinitely reduced, as in the case of a beam section, the position of the centre of gravity will coincide with that of the centroid (centre of area).

The position of the centre of gravity of sheets of metal of various shapes can be obtained by the method shown in Fig. 9.2. The sheet is suspended from a pin and from the same pin is suspended a plumb-line. The line of the string forming the plumb-line can be marked on the sheet behind it with pencil or chalk. The sheet can now be suspended in a different position and

Fig. 9.2 Experimental determination of the centre of gravity of an irregular shape: (a) front view; (b) side view

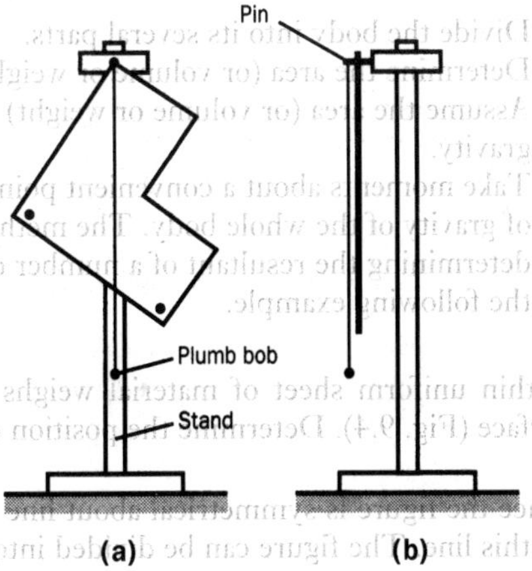

Fig. 9.3 The centre of gravity of simple shapes

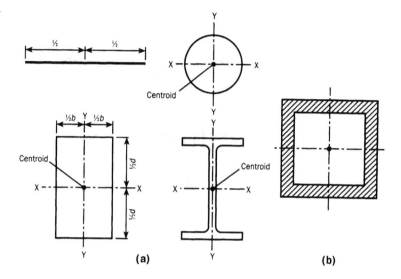

(a)   (b)

another line marked. The intersection of these two lines gives the position of the centre of gravity of the sheet.

The position of the centre of gravity of certain simple shapes is obvious by inspection (Fig. 9.3(a)), e.g. the centre of gravity of a line or uniform rod is halfway along its length and the centre of gravity of a circle is at its centre.

If it were required to balance any of these shapes in a horizontal position by placing a pencil underneath, the point of application of the pencil would be the centre of gravity of the sheet, and the reaction of the pencil would be equal to the weight of the sheet. Each of the shapes shown in Fig. 9.3 has at least two axes of symmetry and whenever a body has two axes of symmetry the centre of gravity is at the intersection of the axes.

The centre of gravity of a body need not necessarily be in the material of the body (see Fig. 9.3(b)). Note that it is impossible to balance this shape on one point so that the sheet lies in a horizontal plane.

To determine the position of the centre of gravity of a compound body or the centroid of a compound section which can be divided into several parts, such that the centres of gravity and weights of the individual parts are known, the following method applies:

1 Divide the body into its several parts.
2 Determine the area (or volume or weight) of each part.
3 Assume the area (or volume or weight) of each part to act at its centre of gravity.
4 Take moments about a convenient point or axis to determine the centre of gravity of the whole body. The method is identical with that of determining the resultant of a number of forces and is explained in the following example.

Example 9.1

A thin uniform sheet of material weighs w newtons for each mm² of its surface (Fig. 9.4). Determine the position of its centre of gravity.

Solution

Since the figure is symmetrical about line AB, its centre of gravity must lie on this line. The figure can be divided into three rectangles.

Fig. 9.4 Example 9.1

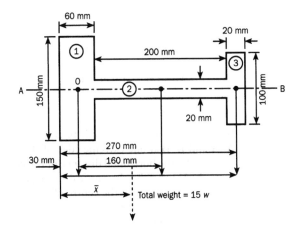

$$\text{Area (1)} = 60 \times 150 = \quad 9000 \text{ mm}^2 \qquad \text{Weight} = \quad 9w \text{ kN}$$

$$\text{Area (2)} = 200 \times 20 = \quad 4000 \text{ mm}^2 \qquad \text{Weight} = \quad 4w \text{ kN}$$

$$\text{Area (3)} = 20 \times 100 = \quad 2000 \text{ mm}^2 \qquad \text{Weight} = \quad 2w \text{ kN}$$

$$\text{Total area} = 15\,000 \text{ mm}^2 \qquad \text{Total Weight} = 15w \text{ kN}$$

Let $\bar{x}$ be the distance of the centre of gravity of the whole figure from O. Take moments about this point.

$$15w\bar{x} = 9w \times 30 + 4w \times 160 + 2w \times 270$$

$$= 270w + 640w + 540w$$

$$= 1450w$$

$$\bar{x} = 1450w/15w = 97 \text{ mm}$$

The centre of gravity is therefore on the line AB at 97 mm from O at the extreme left edge of the figure.

Note that the centre of gravity has been determined without knowing the actual weight of the material, and when a body is of uniform density throughout, its weight may be ignored and moments of areas or moments of volumes can be taken. The position of the centroid of a section is of great importance in beam design for, as will be seen later, the portion of section above this centroid performs a different function to that below the centroid.

Example 9.2

Determine the position of the centre of gravity of the body shown in Fig. 9.5(a). The body has a uniform thickness of 100 mm and weighs 10 N/m3.

Solution

Since the body is homogeneous (i.e. of uniform density) and is of uniform thickness throughout, its weight and thickness may be ignored and moments taken of areas.

$$\text{Area (1)} = \tfrac{1}{2}(3.0 \times 1.5) = 2.25 \text{ m}^2$$

$$\text{Area (2)} = 4.0 \times 1.5 = 6.00 \text{ m}^2$$

$$\text{Area (3)} = 1.0 \times 3.0 = 3.00 \text{ m}^2$$

$$\text{Total area} = \overline{11.25 \text{ m}^2}$$

In the centre of gravity problems it is usually convenient to choose two axes A–A and B–B at one extreme edge of the figure.

Let $\bar{x}$ be the horizontal distance of the centre of gravity of the whole figure from A–A and $\bar{y}$ be the vertical distance of the centre of gravity from axis B–B. The positions of the centre of gravity of the three separate parts of the figure are shown in Fig. 9.5(a).

Taking moments about axis A–A:

$$11.25\,\bar{x} = 3.00 \times 1.5 + 6.00 \times 2.25 + 2.25 \times 2.5$$
$$= 4.5 + 13.5 + 5.6 = 23.6$$
$$\bar{x} = 23.6/11.25 = 2.1 \text{ m}$$

Taking moments about axis B–B:

$$11.25\bar{y} = 3.00 \times 0.5 + 6.00 + 3.0 + 2.25 \times 6.0$$
$$= 1.5 + 18.0 + 13.5 = 33.0$$
$$\bar{y} = 33.0/11.25 = 2.9 \text{ m}$$

Fig. 9.5 Example 9.2

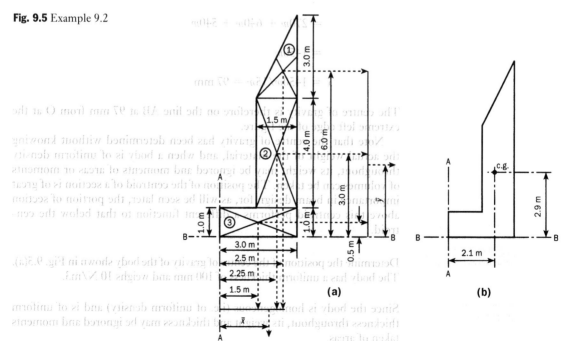

The centre of gravity is in the position indicated by Fig. 9.5(b) and is within the thickness of the body halfway between the front and back faces.

Use of the link polygon

The above problem may also be solved graphically by means of the principle of the link polygon as shown in Fig. 9.6, where the body has to be drawn to scale.

Fig. 9.6 Solving Example 9.2 using the link polygon method: (a) free-body diagram with link polygon; (b) force and polar diagrams (horizontal); (c) force and polar diagrams (vertical)

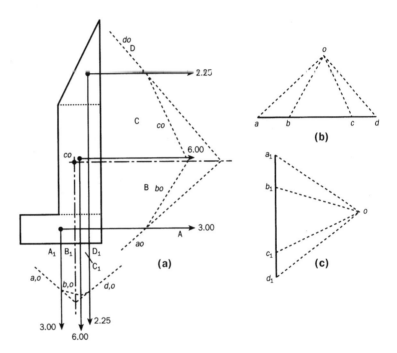

Example 9.3

A compound girder is built up of a 254×146 UB37 steel beam (i.e. a universal beam of 254 mm $\times$ 146 mm at 37 kg/m) with one 250 mm $\times$ 24 mm plate on the top flange only. The area of the steel beam alone is 4720 mm^2. Determine the position of the centroid of the compound girder (Fig. 9.7).

Solution

- The compound section is symmetrical about the vertical axis Y–Y, so the centroid must lie on this line.
- The addition of a plate to the top flange has the effect of moving the centroid (the X–X axis) towards that plate, and taking moments about line B–B at the lower edge of the section,

$$\text{Total area of section} \times \bar{y} = (\text{area of UB} \times 128 \text{ mm})$$

$$+ (\text{area of plate} \times 268 \text{ mm})$$

$$(4720 + 6000)\bar{y} = 4720 \times 128 + 6000 \times 268$$

Fig. 9.7 Example 9.4

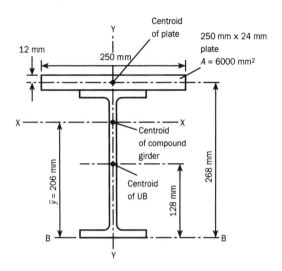

$$10\ 720\bar{y} = 604\ 160 + 1\ 608\ 000$$

$$= 2\ 212\ 160$$

$$\bar{y} = 2\ 212\ 160/10\ 720 = 206\ \text{mm}$$

Moment of inertia

The **moment of inertia** or, more appropriately in the case of a beam section (which, as stated earlier, is a plane figure without weight), the **second moment of area** of a shape, is a property which measures the *efficiency* of that shape in its resistance to bending.

Other factors besides shape enter into the building up of a beam's resistance to bending moment; the material of which a beam is made has a very obvious effect on its strength, but this is allowed for in other ways. The moment of inertia takes no regard of the strength of the material; it measures only the manner in which the geometric properties or shape of a section affect its value as a beam.

A beam such as the one shown in Fig. 9.8 may, in theory, be used with the web vertical or with the web horizontal, and it will offer much more resistance to bending when fixed as in (a) than when as in (b). This is because the moment of inertia about the X–X axis is larger than the moment of inertia about the Y–Y axis.

These axes of symmetry X–X and Y–Y are termed the **principal axes** and they are in fact axes which intersect at the centroid of the section concerned.

It will normally be necessary to calculate the moments of inertia about both these principal axes, and these are usually described as I_{xx} and I_{yy}, the moments of inertia about the X–X and Y–Y axes respectively.

The manner in which the 'build-up' of a shape affects its strength against bending must – to be understood completely – involve the use of calculus, but it may still be made reasonably clear from fairly simple considerations.

The term *moment of inertia* is itself responsible for a certain amount of confusion. Inertia suggests laziness in some ways, whereas in fact the

Fig. 9.8 The universal beam (UB) section.

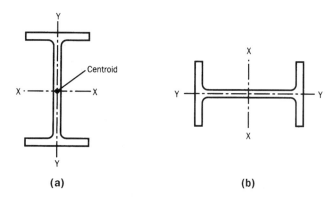

(a)   (b)

true meaning of inertia may be described as the resistance which a body makes to any forces or moments which attempt to change its shape (or its motion in the case of a moving body). A beam tends to change its shape when loaded, and inertia is the internal resistance with which the beam opposes this change of shape. The moment of inertia is a measure of the resistance which the section can supply in terms of its shape alone.

Good and bad beam shapes

In order to make the best use of the beam material, the 'shape' of the section has to be chosen with care. Certain shapes are better able to resist bending than others, and in general it may be stated that a shape will be more efficient when the greater part of its area is as far away as possible from its centroid.

A universal beam (UB) section (Fig. 9.8(a)), for example, has its flanges (which comprise the greater part of its area) well away from the centroid and the X–X axis, and is in consequence an excellent shape for resisting bending.

Figure 9.9(a) shows a shape built up by riveting together four angles. The bulk of its area (shown shaded) is situated *near to* the centroid, and the section is not a good one from the point of view of resistance to bending about the X–X axis.

If the same four angles are arranged as shown in Fig. 9.9(b), however, then the bulk of the area has been moved away from the X–X axis, and the section – which now resembles roughly a UB form – is a good one, with a high resistance to bending.

When the same four angles are used with a plate placed between them as in Fig. 9.9(c), the flanges are moved even further from the axis which passes through the centroid and is called the **neutral axis** or the axis of bending, and the efficiency of the shape is much greater. Large deep girders are built up in this way when very heavy loads have to be supported.

When a shape such as that shown in Fig. 9.9(b) is being designed, it will be found that doubling the area of the *flange* will approximately double the flange's efficiency, but doubling the depth *d* as in Fig. 9.9(c) will increase

Fig. 9.9 Good and bad beam shapes

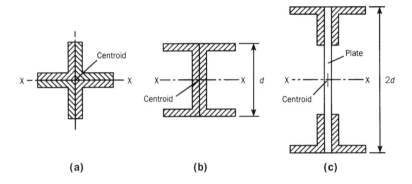

the efficiency by 4, i.e. 2^2. The efficiency of the flange, then, varies directly as its area and as the square of its distance from the neutral axis.

Calculation of the moment of inertia

In order to measure the increase in efficiency of a section as its area and depth increase, the moment of inertia (second moment of area) is calculated, and this property does in fact measure both the direct increase in area and the square of the increase in depth.

To determine the moment of inertia of the rectangle shown in Fig. 9.10(a), it will be necessary to divide the shape into a number of strips of equal area as shown. The area of each strip will be multiplied by the square of the distance of its centroid from the centroid of the whole section. The sum of all such products, $a \times y^2$, will be the moment of inertia (second moment of area) of the whole shape about the X–X axis, i.e. the total I_{xx} of section.

Assume a rectangle to be as shown in Fig. 9.10(b), 180 mm wide and 320 mm in depth, divided into 8 strips each 40 mm deep, each strip having an area of 7200 mm^2.

The sum of all the products

$$\sum ay^2 = \sum 2 \times a(y_1{}^2 + y_2{}^2 + y_3{}^2 + \cdots)$$

$$= 2 \times 7200(20^2 + 60^2 + 100^2 + 140^2)$$

$$= 14\,440 \times 33\,600$$

$$= 483.84 \times 10^6 \text{ mm}^4$$

This is an approximate value of the I_{xx}, or moment of inertia about the X–X axis, but the exact value of this factor will, of course, depend upon the number of strips into which the shape is divided. It can be shown that if b is the width of the rectangle and d is the depth, then the exact value of I_{xx} is given by

$$I_{xx} = \tfrac{1}{12}bd^3$$

Fig. 9.10 Calculating the moment of inertia of a rectangle by division into a number of strips

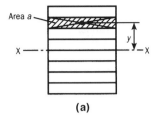

(a)

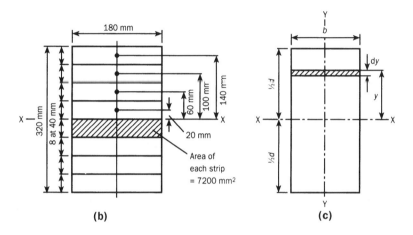

(b)   **(c)**

In the case above this is

$$\tfrac{1}{12} \times 180 \times 320^3 \quad \text{or} \quad 491.52 \times 10^6 \text{ mm}^4$$

The units of the answer are important, and it should be appreciated that, as an area (mm²) has been multiplied by a distance squared (y^2), the answer is the second moment of area, or more simply the moment of inertia of the shape, and is measured in mm⁴.

For those who have an elementary knowledge of calculus, the foregoing approximate derivation of $\tfrac{1}{12} bd^3$ will appear somewhat lengthy, and the exact value is obtained by integrating as follows.

Consider a small strip of area of breadth b and depth $\mathrm{d}y$ at a distance of y from the neutral axis as shown in Fig. 9.10(c). The moment of inertia of this strip (shown shaded) is

$$(\text{its area}) \times (y)^2 = b \times \mathrm{d}y \times y^2 = b \times y^2 \times \mathrm{d}y$$

The second moment of half of the rectangle is the sum of all such quantities $by^2\, \mathrm{d}y$ between the limits of $y = 0$ and $y = \tfrac{1}{2}d$

$$I_{\text{xx}} \text{ of half rectangle} = \int_0^{d/2} by^2\, \mathrm{d}y = \left[\frac{by^3}{3} \right]_0^{d/2} = \frac{bd^3}{24}$$

Thus,

$$I_{\text{xx}} \text{ of the complete rectangle} = \frac{bd^3}{24} \times 2 = \frac{bd^3}{12}$$

Similarly, I_{yy} of the rectangle $= db^3/12$.

The values of the moments of inertia of a number of common shapes are listed below for reference (Fig. 9.11).

Rectangle, about neutral axes:

$$I_{xx} = \tfrac{1}{12}bd^3$$

$$I_{yy} = \tfrac{1}{12}db^3$$

Rectangle, about one edge:

$$I_{uu} = \tfrac{1}{3}bd^3$$

$$I_{vv} = \tfrac{1}{3}db^3$$

Hollow rectangular shape:

$$I_{xx} = \tfrac{1}{12}(BD^3 - bd^3)$$

$$I_{yy} = \tfrac{1}{12}(DB^3 - db^3)$$

Triangle:

$$I_{xx} \text{ about neutral axis } = bd^3/36$$

$$I_{nn} \text{ about base } = bd^3/12$$

Circle:

$$I_{xx} = I_{yy} = \pi d^4/64$$

Fig. 9.11 Moments of inertia of common shapes

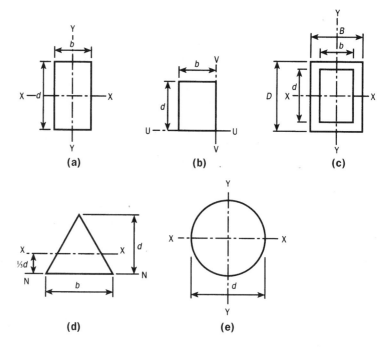

(a)   (b)   (c)

(d)   (e)

Principle of parallel axes

Figure 9.12(a) shows a simple rectangular section of size $b \times d$. The I_{xx} of the rectangle is $\frac{1}{12}bd^3$. It will be seen, however, that there are times when the moment of inertia of the rectangle about some other parallel axis such as Z–Z is required, and the I_{zz} will obviously be greater than the I_{xx}, since larger distances are involved in the summation AH^2.

The rule for use in these cases may be stated as follows: To find the moment of inertia of any shape about an axis Z–Z, parallel to the neutral axis (X–X) and at a perpendicular distance of H away from the neutral axis, the amount AH^2 (area of shape $\times$ distance H squared) must be added to I_{xx}.

For example, in the case of the rectangle shown in Fig. 9.12(b),

$$I_{xx} = \tfrac{1}{12}bd^3 = \frac{150 \times 100^3}{12} = 12.5 \times 10^6 \, \text{mm}^4$$

The I_{zz} about the base (where $H = 50$ mm) is

$$I_{zz} = I_{xx} + AH^2 = 12.5 \times 10^6 + 15\,000 \times 50^2$$

$$= (12.5 + 37.5) \times 10^6$$

$$= 50 \times 10^6 \, \text{mm}^4$$

Note: This particular case could have been derived directly from the formula of $\frac{1}{3}bd^3$ given in Fig. 9.11(b).

$\frac{1}{3}bd^3$ in this example is $\frac{1}{3} \times 150 \times 100^3 = 50 \times 10^6 \, \text{mm}^4$

Fig. 9.12 Principle of parallel axes

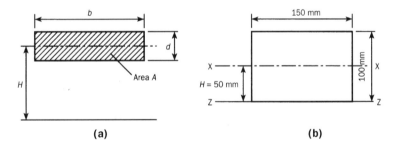

(a)   (b)

| Example 9.4 |
The principle of parallel axes may be used to calculate the values of the moments of inertia of structural sections like the I-beam shown in Fig. 9.13(a).

| Solution |
The web has its centroid on the X–X axis of the beam and the I_{xx} of the web thus equals

$$\tfrac{1}{12} \times 12 \times 400^3 = 64 \times 10^6 \, \text{mm}^4$$

The moment of inertia of one flange about *its own axis* (F–F) is

$$\tfrac{1}{12} \times 200 \times 24^3 = 0.23 \times 10^6 \, \text{mm}^4$$

and from the principle of parallel axes, the I_{xx} of the one flange is

$$I_{xx} = 0.23 \times 10^6 + 200 \times 24 \times 212^2$$

$$= 215.96 \times 10^6 \, \text{mm}^4$$

Fig. 9.13 Example 9.4

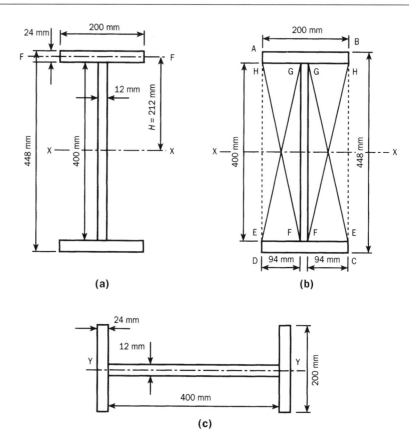

(a)

(b)

(c)

The total I_{xx} of the two flanges plus the web is

$$\text{Total } I_{xx} = (64 + 2 \times 216) \times 10^6 \text{ mm}^4$$

$$= 496 \times 10^6 \text{ mm}^4$$

An alternative method of calculating the moment of inertia about X–X is to calculate the I_{xx} of rectangle ABCD and to subtract the I_{xx} of the two rectangles EFGH (Fig. 9.13(b)). Thus

$$\text{Total } I_{xx} = \tfrac{1}{12} \times 200 \times 448^3 - \tfrac{1}{12} \times 2(94 \times 400^3)$$

$$= (1498.59 - 1002.66) \times 10^6$$

$$= 496 \times 10^6 \text{ mm}^4$$

Note: Particular care should be taken in using this method of the subtraction of moments of inertia to see that all the rectangles concerned have axis X–X as their common neutral axis.

The I_{yy} of the above beam section may most easily be calculated by adding the I_{yy} of the three rectangles of which the joist consists, as shown in Fig. 9.13(c), because axis Y–Y is their common neutral axis.

$$I_{yy} = 2 \times (\tfrac{1}{12} \times 24 \times 200^3) + \tfrac{1}{12} \times 400 \times 12^3$$

$$= (32.00 + 0.06) \times 10^6$$

$$= 32.06 \times 10^6 \text{ mm}^4$$

The method shown above is accurate and can be used for steel girders built up by welding together plates of rectangular cross-section. It should be noted, however, that for structural rolled steel sections the moments of inertia can be found tabulated in handbooks. An example of such a table is given on pages 224–7 and the given moments of inertia take into account the root radius, fillets, etc.

Example 9.5

A T-section measures 140 mm $\times$ 140 mm $\times$ 20 mm as shown in Fig. 9.14. Calculate the I_{xx}.

Solution

Taking moments of areas about the base to determine the position of the neutral axis X–X:

$$(2400 + 2800)\bar{x} = 2400 \times 80 + 2800 \times 10$$

$$\bar{x} = \frac{(192 + 28) \times 10^3}{5.2 \times 10^3} = 42.3 \text{ mm}$$

I_{xx} of vertical rectangle:

$$\tfrac{1}{12} \times 20 \times 120^3 + 2400 \times 37.7^2$$

$$= (2.88 + 3.41) \times 10^6$$

$$= 6.29 \times 10^6 \text{ mm}^4$$

I_{xx} of horizontal rectangle:

$$\tfrac{1}{12} \times 140 \times 20^3 + 2800 \times 32.2^2$$

$$= (0.09 + 2.92) \times 10^6$$

$$= 3.01 \times 10^6 \text{ mm}^4$$

Fig. 9.14 Example 9.5

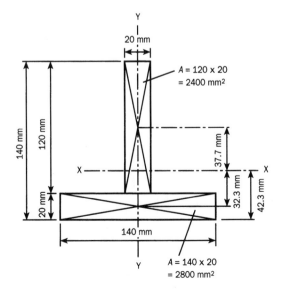

Total I_{xx}:

$$(6.29 + 3.01) \times 10^6 = 9.30 \times 10^6 \, mm$$

Calculate the I_{xx} and I_{yy} of the section shown in Fig. 9.15.

Taking moments of area about the base to find the distance $\bar{x}$ to the neutral axis X–X:

$$9600\bar{x} = 4800 \times 12 + 2400 \times 124 + 2400 \times 236$$

$$= 57\,600 + 297\,600 + 566\,400$$

$$\bar{x} = 921\,600/9600 = 96 \, mm$$

The figure is divided into three rectangles as shown in Fig. 9.15 and the distances of the centroids of each from the centroid of the whole section are also given.

$$I_{xx}\,(\text{top flange}) = \tfrac{1}{12} \times 100 \times 24^3 + 2400 \times 140^2$$

$$= (0.11 + 47.04) \times 10^6$$

$$= 47.15 \times 10^6 \, mm^4$$

$$I_{xx}\,(\text{web}) = \tfrac{1}{12} \times 12 \times 200^3 + 2400 \times 28^2$$

$$= (8.00 + 1.88) \times 10^6$$

$$= 9.88 \times 10^6 \, mm^4$$

Fig. 9.15 Example 9.6

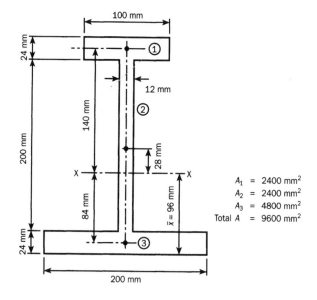

$A_1 = 2400 \, mm^2$
$A_2 = 2400 \, mm^2$
$A_3 = 4800 \, mm^2$
Total $A = 9600 \, mm^2$

$$I_{xx} \text{(bottom flange)} = \tfrac{1}{12} \times 200 \times 24^3 + 4800 \times 84^2$$

$$= (0.23 + 53.87) \times 10^6$$

$$= 34.10 \times 10^6 \, \text{mm}^4$$

Total I_{xx} of the whole section

$$= (47.15 + 9.88 + 34.10) \times 10^6$$

$$= 91.13 \times 10^6 \, \text{mm}^4$$

$$\text{Total } I_{yy} = \tfrac{1}{12} \times 24 \times 100^3 + \tfrac{1}{12} \times 24 \times 200^3 + \tfrac{1}{12} \times 200 \times 12^3$$

$$= (2.00 + 16.00 + 0.03) \times 10^6$$

$$= 18.03 \times 10^6 \, \text{mm}^4$$

Example 9.7

Calculate the I_{xx} and I_{yy} of the channel section shown in Fig. 9.16.

Solution

X–X axis:

$$I_{xx} = \tfrac{1}{2} \times 100 \times 300^3 - \tfrac{1}{12} \times 88 \times 252^3$$

$$= (225.00 - 117.36) \times 10^6$$

$$= 107.64 \times 10^6 \, \text{mm}^4$$

Check by addition:

$$I_{xx} = \tfrac{1}{12} \times 12 \times 300^3 + 2 \times (\tfrac{1}{12} \times 88 \times 24^3 + 88 \times 24 \times 138^2)$$

$$= 27 \times 10^6 + 2(0.10 + 40.22) \times 10^6$$

$$= 107.64 \times 10^6 \, \text{mm}^4 \text{ (as before)}$$

Fig. 9.16 Example 9.7

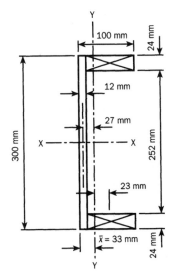

Y–Y axis:

Taking moments of areas about back of channel:

$$\text{Area} \times \text{distance } \bar{x} = 300 \times 12 \times 6 + 2(24 \times 88 \times 56)$$

$$= 21\ 600 + 236\ 544$$

$$= 258\ 144 \text{ mm}^3$$

$$\text{Distance } \bar{x} = \frac{258\ 144}{3600 + 2 \times 2112} = 33 \text{ mm}$$

$$I_{yy} = \tfrac{1}{12} \times 300 \times 12^3 + 3600 \times 27^2$$

$$+ 2(\tfrac{1}{12} \times 24 \times 88^3 + 2112 \times 23^2)$$

$$= (0.04 + 2.63 + 2(1.36 + 1.12)) \times 10^6$$

$$= 7.63 \times 10^6 \text{ mm}^4$$

Check by subtraction:

$$I_{yy} = (\tfrac{1}{12} \times 300 \times 100^3 + 30\ 000 \times 17^2)$$

$$- (\tfrac{1}{12} \times 252 \times 88^3 + 22\ 176 \times 23^2)$$

$$= (25.00 + 8.67 - 14.31 - 11.73) \times 10^6$$

$$= 7.63 \times 10^6 \text{ mm}^4 \text{ (as before)}$$

Example 9.8

Calculate the I_{xx} and I_{yy} of the compound girder shown in Fig. 9.17. The properties of the UB alone are:

$$\text{Area} = 2800 \text{ mm}^2$$

$$I_{xx} = 28.41 \times 10^6 \text{ mm}^4$$

$$I_{yy} = 1.19 \times 10^6 \text{ mm}^4$$

Fig. 9.17 Example 9.8

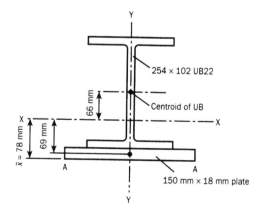

254 × 102 UB22

66 mm

Centroid of UB

$\bar{x} = 78$ mm

69 mm

150 mm × 18 mm plate

Solution

Note: The moment of inertia given above for the UB (without the plate) is of course about the X–X axis of the UB.

The addition of a single plate renders the compound section unsymmetrical about the X–X axis of the compound, and the AH^2 of the UB must be added to its own I_{xx}.

Taking moments about the line A–A:

$$\text{Distance } \bar{x} = \frac{2800 \times 145 + 2700 \times 9}{2800 + 2700}$$

$$= 430\ 300/5500$$

$$= 78 \text{ mm}$$

$$I_{xx} = 28.41 \times 10^6 + 2800 \times 66^2$$

$$+ \tfrac{1}{12} \times 150 \times 18^3 + 2700 \times 69^2$$

$$= (28.41 + 12.20 + 0.07 + 12.85) \times 10^6$$

$$= 53.53 \times 10^6 \text{ mm}^4$$

$$I_{yy} = 1.19 \times 10^6 + \tfrac{1}{12} \times 18 \times 150^3$$

$$= (1.19 + 5.06) \times 10^6$$

$$= 6.25 \times 10^6 \text{ mm}^4$$

Example 9.9

Calculate the I_{xx} and I_{yy} of the compound girder shown in Fig. 9.18. The properties of the UB alone are:

$$\text{Area} = 2800 \text{ mm}^2$$

$$I_{xx} = 28.41 \times 10^6 \text{ mm}^4$$

$$I_{yy} = 1.19 \times 10^6 \text{ mm}^4$$

Fig. 9.18 Example 9.9

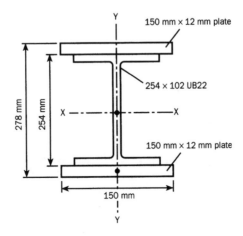

Solution *Note*: The addition of a 150 mm × 12 mm plate to each flange does not move the neutral axis of the UB from its original position, so that

$$I_{xx} = 28.41 \times 10^6 + \tfrac{1}{12} \times 150 \times 278^3 - \tfrac{1}{12} \times 150 \times 254^3$$

$$= (28.41 + 268.56 - 204.84) \times 10^6$$

$$= 92.13 \times 10^6 \text{ mm}^4$$

Check I_{xx} by addition of the I_{xx} of plates:

$$I_{xx} = 28.41 \times 10^6 + 2(\tfrac{1}{12} \times 150 \times 12^3 + 150 \times 12 \times 133^2)$$

$$= (28.41 + 2 \times 31.86) \times 10^6$$

$$= 92.13 \times 10^6 \text{ mm}^4 \text{ (as before)}$$

$$I_{yy} = 1.19 \times 10^6 + 2(\tfrac{1}{12} \times 12 \times 150^3)$$

$$= (1.19 + 6.75) \times 10^6$$

$$= 7.94 \times 10^6 \text{ mm}^4$$

Summary

The *centre of gravity* of a body is a point in or near the body through which the resultant weight of the body acts. When dealing with areas, the terms *centroid*, *centre of area* and *neutral axis* are frequently used.

With respect to beam sections, which are plane figures without weight, the *moment of inertia* about an axis is the sum of the *second moments of area* about that axis, i.e.

$$I = \sum ay^2$$

If a section is divided into an infinite number of strips parallel to the axis in question, and the area of each strip is multiplied by the square of its distance from the axis, and then all these quantities are added together, the result is the moment of inertia of the section.

For geometrical shapes, formulae for the moments of inertia can be obtained with the aid of calculus.

The *principle of parallel axes* is used to determine the moment of inertia of any shape about an axis Z–Z parallel to another axis X–X at a perpendicular distance of *H*; the amount AH^2 (i.e. area of shape multiplied by the square of the distance *H*) must be added to the moments of inertia about X–X.

Exercises

1 Figure 9.Q1 shows a system of three weights connected together by rigid bars. Determine the position of the centre of gravity of the system with respect to point A, ignoring the weight of the bars.

2 Each of the shapes shown in Fig. 9.Q2(a), (b), (c), (d) is symmetrical about a vertical axis. Calculate the distance of the centre of gravity from the base. (All dimensions are in millimetres.)

3 In Fig. 9.Q3(a) to (e) take point P as the intersection of the A–A and B–B axes and calculate the position of the centre of gravity with respect to these axes. ($\bar{x}$ is the distance of the c.g. from the vertical axis A–A and $\bar{y}$ is the distance of the c.g. from the horizontal axis B–B. All dimensions are in millimetres.)

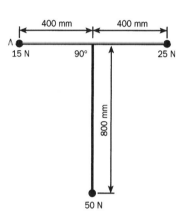

Fig. 9.Q1

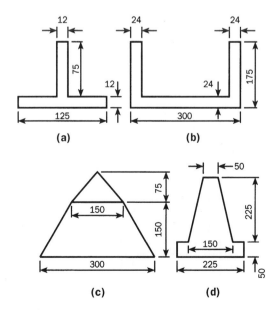

(a)   (b)

(c)   (d)

Fig. 9.Q2

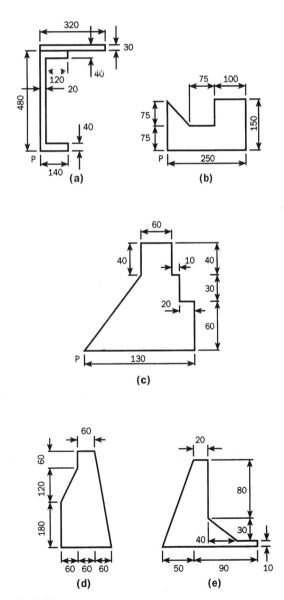

Fig. 9.Q3

4–9　Calculate the I_{xx} and the I_{yy} of the shapes shown in Figs. 9.Q4–9.Q9.

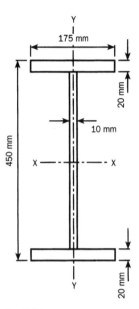

Fig. 9.Q4

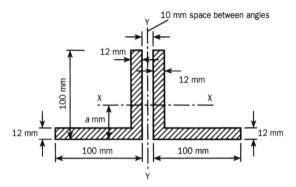

Fig. 9.Q6

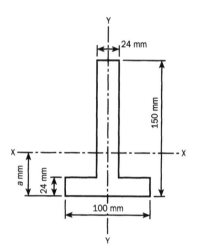

Fig. 9.Q7

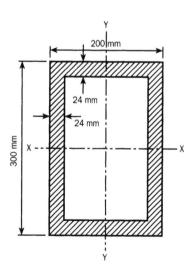

Fig. 9.Q5

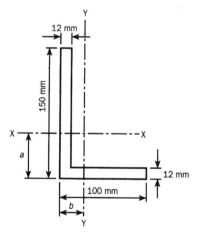

Fig. 9.Q8

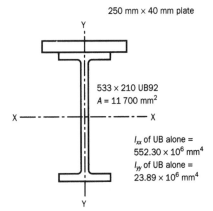

250 mm × 40 mm plate

533 × 210 UB92
A = 11 700 mm²

I_{xx} of UB alone =
552.30 × 10⁶ mm⁴
I_{yy} of UB alone =
23.89 × 10⁶ mm⁴

Fig 9.Q9

10 Two steel channels, 229 mm × 76 mm, are to be arranged as shown in Fig. 9.Q10, so that the I_{xx} and the I_{yy} of the compound section are equal.

The properties of one single channel are:

Area = 3320 mm²

I_{xx} = 26.15 × 10⁶ mm⁴

Distance b = 20 mm

I_{yy} = 15.90 × 10⁶ mm⁴

(about axis shown dotted)

What should be the distance a?

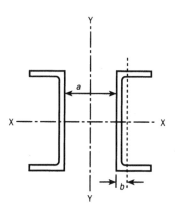

Fig. 9.Q10

11 The properties of a single 762 × 267 UB173 are

Area = 22 000 mm²

I_{xx} = 2053.0 × 10⁶ mm⁴

I_{yy} = 68.5 × 10⁶ mm⁴

Calculate the I_{xx} and I_{yy} of a compound girder which consists of two such beams at 300 mm centres and two 700 mm × 36 mm steel plates attached to the flanges of the beams as shown in Fig. 9.Q11.

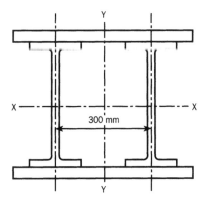

300 mm

Fig. 9.Q11

12–14 Determine the position of the centre of area of the shapes shown in Figs. 8.Q12–8.Q14 and calculate the values of their I_{xx}.

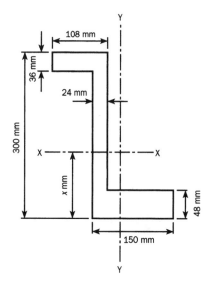

108 mm

36 mm

24 mm

300 mm

x mm

150 mm

48 mm

Fig. 9.Q12

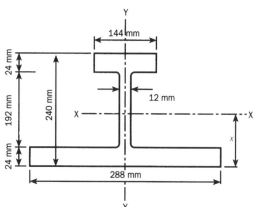

Fig. 9.Q13

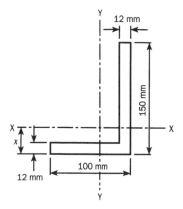

Fig. 9.Q14

15   A strut is made up by inserting a 178 mm × 24 mm steel plate between two channels as shown in Fig. 9.Q15. The properties of a single 178 mm × 76 mm steel channel are

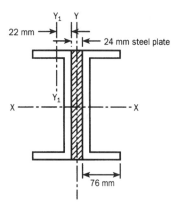

Fig. 9.Q15

Area = 2660 mm²

$I_{yy} = 13.38 \times 10^6$ mm⁴

$I_{yy} = 1.34 \times 10^6$ mm⁴

Calculate the values of I_{xx} and I_{yy} for the strut.

16   Calculate the I_{xx} and I_{yy} of the square section shown in Fig. 9.Q16.

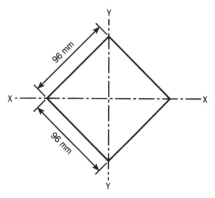

Fig. 9.Q16

17   A special stanchion section is built up using two UBs fixed together as shown in Fig. 9.Q17.

The properties of each individual UB are

356 × 171 UB 57	254 × 102 UB 22
Area 7260 mm²	2800 mm²
I_{xx} 160.40 × 10⁶ mm⁴	28.41 × 10⁶ mm⁴
I_{yy} 11.08 × 10⁶ mm⁴	1.19 × 10⁶ mm⁴

Calculate the dimension x to the centroid of the compound section and the values of I_{xx} and I_{yy}.

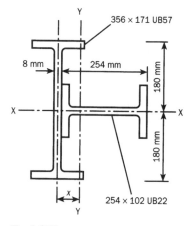

Fig. 9.Q17

18 A timber I-beam consists of two 120 mm × 36 mm flanges and a 48 mm thick web. The overall depth of the beam is 276 mm. Determine its I_{xx} and I_{yy}.

19 A hollow circular section has an external diameter of 200 mm. The I_{xx} of the section is 52.34 × 10⁶ mm⁴. What is the thickness of the material?

20 The overall depth of a hollow rectangular section is 300 mm and the thickness of the material is 20 mm throughout. The I_{xx} about the horizontal axis is 184.24 × 10⁶ mm⁴. What is the overall width of the section?

21 The I_{xx} of a 352 mm deep UB is 121 × 10⁶ mm⁴. The beam has to be increased in strength by the addition of a 220 mm wide plate to each flange so that the total I_{xx} of the compound girder is brought up to 495 × 10⁶ mm⁴. What thickness plate will be required? Bolt or rivet holes need not be deducted.

22 The UB mentioned in Question 21 (I_{xx} = 121 × 10⁶ mm⁴) is to be strengthened by the addition of 30 mm thick plate to each flange so that the I_{xx} is increased to 605 × 10⁶ mm⁴. What is the required width of the plates? Bolt or rivet holes may be ignored.

23 Calculate the second moment of area about the X–X (horizontal) axis of the crane gantry girder shown in Fig. 9.Q23.

The properties of the individual components are

	610 × 305 UB238	432 × 102 [
Area	30 300 mm²	8340 mm²
I_{xx}	2095.0 × 10⁶ mm⁴	213.7 × 10⁴ mm⁴
I_{yy}	158.4 × 10⁶ mm⁴	6.27 × 10⁶ mm⁴
Web thickness	18.4 mm	12.2 mm

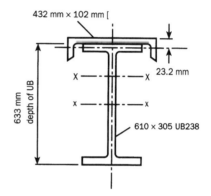

Fig. 9.Q23

Part two Understanding structural design

Chapter ten   Introduction to structural stability, durability and environmental conditions

General

Structures should be designed for robustness, long life, and with sufficient provision to resist the effects both of accidental actions, such as blast and impact forces from gas pipe explosions or traffic accidents, and of misuse which leads to loss of stability and the risk of progressive failure. When actual damage (failure) of a structural element such as a beam, wall or roof truss occurs because of an accident or because localized damage destroys or removes that structural element, the adjoining structural elements should be designed to carry without failure the additional load impacts associated with the removal of the damaged member of the structure. Thus the replacement and/or the repairs of the damaged part can be completed without causing further damage to the rest of the structure.

British and European codes of practice recommend that overall structural stability should be ensured. This means that the design details, connections and fixing details of the parts of the structure should be adequate and compatible; in other words a stable and robust design. Structural robustness means a structure's ability to remain stable (not suffer major collapse) when subjected to minor accidental damage.

The interaction between structural parts such as roof trusses, floors, walls, columns or beams affects the robustness of the structure. Figure 10.1 shows structures with different degrees of robustness.

Structures and environmental conditions

Building and civil engineering structures may become unsafe and unfit for use due to extremes of climate, man-made events and the deterioration of the materials used. The loss of structural safety may be caused by factors such as changes in the requirements of clients, materials and construction technology, and the processes of design, construction, maintenance and use. For a safer structure the following topics should be considered.

Assessment of structural safety and risk at the design stage

Designers should consider means to protect structural parts and the structure as a whole from hazards, which are based on systematic identification and analysis of critical situations that may arise during the life of the structure. The European code includes further information and specific advice on identification of hazards and critical design situations.

Structures should be safe in normal use, whatever materials are used in them, and should have high resistance to disproportionate damage in the

Fig. 10.1 Structures with different degrees of robustness: (a) unstable structure; (b) stable structure. Walls need to be tied properly to roof; (c) stable steel building structure; (d) stable steel or concrete building ; (e) and (f) ideal core locations; (g) and (h) not recommended

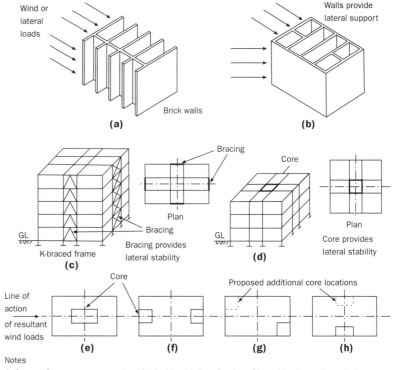

Notes
1. Centre of core arrangement should coincide with line of action of lateral loads, such as wind load, to avoid eccentricity.
2. Common core structures are reinforced concrete walls and brickwork to enclose lift, stairs, etc.

event of accidental misuse or exceptional circumstances. Structural buildings or forms which have a low sensitivity to damage and a high resistance to local and/or disproportionate collapse should be constructed. The designer should consult the Eurocode on accidental actions due to impact and explosions.

Design for durability and sustainability

Structural fabrics and major structural elements are often exposed to the weather and therefore are vulnerable to corrosion or other degradation processes. In-service performance is influenced not only by the indoor and outdoor environmental conditions surrounding the materials but also by the design of the structures, and the properties and behaviour of its individual parts. Structural movement, thermal movement, wind effects and contact with water or chemical solutions may over a period of years weaken fixings, materials and members. Both the British Standards and the European codes of practice recommend that, to ensure protection of materials during the useful life of a structure, the environmental conditions should be considered at both the design and construction stages. These include climatic conditions such as temperature, humidity, precipitation, wind, atmospheric pollution, smoke, gases, and vehicle and aeroplane exhausts. These environmental conditions will influence the choice of building materials, surface treatments and structural forms, and will also affect maintenance costs. For example, the durability of reinforced concrete elements such as beams, slabs, foundations, columns and walls is likely to be poorer with a higher water/ cement ratio and/or with a lower cement content in the concrete mixture used in making such elements. This is because a higher water/cement ratio and/or a lower cement content will increase the permeability of the concrete and hence reduce its durability. Means of protection of structural steel works are discussed in Chapter 7. The following section is intended to give an insight into the reason why steel reinforcement or steel bars in concrete elements corrode.

Corrosion of steel bars in reinforced concrete elements and members

BS 8110 and EC2 recommend that concrete cover to all steel reinforcement, including links, should never be less than the nominal cover minus 5 mm. Table 3.3 of BS 8110: Part 1: 1997 shows the nominal cover to all reinforcement, including links, needed to meet durability requirements. This nominal cover of concrete protects the steel bars from corrosion and is determined by the following factors:

- exposure conditions
- size of aggregate in the concrete
- water/cement ratio
- minimum content of cement
- grade of concrete or the characteristic strength of concrete in N/mm^2.

Why corrosion takes place and what it means in terms of chemical reactions

Exposure to environmental conditions covers the following:

(a) the ambient temperature and humidity, i.e. the presence of moisture (H_2O) and oxygen (O_2)
(b) the presence of chloride ions in the cement, and from seawater, contaminated soil, de-icing salts, contaminated water used in mixing concrete at the plant or on site, and contaminated air or atmosphere
(c) the presence of carbon dioxide (CO_2) in the atmosphere.

From the start, i.e. from mixing the constituents of concrete with water, a major product of the chemical reaction (usually called the hydration reaction) of concrete is calcium silicate hydroxide ($C_3S_2H_3$) and calcium hydroxide gel ($Ca(OH)_2$). The calcium hydroxide gel ($Ca(OH)_2$) forms an alkaline environment around the steel reinforcement, which protects the steel against corrosion.

$$2C_2S + 4H_2O \Rightarrow C_3S_2H_3 + Ca(OH)_2$$

$$2C_3S + 6H_2O \Rightarrow C_3S_2H_3 + 3Ca(OH)_2$$

Note: C_2S is an abbreviation of $2CaOSiO_2$ (dicalcium silicate) and C_3S is an abbreviation of $3CaOSiO_2$ (tricalcium silicate).

$C_3S_2H_3$ is calcium silicate hydroxide, which provides most of the strength of the concrete (although the calcium hydroxide gel $Ca(OH)_2$ also contributes to the strength of concrete), *and it provides an alkaline environment to protect steel against corrosion.*

In the presence of moisture (H_2O) and oxygen (O_2), steel reinforcement will corrode and form hydrated iron oxides known as rust. The rust occupies a much greater volume than the steel metal from which it was formed. The rust sets up bursting forces in the concrete surrounding the corroded area of the steel. The bursting forces are capable of causing concrete to crack and spall.

The mechanism of corrosion is explained as follows: the carbon dioxide in the atmosphere surrounding a reinforced concrete element penetrates through the pores in the concrete cover to the steel bars. It then reacts with the alkaline environment, $Ca(OH)_2$, to form $CaCO_3$ and water as shown in the following chemical reaction, which is known as the concrete carbonation reaction:

$$Ca(OH)_2 + CO_2 \Longrightarrow CaCO_3 + H_2O$$

With time, the alkaline environment $Ca(OH)_2$ will be destroyed and the alkalinity of concrete will be altered. When the alkalinity of concrete rises to a pH value greater than 12.5, an oxide layer (very thin) is formed on the surface of the steel bars which impedes the dissolution of the ions. This dissolution of ions is accelerated in the presence of moisture and water. The rate of carbonation or depth of carbonation (D) is described by the following formula:

$$D = k\sqrt{t}$$

where t = time in years, and K = a constant, its value depending on the following:

a) Water/cement ratio, W/C

The lower the W/C the higher the carbonation time in years. This means that concrete with low W/C has fewer pores and higher resistance to carbonation, i.e. protected steel bars take a longer time to corrode (see Fig. 10.2).

b) Quality of concrete constituents, concrete curing and concrete compaction

The chemical components of the cement, aggregate, water and any additional materials used to produce the concrete mass have to be selected carefully before constructing a concrete structure or a concrete structural element. Some aggregates containing silica compounds may be susceptible to attack by alkalis Na_2O and K_2O originating from the cement, contaminated water, contaminated soil, etc. The attack by alkalis on the aggregate, which contains silica compounds, produces an exposure action that can damage the concrete, especially when the level of moisture is high within the concrete and the concrete contains a high alkali content.

Curing of concrete is very important to the achievement of a high level of chemical reaction between cement and water. This of course produces more of the hydration products required to reduce the number and size of concrete capillaries.

Thorough compaction of concrete produces a homogenous mass of concrete with few voids. Both BS8110 and ENV206 stress the importance of thorough curing and compaction of concrete, and the time required for curing depends on the rate at which a certain resistance to the penetration of gases or liquids beyond the nominal cover is reached.

Fig. 10.2 Carbonation versus concrete nominal cover for concrete containing ordinary portland cement, gravel and sand, with different water/cement ratios (Also see Table 2.5, *Introduction to Euro-Code 2*, Deckett and Alexander, 1997)

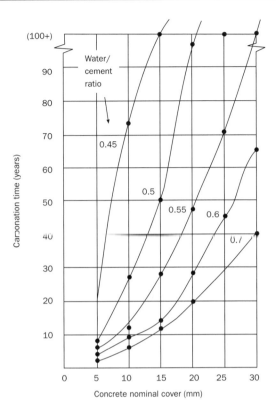

High resistance to the penetration of gases or liquids is important. Research work (see references) has shown that better curing improves the resistance of the concrete nominal cover to the penetration of chlorides and CO_2. The potential of corrosion to steel bars is increased if chloride ions are present in the concrete. Chloride ions first penetrate the surface layer (approximately 18 mm) of the concrete nominal cover by means of absorption. Any further penetration is by diffusion. Therefore the level of curing of the concrete and the compaction of the mix should be sufficient to increase the ability of the concrete nominal cover to prevent penetration of the concrete and protect the steel bars from corrosion.

Exercises

1 Describe the measures proposed in BS 8110 and EC2 regarding designing for stability, durability and the environmental conditions of structures.

2 Discuss how the corrosion of steel bars arises in re-inforced concrete members and how such failures can be avoided at both design and construction stages.

3 Discuss how the water/cement ratio, curing and compaction of concrete influence the durability of re-inforced concrete members. Support your answer with evidence from BS 8110, EC2 and other sources.

4 Explain how concrete carbonation takes place. Also explain what the formula $D = k\sqrt{t}$ means. Does a high value of k give a concrete with a high resistance to carbonation? Comment on your answer.

In the previous two chapters it has been shown that

- When a beam is loaded, the beam has a bending tendency which is measured in newton millimetres (N mm), etc., and is known as the *bending moment* on the beam. There is, for every loaded beam, a certain critical point at which this bending moment has a maximum value (M_{max}).
- The shape of the beam's cross-section has an effect upon its strength and this shape effect is measured in mm⁴ units and is called the *moment of inertia* (second moment of area).
- The material of which the beam is constructed also affects the beam's strength and this factor is measured in terms of the material's *safe allowable stress* (in tension or compression) and is measured in newtons per square millimetre (N/mm^2), etc.

The maximum bending moment, M_{max}, depends on the length of the beam and on the nature and disposition of the applied loads.

The factor measuring the strength of the section by virtue of its shape may be varied by the designer so that, when the strength of the material is also taken into account, a beam may be designed of just sufficient strength to take the calculated bending moment.

The bending moment may be thought of as the *punishment factor*.

A suitable combination of the moment of inertia and the safe allowable stress may be described as the beam's *resistance factor* and, indeed, this factor is termed the *moment of resistance M_r* of the beam.

The general theory of bending

It has been seen that, in general, most beams tend to bend in the form shown in Fig. 11.1(a), and if X–X is the neutral axis of such a beam, the fibres above X–X are stressed in compression and those below X–X in tension.

Furthermore, fibres far away from the neutral axis are stressed more heavily than those near to the neutral axis. Fibres lying on the neutral axis are neither in tension nor in compression, and are, in fact, quite unstressed.

The distribution of stress across a plane section such as A–A in Fig. 11.1(a) may thus be seen to vary from zero at the neutral axis to maximum tension and maximum compression at the extreme fibres, as shown in the stress distribution diagram (Fig. 11.1(b)).

The stress varies directly as the distance from the neutral axis, i.e. the stress at C is twice that at B.

It should be appreciated by now that material far from the neutral axis is more useful than material near the neutral axis. This is why the universal

Fig. 11.1 General theory of bending

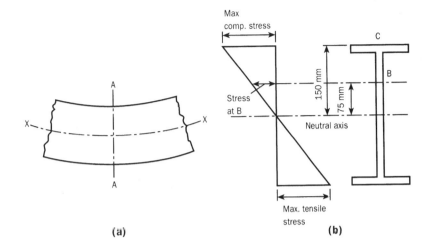

(a)

(b)

beam (UB) is such a popular beam section; the mass of its material (i.e. the two flanges) is positioned as far as possible from the centroid, that is, from the neutral axis (NA).

A simple analogy will help to show how the compression and tension generated within the beam combine to resist the external bending moment.

Consider, for example, the cantilever shown in Fig. 11.2(a), and let a load of W kN be suspended from the free end B, the self-weight of the beam being neglected. If the beam were to be sawn through at section C–C then the portion to the right of the cut would fall, but it could be prevented from falling by an arrangement as shown in Fig. 11.2(b).

A load of W kN, represented by the weight W, is attached to a wire which, in turn, passes over a pulley P and is connected to the small portion

Fig. 11.2 Bending of a cantilever: (a) load diagram; (b) theoretical representation at section C–C; (c) representation preventing rotation

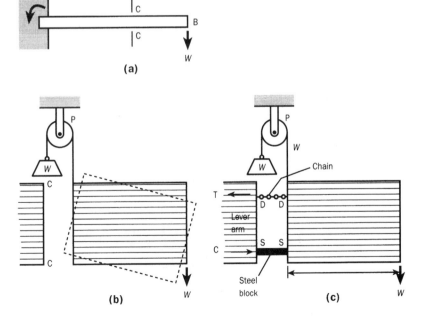

(a)

(b)

(c)

of cantilever. (The weight of the cantilever itself is ignored for the purpose of this discussion.)

However, although this arrangement would prevent the small portion from falling, the portion would still not be in equilibrium, for it would now twist or rotate as shown by the broken lines. This rotation could also be prevented by a small length of chain D–D and a steel block at S–S, as shown in Fig. 11.2(c).

This is because the chain exerts a pull (tension) on the portion of canti-lever to the right, and the steel block exerts a push (compression) on the portion.

These forces T and C are equal to each other, and they form an *internal couple* whose moment is equal to and resists the bending moment Wx. This couple provides the moment of resistance of the section which equals $T \times$ lever arm or $C \times$ lever arm. In a real beam these forces T and C are, of course, provided by the actual tension and compression in the many fibres of the cross-section.

Similarly, in a real beam, the force W which prevents the portion from falling is provided by the shear stress generated within the beam itself.

A cantilever has been chosen to illustrate the nature of these internal forces and moments, but the principle remains the same in the case of the simply supported beam, except that the tension will now occur in the layers below the NA and compression in those above the NA.

Figure 11.3(a) shows a short portion of beam (of any shape) before loading, with two plane sections A–B and C–D taken very close together. The layer E–F lies on the NA and layer G–H is one at a distance of y below the NA and having a small cross-sectional area of a.

Fig. 11.3 Bending of a simple beam: (a) cross–section and longitudinal section; (b) the beam after bending

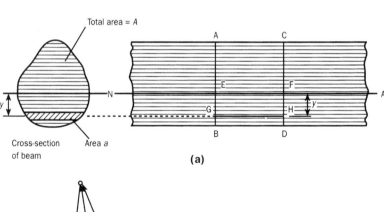

(a)

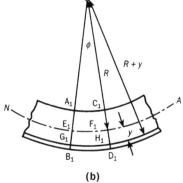

(b)

Figure 11.3(b) shows the same length of beam after bending. The sections A–B and C–D remain straight but are rotated to the positions shown, $A_1 – B_1$ and $C_1 – D_1$. The small portion has bent through an angle of ϕ radians, at a radius of curvature (measured to the NA) of R.

The fibre AC has decreased in length to $A_1–C_1$, whilst the fibre BD has increased in length to $B_1–D_1$. G–H has also increased to $G_1–H_1$ but the increase is less than in the case of the fibre B–D. The fibre E–F, being on the NA, neither increases nor decreases and E–F = $E_1–F_1$

This straining (alteration in length) of the various fibres in the depth of the section is shown in Fig. 11.4(b), and as stress and strain are proportional to each other (stress/strain = E, see Chapter 7) a similar diagram, Fig. 11.4(c), shows the distribution of stress over the depth of the section.

Referring again to Fig. 11.3(b), the fibre G–H increases in length to $G_1–H_1$. Therefore

$$\text{Strain in fibre } G-H = \frac{\text{increase in length}}{\text{original length}}$$

$$= \frac{(G_1–H_1) - (G–H)}{G–H}$$

But length of arc = radius × angle in radians, therefore

$$G_1–H_1 = (R + y)\phi$$

Now $E_1–F_1$ remains the same length and thus $E_1–F_1 = G–H$. Also $E_1–F_1 = R\phi$.

Thus

$$\frac{(G_1–H_1) - (G–H)}{G–H} = \text{strain in fibres } G–H = \frac{(R + y)\phi - R\phi}{R\phi}$$

$$= \frac{R\phi + y\phi - R\phi}{R\phi} = \frac{y}{R} \tag{1}$$

and, since stress/strain = E, then stress = strain × E, therefore,

$$\text{Stress in fibres } G–H = f = \frac{y}{R} \times E = \frac{Ey}{R} \tag{2}$$

Fig. 11.4 Stress and strain within the beam: (a) cross-section; (b) strain profile; (c) stress profile

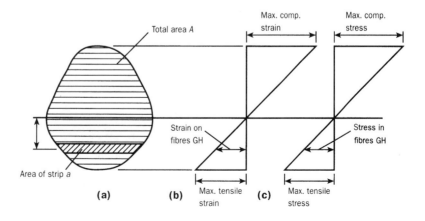

i.e.

$$\frac{f}{y} = \frac{E}{R} \tag{3}$$

The total load over the area a = stress on the area $\times$ the area

$$= \frac{Ey}{R} \times a = \frac{Eay}{R}$$

The first moment of the load about the NA is

$$\text{Load} \times y = \frac{Eay^2}{R}$$

The summation of all such moments over the total area A (i.e. the sum of the moments of all such areas as a), including areas above the neutral axis as well as below, is

$$\sum \frac{Eay^2}{R}$$

and this is the total moment of resistance of the section.

But E and R are constants (i.e. only a and y^2 vary) and, since in design the moment of resistance is made equal to the bending moment (usually the maximum bending moment, M_{max}),

$$M_{max} = M_r = \frac{E}{R}\sum ay^2$$

i.e.

$$\text{Safe bending moment} = \frac{E}{R} \times \text{the sum of all the } ay^2$$

But as was already stated in Chapter 9, the sum of all the ay^2, i.e. $\sum ay^2$, is the moment of inertia or second moment of area of the shape, I. So it may now be written

$$M_r = \frac{EI}{R} \tag{4}$$

Referring again to equation (3), $f/y = E/R$ and substituting f/y for E/R in equation (4),

$$M_r = \frac{fI}{y} = f\left(\frac{I}{y}\right) \tag{5}$$

Finally, the main formulae arising from this theory may be stated as follows:

$$\frac{M_r}{I} = \frac{f}{y} = \frac{E}{R} \tag{6}$$

and the most important portion consists of the first two terms

$$\frac{M_r}{I} = \frac{f}{y} \quad \text{or} \quad M_r = f\left(\frac{I}{y}\right) \quad \text{i.e. equation (5)}$$

In the above formula f is the permissible bending stress for the material of the beam. For the most commonly used steel (Grade S275), f may be taken as 165 N/mm^2, whereas the value of f for timber ranges between 3 N/mm^2 and 12 N/mm^2 depending on the species.

These values, however, apply only in those cases where the compression flange of the beam is adequately restrained laterally to prevent sideways buckling. For beams which are not restrained laterally, the above values have to be appropriately reduced. This is outside the scope of this textbook and it can be assumed that all the beams dealt with in this chapter are adequately supported laterally.

Elastic section modulus – symmetrical sections

In equation (5) the factor I/y is called the **section modulus** (or modulus of section) and is usually denoted by the letter Z,

$$\frac{I}{y} = Z$$

Where a beam is symmetrical about its X–X axis, as in the case of those shown in Fig. 11.5, the X–X axis is also the neutral axis (NA) of the beam section, and so the distance y to the top flange is the same as that to the bottom flange, hence

$$Z = \frac{I_{xx}}{y} = \frac{I_{xx}}{\frac{1}{2}d}$$

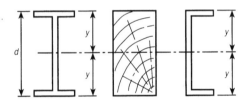

Fig. 11.5 Beams with symmetrical sections about the X–X axis

Now, for a rectangular section, $I_{xx} = \frac{1}{12}bd^3$, and so

$$Z = \frac{I}{y} = \frac{bd^3}{12} \div \frac{d}{2} = \frac{bd^2}{6}$$

In the case of rolled steel sections the values of Z (i.e. I/y) are normally taken directly from tables similar to those on pages 224–7 (elastic modulus column) but the following examples include some in which the values are calculated.

Example 11.1

A timber beam of rectangular cross-section is 150 mm wide and 300 mm deep. The maximum allowable bending stress in tension and compression must not exceed 6 N/mm^2. What maximum bending moment in N mm can the beam safely carry?

Solution

Section modulus $Z = \frac{1}{6}bd^2 = \frac{1}{6} \times 150 \times 300^2$

$$= 2.25 \times 10^6 \text{ mm}^3 \text{ units}$$

Safe allowable bending moment $M_{max} = fZ = 6 \times 2.25 \times 10^6$

$$= 13.5 \times 10^6 \text{ N mm}$$

Example 11.2

A timber beam has to support loading which will cause a maximum bending moment of 25×10^6 N mm. The safe bending stress must not exceed 7 N/mm^2. What section modulus will be required in choosing a suitable size section?

Solution

$M_r = fZ$ and, therefore, the required $Z = M_r/f$, i.e.

$$Z = \frac{25 \times 10^6}{7} = 3.57 \times 10^6 \text{ mm}^3$$

Example 11.3

A timber beam is required to span 4 m carrying a total uniform load (inclusive of the beam's self-weight) of 40 kN. The safe allowable bending stress is 8 N/mm^2. Choose a suitable depth for the beam if the width is to be 120 mm.

Solution

$$M_{max} = \tfrac{1}{8}Wl = \tfrac{1}{8} \times 40 \times 4 \times 10^6 = 20 \times 10^6 \text{ N mm}$$

but $M_{max} = M_r = f \times Z$, and it follows that

$$\text{required } Z = \frac{M_{max}}{f} = \frac{20 \times 10^6}{8} = 2.5 \times 10^6 \text{ mm}^3$$

The Z of the section is $\tfrac{1}{6}bd^2$, where b is given as 120 mm, therefore

$$\tfrac{1}{6} \times 120 \times d^2 = 2.5 \times 10^6 \text{ mm}^3$$

$$d^2 = \frac{25 \times 10^6 \times 6}{120}$$

$$d = \sqrt{125\,000} = 353 \text{ mm, say 360 mm}$$

Example 11.4

The section of floor shown in Fig. 11.6 is to be carried by 125 mm $\times$ 75 mm timber joists spanning the 3 m length. The bending stress must not exceed 4.6 N/mm^2 and the total inclusive load per m^2 of floor is estimated to be 2.0 kN. At what cross-centres x in mm must the timber beams be fixed?

Solution

Total load carried by *one* timber joist is

$$3 \times x \times 2.0 = 6x \text{ kN}$$

$$Z \text{ of joist} = \tfrac{1}{6}bd^2 = \tfrac{1}{6} \times 75 \times 125^2 = 195 \times 10^3 \text{ mm}^3$$

Moment of resistance of one joist = bending moment on one joist

$$M_r = fZ = 4.6 \times 195 \times 10^3 = 0.9 \times 10^6 \text{ N mm}$$

Fig. 11.6 Example 11.4

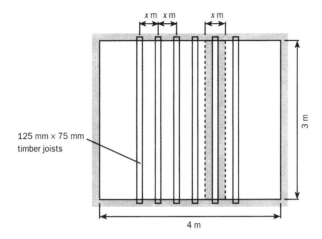

Thus

$$M_{max} = \tfrac{1}{8}Wl = \tfrac{1}{8} \times 6x \times 3 \times 10^6 = 0.9 \times 10^6 \text{ N mm}$$

and

$$x = \frac{0.9 \times 10^6 \times 8}{6 \times 3 \times 10^6} = 0.4 \text{ m}$$

Example 11.5

If the floor mentioned in Example 11.4 has its 125 mm $\times$ 75 mm timber joists at 450 mm centres and the span of the timber joists is halved by the introduction of a main timber beam 150 mm wide, as shown in Fig. 11.7, what load in kN/m² will the floor now safely carry? And what will be the required depth of the 150 mm wide main timber beam if $f = 5.25$ N/mm²?

Fig. 11.7 Example 11.5

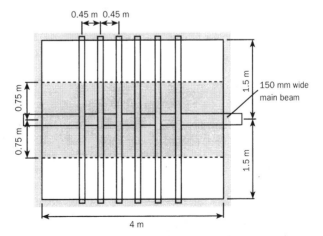

Solution

Let w be the safe load in kN/m². Then load carried by one secondary timber joist is

$$1.5 \times 0.45 \times w = 0.675w \text{ kN}$$

M_r of the joists is unchanged at 0.9×10^6 N mm, therefore

$$M_{\text{max}} = \tfrac{1}{8} \times 0.675w \times 1.5 \times 10^6 = 0.9 \times 10^6 \text{ N mm}$$

and

$$w = \frac{0.9 \times 10^6 \times 8}{0.675 \times 1.5 \times 10^6} = 7 \text{ kN/m}^2$$

Total load on the main beam

Each secondary timber joist transfers half of its load to the main beam in the form of an end reaction, and the other half to the wall.

Thus, the total area of floor load carried by the main beam is the area shown shaded in Fig. 11.7.

The total load carried by the main beam is thus $4.0 \times 1.5 \times 7 = 42 \text{ kN}$ and, although this in fact consists of a large number of small point loads from the secondary beams, it is normal practice to treat this in design as a uniformly distributed load. Therefore

$$M_{\text{max}} \text{ on main beam} = \tfrac{1}{8} \times 42 \times 4 \times 10^6$$
$$= 21 \times 10^6 \text{ N mm}$$

and

$$Z \text{ required} = \frac{M_{\text{r}}}{f} = \frac{21 \times 10^6}{5.25} = 4 \times 10^6 \text{ mm}^3$$

but for a rectangular section, $Z = \tfrac{1}{6}bd^2$, so,

$$d = \sqrt{\left(\frac{4 \times 10^6 \times 6}{150}\right)} = 400 \text{ mm}$$

Example 11.6   A welded steel girder is made up of plates as shown in Fig. 11.8.

Fig. 11.8 Example 11.6

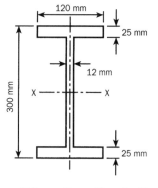

What safe, uniformly distributed load can this girder carry on a simply supported span of 4.0 m if the permissible bending stress is limited to 165 N/mm²?

Solution

$$I_{\text{xx}} \text{ of section} = \tfrac{1}{12} \times 120 \times 300^3 - \tfrac{1}{12} \times 108 \times 250^3$$
$$= 270 \times 10^6 - 140.625 \times 10^6$$
$$= 129 \times 10^6 \text{ mm}^4$$

$$Z_{xx} \text{ of section} = \frac{I}{y} = \frac{129 \times 10^6}{150} = 0.86 \times 10^6 \text{ mm}^3$$

Hence

$$M_r = 165 \times 0.86 \times 10^6 = 142 \times 10^6 \text{ N mm}$$

Let W be the total safe, uniformly distributed load the girder can carry (in kN), then

$$\tfrac{1}{8}W \times 4 \times 10^6 = 142 \times 10^6 \text{ N mm}$$

$$W = \frac{142 \times 8}{4} = 284 \text{ kN}$$

Example 11.7

A steel beam is required to span **5.5 m** between centres of simple supports carrying a 220 mm thick brick wall as detailed in Fig. 11.9.

Choose from the table of properties (pages 224–7) a suitable beam section given that the permissible stress in bending is 165 mm².

Fig. 11.9 Example 11.7

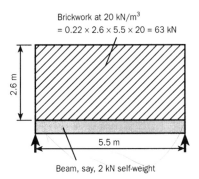

Brickwork at 20 kN/m³
= 0.22 × 2.6 × 5.5 × 20 = 63 kN

2.6 m

5.5 m

Beam, say, 2 kN self-weight

Solution

Total uniformly distributed load $W = 63 + 2 = 65$ kN.
Therefore,

$$M_{max} = \tfrac{1}{8} \times 65 \times 5.5 \times 10^6 = 44.69 \times 10^6 \text{ N mm}$$

and

$$\text{required } Z = 44.69 \times 10^6 / 165 = 270\,800 \text{ mm}^3$$

$$= 270.8 \text{ cm}^3 \text{ (as shown in tables)}$$

Z: elastic
modulus

A suitable section would be the 305 × 102 UB25 (i.e. 305 mm × 102 mm at 25 kg/m universal beam) with an actual value of Z 292 cm³.

The 203 × 133 UB30 with a Z value of 280 cm³ is nearer to the required 270.5 cm³, but the former section was chosen because of its smaller weight.

Example 11.8

Figure 11.10 shows a portion of floor of a steel-framed building. The floor slab, which weighs 3 kN/m², spans in the direction of the arrows, carrying a super (live) load of 5 kN/m² and transferring it to the secondary beams marked A. These in turn pass the loads to the main beams marked B and to the columns.

Fig. 11.10 Example 11.8

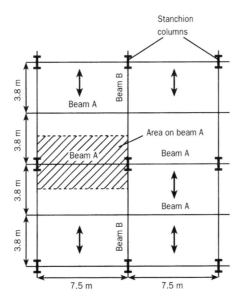

Choose suitable sizes for beams A and B, using a permissible bending stress of 165 N/mm².

Solution

Secondary beam, A

$$\text{Total load } W = 3.8 \times 7.5 \times (3 + 5) = 228 \text{ kN}$$

$$M_{\max} = \tfrac{1}{8} \times 228 \times 7.5 \times 10^6 = 213.75 \times 10^6 \text{ N mm}$$

and

$$\text{required } Z = \frac{213.75 \times 10^6}{165} = 1.295 \times 10^6 \text{ mm}^3$$

$$= 1295 \text{ cm}^3$$

Use 457×191 UB67 with a Z value of 1296 cm³.

Main beam, B
Reaction from beams A at midspan = 228 kN (point load). Assumed weight of beam (UDL), say, 3 kN/m

$$M_{\max} = (\tfrac{1}{4} \times 228 + \tfrac{1}{8} \times 3 \times 7.6) \times 7.6 \times 10^6$$

$$= (57 + 2.85) \times 7.6 \times 10^6$$

$$= 454.86 \times 10^6 \text{ N mm}$$

and

$$\text{required } Z = \frac{454.86 \times 10^6}{165} = 2.757 \times 10^6 \text{ mm}^3$$

$$= 2757 \text{cm}^3$$

Use 610×229 UB113 with a Z value of 2874 cm³.

Example 11.9 The timber box beam shown in Fig. 11.11 spans 6 m and carries a uniformly distributed load of 2 kN/m and a centrally placed point load of 3 kN. Determine the maximum stresses due to bending.

Fig. 11.11 Example 11.9

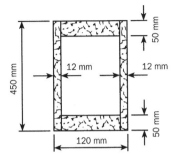

Solution Because of the symmetry of the loading, the maximum bending moment due to the combined action of the two loads will occur at midspan.

$$M_{max} = (\tfrac{1}{8} \times 2 \times 6 + \tfrac{1}{4} \times 3) \times 6 \times 10^6$$

$$= 13.5 \times 10^6 \text{ N mm}$$

$$I_{xx} = \tfrac{1}{12} \times 120 \times 450^3 - \tfrac{1}{12} \times 96 \times 350^3$$

$$= (911.25 - 343.00) \times 10^6 = 568.25 \times 10^6 \text{ mm}^4$$

$$Z_{xx} = \frac{568.25 \times 10^6}{225} = 2.525 \times 10^6 \text{ mm}^3$$

Now, from $M = f \times Z, f = M/Z$, therefore

$$f = \frac{13.5 \times 10^6}{2.525 \times 10^6} = 5.35 \text{ N/mm}^2$$

Elastic section modulus – non-symmetrical sections

Figure 11.12 shows two examples of beam sections which are not symmetrical about the X–X axis. In such cases the distance y between the neutral axis and

Fig. 11.12 Beams with non-symmetrical sections about the X–X axis

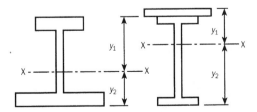

the top and bottom fibres will have two values, y_1 and y_2, and, consequently, there will also be two values of Z_{xx} – one obtained by dividing I_{xx} by the distance y_1, and the other obtained by dividing I_{xx} by the value of y_2.

If in these unsymmetrical shapes the permissible stress is the same for both top and bottom flanges, as it is, for example, in the case of steel, then

$$\text{Safe bending moment} = f \times \text{least } Z$$

If, as in the case of the now obsolete cast iron joist shown in Fig. 11.12, the permissible stresses in tension and compression are not the same, then the safe bending moment is the lesser of the following:

$$M_{\text{rc}} = \frac{I}{y_1} \times \text{compression stress}$$

$$M_{\text{rt}} = \frac{I}{y_2} \times \text{tension stress}$$

Example 11.10

A 254 × 102 UB22 has one 200 mm × 12 mm steel plate welded to the top flange only as shown in Fig. 11.13. The area of the universal beam alone is 2800 mm² and $I_{\text{xx}} = 28.41 \times 10^6$ mm⁴. What safe uniform load can such a beam carry on a 5.0 m simply supported span if $f = 165$ N/mm² in tension or compression?

Fig. 11.13 Example 11.10

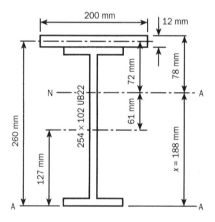

Solution

Taking moments about line A–A to find the neutral axis which passes through the centroid of the section:

$$(2800 + 2400)x = 2800 \times 127 + 2400 \times 260 = 979\,600$$

$$x = 979\,600/5200 = 188 \text{ mm}$$

Applying the principle of parallel axes (Chapter 9) to determine the second moment of area of the section about NA:

$$I_{\text{NA}} = 28.41 \times 10^6 + 2800 \times 61^2$$

$$+ \tfrac{1}{12} \times 200 \times 12^3 + 2400 \times 72^2$$

$$= (28.41 + 10.45 + 0.03 + 12.44) \times 10^6$$

$$= 51.30 \times 10^6 \text{ mm}^4$$

$$Z_{\text{top}} = (51.30 \times 10^6)/78 = 657\ 700 \text{ mm}^3$$

$$Z_{\text{bot}} = (51.30 \times 10^6)/188 = 272\ 800 \text{ mm}^3$$

Therefore

$$\text{Critical } M_r = 165 \times 272\ 800 = 45.01 \times 10^6 \text{ N mm}$$

For a UDL, $M_{\text{max}} = \frac{1}{8}Wl$

$$W = 45.01 \times 8/5 = 72.0 \text{ kN}$$

This load will cause the bottom fibres of the beam to be stressed in tension at 165 N/mm², while the maximum compressive stress in the top fibres will be much less, namely

$$165 \times \frac{272\ 800}{657\ 700} \text{ or, more simply,}$$

$$165 \times \frac{78}{188} = 68.5 \text{ N/mm}^2$$

Example 11.11

Figure 11.14 shows an old type cast iron joist (dimensions in mm) with a tension flange of 9600 mm², a compression flange of 2880 mm² and a web of 7200 mm².

The safe stress in compression is 5 N/mm², and in tension 2.5 N/mm². What is the safe bending moment for the section? What safe uniform load will the beam carry on a 4.8 m span?

Fig. 11.14 Example 11.11

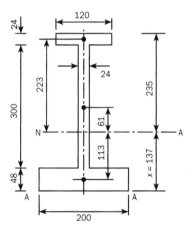

Solution

Taking moments about the lower edge A–A to find the neutral axis,

$$2880 \times 360 = 1\ 036\ 800$$

$$7200 \times 198 = 1\ 425\ 600$$

$$\underline{9600 \times 24 =\ \ \underline{230\ 400}}$$

$$19\ 680 \qquad 2\ 692\ 800$$

Therefore $x = 2\ 692\ 800/19\ 680 = 137$ mm

Total I_{NA} is

$$\tfrac{1}{12} \times 120 \times 24^3 + 2880 \times 223^2 = 143.36 \times 10^6$$

$$\tfrac{1}{12} \times 24 \times 300^3 + 7200 \times 61^2 = 80.79 \times 10^6$$

$$\tfrac{1}{12} \times 200 \times 48^3 + 960 \times 113^2 = 124.42 \times 10^6$$

$$348.57 \times 10^6 \text{ mm}^4$$

Safe bending moment in tension $=$ tension $f \times$ tension Z

$$= 2.5 \times \frac{348.57 \times 10^6}{137}$$

$$= 6.36 \times 10^6 \text{ N mm}$$

Safe bending moment in compression $=$ compression $f \times$ compression Z

$$= 5.0 \times \frac{348.57 \times 10^6}{235}$$

$$= 7.42 \times 10^6 \text{ N mm}$$

- Thus tension is the critical stress, and safe bending moment $= 6.36 \times 10^6$ N mm
- Therefore, since $\tfrac{1}{8}Wl = 6.36 \times 10^6$ N mm the safe uniformly distributed load, $W = (6.36 \times 8)/4.8 = 10.6$ kN

Example 11.12

A timber joist has a 48 mm thick flange of the same timber fixed to it as shown below in Fig. 11.15, so that the resulting section may be considered to act as a tee beam.

Fig. 11.15 Example 11.12

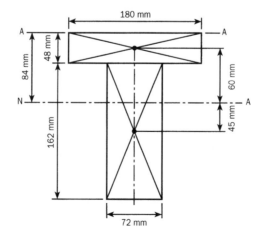

Calculate the safe distributed load on a span of 3.6 m with an extreme fibre stress of 7 N/mm².

Taking moments about line A–A to find the NA:

$$8640 \times 24 \qquad\qquad\qquad = \quad 207\,360$$

$$\underline{11\,664 \times (48 + \tfrac{1}{2} \times 162)} \; = \underline{1\,504\,656}$$

$$20\,304 \qquad\qquad\qquad\qquad\quad 1\,712\,016$$

Therefore $x = 1\,712\,016/20\,304 = 84$ mm

Total I_{NA} is

$$\tfrac{1}{12} \times 180 \times 48^3 + 8640 \times 60^2$$

$$+ \tfrac{1}{12} \times 72 \times 162^3 + 11\,664 \times 45^2$$

$$= (1.66 + 31.10 + 25.51 + 23.63) \times 10^6$$

$$= 81.9 \times 10^6 \text{ mm}^4$$

$$\text{Least } Z_{\text{xx}} = \frac{I_{\text{xx}}}{y} = \frac{81.9 \times 10^6}{210 - 84} = 650\,000 \text{ mm}^3$$

Therefore

$$M_{\text{r}} = fZ = 7 \times 650\,000 = 4.55 \times 10^6 \text{ N mm}$$

If W = safe distributed load in kN

$$\tfrac{1}{8}Wl = 4.55 \times 10^6 \text{ N mm}$$

$$W = \frac{4.55 \times 10^6 \times 8}{3600} = 10\,000 \text{ N} = 10 \text{ kN}$$

A timber cantilever beam projects 2 m and carries a 6 kN point load at the free end. The beam is 150 mm wide throughout, but varies in depth from 150 mm to 250 mm, as shown in Fig. 11.16. Calculate the stress in the extreme fibres at the support and at a point 1 m from the support. Ignore the weight of the beam.

Fig. 11.16 Example 11.13

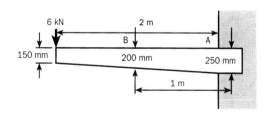

| Solution | To find stress at the support (point A): |

$$M_{max} = 6 \times 2 \times 10^6 = 12 \times 10^6 \,\text{N mm}$$

$$Z \text{ at point } A = \tfrac{1}{6}bd^2 = \tfrac{1}{6} \times 150 \times 250^2 = 1.56 \times 10^6 \,\text{mm}^3$$

$$\text{Stress } f \text{ at } A = \frac{M}{Z} = \frac{12 \times 10^6}{1.56 \times 10^6} = 7.68 \,\text{N/mm}^2$$

To find stress at point B:

$$M_B = 6 \times 1 \times 10^6 = 6 \times 10^6 \,\text{N mm}$$

$$Z \text{ at point } B = \tfrac{1}{6}bd^2 = \tfrac{1}{6} \times 150 \times 200^2 = 1.0 \times 10^6 \,\text{mm}^3$$

$$\text{Stress } f \text{ at } B = \frac{M}{Z} = \frac{6 \times 10^6}{1 \times 10^6} = 6.0 \,\text{N/mm}^2$$

| Example 11.14 | A hollow steel pipe of 150 mm external and 100 mm internal diameter, as shown in Fig. 11.17 is to span between two buildings. What is the greatest permissible span in metres if the stresses in tension and compression must not exceed 150 N/mm²? |

Fig. 11.17 Example 11.14

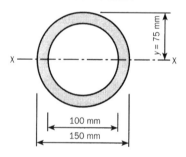

| Solution | *Note*: The unit weight of steel is 78 kN/m³ |

$$\text{Area} = \tfrac{1}{4}\pi(150^2 - 100^2) = 0.7854 \times (22\,500 - 10\,000)$$

$$= 9.8 \times 10^3 \,\text{mm}^2$$

If l = length of span in millimetres, then

$$\text{Total volume of pipe} = l \times 9800 \,\text{mm}^3$$

$$\text{Total weight of pipe} = \frac{l \times 9800 \times 78\,000}{1\,000\,000\,000} = 0.77 \times l \,\text{N}$$

But I_{xx} of pipe $= \tfrac{1}{64}\pi(150^4 - 100^4) = \tfrac{1}{64}\pi(406.25 \times 10^6) \,\text{mm}^4 = 19.94 \times 10^6 \,\text{mm}^4$, therefore

$$Z_{xx} \text{ of pipe} = \frac{I}{y} = \frac{19.94 \times 10^6}{75} = 0.27 \times 10^6 \,\text{mm}^3$$

$$\text{Permissible } M_{max} = M_r = \tfrac{1}{8}Wl = fZ = 150 \times 0.27 \times 10^6$$

$$= 40.5 \times 10^6 \,\text{N mm}$$

therefore

$$\tfrac{1}{8} \times 0.77l \times l = 40.5 \times 10^6 \, \text{N mm}$$

$$l = \sqrt{\left(\frac{40.5 \times 10^6 \times 8}{0.77 \times 10^6} \right)} = 20.5 \, \text{m}$$

Limit state design

A limit state may be defined as that state of a structure at which it becomes unfit for the use for which it was designed.

To satisfy the object of this method of design, all relevant limit states should be considered in order to ensure an adequate degree of safety and serviceability. The codes therefore distinguish between the **ultimate limit states**, which apply to the safety of the structure, and the **serviceability limit states**, which deal with factors such as deflection, vibration, local damage and cracking of concrete.

Whereas the permissible stress design methods relied on a single factor of safety (or load factor), the limit state design codes introduce two partial safety factors, one applying to the strength of the materials, γ_m, and the other to loads, γ_f. This method, therefore, enables the designer to vary the degree of risk by choosing different partial safety factors. In this way the limit state design ensures an acceptable probability that the limit states will not be reached, and so provides a safe and serviceable structure economically.

Part 1 of BS 5950 *Structural use of steelwork in building*, 'Code of practice for design in simple and continuous construction: hot rolled sections', first published in 1985, current publication dated 2000, gives some examples of limit states. These are listed in Table 11.1.

Table 11.1 Limit states

Ultimate	Serviceability
1 Strength (including general yielding, rupture, buckling and forming a mechanism)	5 Deflection
	6 Vibration
	7 Wind-induced oscillation
2 Stability against overturning and sway	8 Durability
3 Fracture due to fatigue	
4 Brittle fracture	

In the design of a steel beam for the ultimate limit state of strength, the following points should be considered.

Loads

For the purpose of checking the strength and stability of a structure the specified loads, which are based on the values given in BS 6399: Part 1: 1996 for dead and imposed loads and in BS 399–2: 1997 for wind loads, have to be multiplied by the partial factor γ_f. The value of γ_f depends on the combination of loads: for dead loads only it is 1.4, for imposed (or live) loads it is 1.6 and for wind loads it is 1.4. However, for dead loads combined with wind and live loads it is 1.2.

So, the **factored load**, also called the **design load**, for a combination of dead and live loads will be 1.4 × dead + 1.6 × live load and for the combination of dead, live and wind loads it will be 1.2 × (dead + live + wind load).

Strength of steel

The design strength, p_y, is based on the minimum yield strength as specified in BS 4360: 1990 *Specification for weldable steels* (see Table 11.2). The material factor, γ_m, for steel is 1.0 and the modulus of elasticity, E, is 205 000 N/mm².

Table 11.2 Design strength, p_y, for sections, plates and hollow sections

Thickness ⩽ (mm)	Design strength (N/mm²) for BS 5950–2		
	S275	S355	S460
16	275	355	460
40	265	345	440
63	255	335	430
80	245	325	410
100	235	315	400
150	225	295	

(*Source*: Adapted from Table 9, BS 5950: Part 1: 2000)
Note: For rolled sections, the thickness is that specified for the thickest element of the cross-section.

Limit state of strength

The **moment capacity** (or moment of resistance), M_c, of a steel section depends on lateral restraint and the cross-section of the beam. Clause 3.5.2 of BS 5950: Part 1: 2000 distinguishes four classes of cross-section, according to their width-to-thickness ratio. The universal beam sections are Class 1 or plastic cross-sections, and as such may be used for plastic (limit state) design.

For these sections the moment capacity should be taken as follows (see Clause 4.2.5 of BS 5950):

$$M_c = p_y S \leq 1.2 p_y Z \text{ (in the case of a simply supported beam or cantilever)}$$

$$\leq 1.5 p_y Z \text{ (generally)}$$

provided that the shear force, F_v, does not exceed

$$0.36 \, p_y s h$$

where

S is the plastic modulus of section
s is the web thickness of the section
h is the depth of the section

Example 11.15

Using the data given in Example 11.8, choose suitable sections for beams A and B, given that $p_y = 265 \, \text{N/mm}^2$ and the beams have full lateral restraint.

Solution

Secondary beam, A

$$\text{Loading} = 3.8 \times 7.5 \times (1.4 \times 3 + 1.6 \times 5) = 347.7 \, \text{kN}$$

$$\text{Applied moment, } M_{\text{max}} = \tfrac{1}{8} \times 347.7 \times 7.5 \times 10^6$$

$$= 325.97 \times 10^6 \, \text{N mm}$$

$$\text{So required } S = \frac{325.97 \times 10^6}{265}$$

$$= 1.23 \times 10^6 \, \text{mm}^3 = 1230 \, \text{cm}^3$$

Use a 457 × 152 UB60 ($S = 1287 \, \text{cm}^3 < 1.2 \times 1122 \, \text{cm}^2$)

Note: For this beam, $p_y = 275 \, \text{N/mm}^2$ (as $t_{\text{flange}} = 13.3$ mm). Therefore, $S = (325.97 \times 10^6)/275 = 1.185 \times 10^6 \, \text{mm}^3 = 1185 \, \text{cm}^3$, which is still less than $S = 1287 \, \text{cm}^3$ provided by the beam. Therefore the beam size is still satisfactory.
Check for shear:

$$F_v = \frac{347.7}{2} \, \text{kN} < \frac{0.36 \times 275 \times 8.1 \times 454.6}{1000} \, \text{kN}$$

i.e. the chosen section is satisfactory.

Main beam, B

Loading: own weight of beam, say, 3 kN/m UDL, point load at midspan from beams A = 347.7 kN.

$$\text{So } M_{\text{max}} = (\tfrac{1}{4} \times 347.7 + \tfrac{1}{8} \times 1.4 \times 3 \times 7.6) \times 7.6 \times 10^6$$

$$= 690.95 \times 10^6 \, \text{N mm}$$

$$\text{and required } S = \frac{690.95 \times 10^6}{265}$$

$$= 2.607 \times 10^6 \, \text{mm}^3 = 2607 \, \text{cm}^3$$

Use a 533 × 210 UB101 ($S = 2612 \, \text{cm}^3 < 1.2 \times 2292 \, \text{cm}^3$)

Check for shear:

$$F_v = \frac{347.7 + 31.92}{2} \text{ kN} < \frac{0.36 \times 265 \times 10.8 \times 536.7}{1000} \text{ kN}$$

It has to be pointed out that this very brief introduction to limit state design of steelwork treats the topic in an extremely simplified manner. For more detailed application of the principles it is imperative that specialist textbooks be consulted.

Summary

The following variables are used in the design of timber and steel beams by the *elastic method*:

$$M_r = f \times \frac{I}{y} = fZ$$

= moment of resistance

= maximum bending moment, M_{max} in N mm, kN m, etc.

f = maximum permissible bending stress in N/mm^2

I = moment of inertia about the axis of bending (mm^4)

y = the distance from the neutral axis to the extreme beam fibres

Z = elastic section modulus (mm^3) = $\dfrac{I}{y}$

For rectangular cross-section beams, $Z = \frac{1}{6}bd^2$

For circular cross-section beams, $Z = \pi D^3 / 32$

For rolled steel sections, values of Z can be found from Table 11.3 (pages 224–7) under the heading Elastic modulus

For sections built up with rectangular plates, values of Z can be calculated from first principles, havimg first obtained the moments of inertia (see Chapter 9)

For limit state design of steel beams see BS 5950: Part 1: 2000

The code for the limit state design in timber is BS 5268: Part 1: 1996

Exercises

1 A 250 mm × 75 mm timber beam with its longer edge vertical spans 2 m between simple supports. What safe uniformly distributed load W can the beam carry if the permissible bending stress is 8 N/mm^2?

2 A timber joist 75 mm wide has to carry a uniform load of 10 kN on a span of 4 m. The bending stress is to be 6 N/mm^2. What depth should the joist be?

3 A girder shaped as shown in Fig. 11.Q3 spans 3 m carrying a uniform load over the entire span. The bending stress is to be 150 N/mm^2. What may be the value of the uniform load?

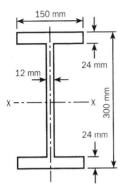

Fig. 11.Q3

4 The properties of a 356 × 171 UB45 are as follows:

$$I_{xx} = 120.7 \times 10^6 \text{ mm}^4$$

$$I_{yy} = 8.11 \times 10^6 \text{ mm}^4$$

What safe uniform load will the beam carry on a span of 4 m, if the stress is to be 125 N/mm²?

5 The universal beam used in Question 4, now spans 4.8 m carrying a uniform load of 170 kN. What is the maximum bending stress in N/mm²?

6 Two 254 mm × 89 mm steel channels, arranged back to back with 20 mm space between them, act as a beam on a span of 4.8 m. Each channel has a section modulus (Z) about axis X–X of 350 000 mm³. Calculate the maximum point load that the beam can carry at midspan if the safe stress is 165 N/mm² and the beam's self-weight is ignored.

7 A timber T-beam is formed by rigidly fixing together two joists, as shown in Fig. 11.Q7. The resulting

section is used (with the 216 mm rectangle placed vertically) as a beam spanning 3.6 m between simple supports. Calculate the safe uniformly distributed load if the permissible fibre stress is 5.6 N/mm².

8 A 152 mm × 76 mm @ 19 kg/m steel tee section, as shown in Fig. 11.Q8, may be stressed to not more than 155 N/mm². What safe inclusive uniform load can the section carry as a beam spanning 2.0 m between simple supports?

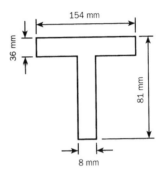

Fig. 11.Q8

9 Timber floor joists 200 mm × 75 mm at 300 mm centres span 3 m between centres of simple supports (Fig. 11.Q9). What safe inclusive load, in kN/m², may the floor carry if the timber stress is not to exceed 6.0 N/mm²?

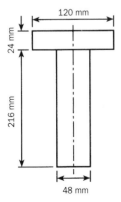

Fig. 11.Q7

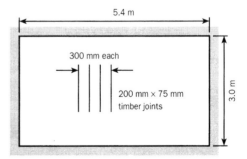

Fig. 11.Q9

10 A small floor 4.8 m × 4.2 m is to be supported by one main beam and 150 mm × 48 mm joists spanning

Table 11.3 Dimensions and properties of universal beams

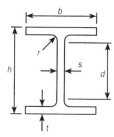

Designation	Area of section (cm²)	Mass per metre (kg/m)	Dimensions (mm)						Ratios for local buckling	
			h Section depth	b Section width	s Web thickness	t Flange thickness	r Root radius	d Depth between fillets	b/2t Flange	d/s Web
1016 × 305 × 487	619	487.0	1036.1	308.5	30.0	54.1	30.0	867.8	2.85	29.0
1016 × 305 × 438	556	438.0	1025.9	305.4	26.9	49.0	30.0	868.0	3.12	32.3
1016 × 305 × 393	500	393.0	1016.0	303.0	24.4	43.9	30.0	868.2	3.45	35.7
1016 × 305 × 349	445	349.0	1008.1	302.0	21.1	40.0	30.0	868.0	3.78	41.2
1016 × 305 × 314	400	314.0	1000.0	300.0	19.1	35.9	30.0	868.2	4.18	45.6
1016 × 305 × 272	346	272.0	990.1	300.0	16.5	31.0	30.0	868.0	4.84	52.7
1016 × 305 × 249	316	249.0	980.2	300.0	16.5	26.0	30.0	868.0	5.77	52.7
1016 × 305 × 222	282	222.0	970.3	300.0	16.0	21.1	30.0	867.8	7.11	54.4
914 × 419 × 388	494	388.0	921.0	420.5	21.4	36.6	24.1	799.6	5.74	37.4
914 × 419 × 343	437	343.3	911.8	418.5	19.4	32.0	24.1	799.6	6.54	41.2
914 × 305 × 289	368	289.1	926.6	307.7	19.5	32.0	19.1	824.4	4.81	42.3
914 × 305 × 253	323	253.4	918.4	305.5	17.3	27.9	19.1	824.4	5.47	47.7
914 × 305 × 224	286	224.2	910.4	304.1	15.9	23.9	19.1	824.4	6.36	51.8
914 × 305 × 201	256	200.9	903.0	303.3	15.1	20.2	19.1	824.4	7.51	54.6
838 × 292 × 226	289	226.5	850.9	293.8	16.1	26.8	17.8	761.7	5.48	47.3
838 × 292 × 194	247	193.8	840.7	292.4	14.7	21.7	17.8	761.7	6.74	51.8
838 × 292 × 176	224	175.9	834.9	291.7	14.0	18.8	17.8	761.7	7.76	54.4
762 × 267 × 197	251	196.8	769.8	268.0	15.6	25.4	16.5	686.0	5.28	44.0
762 × 267 × 173	220	173.0	762.2	266.7	14.3	21.6	16.5	686.0	6.17	48.0
762 × 267 × 147	187	146.9	754.0	265.2	12.8	17.5	16.5	686.0	7.58	53.6
762 × 267 × 134	171	133.9	750.0	264.4	12.0	15.5	16.5	686.0	8.53	57.2
686 × 254 × 170	217	170.2	692.9	255.8	14.5	23.7	15.2	615.1	5.40	42.4
686 × 254 × 152	194	152.4	687.5	254.5	13.2	21.0	15.2	615.1	6.06	46.6
686 × 254 × 140	178	140.1	683.5	253.7	12.4	19.0	15.2	615.1	6.68	49.6
686 × 254 × 125	159	125.2	677.9	253.0	11.7	16.2	15.2	615.1	7.81	52.6
610 × 305 × 238	303	238.1	635.8	311.4	18.4	31.4	16.5	540.0	4.96	29.3
610 × 305 × 179	228	179.0	620.2	307.1	14.1	23.6	16.5	540.0	6.51	38.3
610 × 305 × 149	190	149.1	612.4	304.8	11.8	19.7	16.5	540.0	7.74	45.8
610 × 229 × 140	178	139.9	617.2	230.2	13.1	22.1	12.7	547.6	5.21	41.8
610 × 229 × 125	159	125.1	612.2	229.0	11.9	19.6	12.7	547.6	5.84	46.0
610 × 229 × 113	144	113.0	607.6	228.2	11.1	17.3	12.7	547.6	6.60	49.3
610 × 229 × 101	129	101.2	602.6	227.6	10.5	14.8	12.7	547.6	7.69	52.2
533 × 210 × 122	155	122.0	544.5	211.9	12.7	21.3	12.7	476.5	4.97	37.5
533 × 210 × 109	139	109.0	539.5	210.8	11.6	18.8	12.7	476.5	5.61	41.1
533 × 210 × 101	129	101.0	536.7	210.0	10.8	17.4	12.7	476.5	6.03	44.1
533 × 210 × 92	117	92.1	533.1	209.3	10.1	15.6	12.7	476.5	6.71	47.2
533 × 210 × 82	105	82.2	528.3	208.8	9.6	13.2	12.7	476.5	7.91	49.6

Table 11.3 (*Continued*)

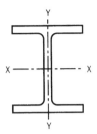

Second moment of area (cm⁴)		Radius of gyration (cm)		Elastic modulus (cm³)		Plastic modulus (cm³)		Parameters and constants				Designation
								u Buckling parameter	x Torsional index	H Warping constant (dm⁶)	J Torsional constant (cm⁴)	
X–X axis	Y–Y axis	X–X axis	Y–Y axis	X–X axis	Y–Y axis	X–X axis	Y–Y axis					
1020400	26720	40.6	6.57	19700	1732	23180	2799	0.867	21.2	64.4	4276	1016 × 305 × 487
908900	23440	40.4	6.49	17720	1535	20740	2467	0.868	23.2	55.9	3166	1016 × 305 × 438
806600	20490	40.2	6.40	15880	1353	18520	2167	0.868	25.6	48.4	2314	1016 × 305 × 393
722100	18460	40.3	6.44	14330	1222	16570	1940	0.872	28.0	43.2	1706	1016 × 305 × 349
643200	16230	40.1	6.37	12860	1082	14830	1712	0.871	30.8	37.7	1253	1016 × 305 × 314
552900	14000	40.0	6.36	11170	934	12800	1469	0.872	35.1	32.2	826	1016 × 305 × 272
480300	11750	39.0	6.09	9799	784	11330	1244	0.861	40.1	26.8	575	1016 × 305 × 249
406900	9544	38.0	5.81	8387	636	9784	1019	0.849	46.0	21.5	384	1016 × 305 × 222
719600	45440	38.2	9.59	15630	2161	17607	3341	0.885	26.7	88.9	1734	914 × 419 × 388
625800	39160	37.8	9.46	13730	1871	15480	2890	0.883	30.1	75.8	1193	914 × 419 × 343
504200	15600	37.0	6.51	10880	1014	12570	1601	0.867	31.9	31.2	926	914 × 305 × 289
436300	13300	36.8	6.42	9501	871	10940	1371	0.866	36.2	26.4	626	914 × 305 × 253
376400	11240	36.3	6.27	8269	739	9535	1163	0.861	41.3	22.1	422	914 × 305 × 224
325300	9423	35.7	6.07	7204	621	8351	982	0.854	46.8	18.4	291	914 × 305 × 201
339700	11360	34.3	6.27	7985	773	9155	1212	0.870	35.0	19.3	514	838 × 292 × 226
279200	9066	33.6	6.06	6641	620	7640	974	0.862	41.6	15.2	306	838 × 292 × 194
246000	7799	33.1	5.90	5893	535	6808	842	0.856	46.5	13.0	221	838 × 292 × 176
240000	8175	30.9	5.71	6234	610	7176	959	0.869	33.2	11.3	404	762 × 267 × 197
205300	6850	30.5	5.58	5387	514	6198	807	0.864	38.1	9.39	267	762 × 267 × 173
168500	5455	30.0	5.40	4470	411	5156	647	0.858	45.2	7.40	159	762 × 267 × 147
150700	4788	29.7	5.30	4018	362	4644	570	0.854	49.8	6.46	119	762 × 267 × 134
170300	6630	28.0	5.53	4916	518	5631	811	0.872	31.8	7.42	308	686 × 254 × 170
150400	5784	27.8	5.46	4374	455	5000	710	0.871	35.5	6.42	220	686 × 254 × 152
136300	5183	27.6	5.39	3987	409	4558	638	0.868	38.7	5.72	169	686 × 254 × 140
118000	4383	27.2	5.24	3481	346	3994	542	0.862	43.9	4.80	116	686 × 254 × 125
209500	15840	26.3	7.23	6589	1017	7486	1574	0.886	21.3	14.5	875	610 × 305 × 238
153000	11410	25.9	7.07	4935	743	5547	1144	0.886	27.7	10.2	340	610 × 305 × 179
125900	9308	25.7	7.00	4111	611	4594	937	0.886	32.7	8.17	200	610 × 305 × 149
111800	4505	25.0	5.03	3622	391	4142	611	0.875	30.6	3.99	216	610 × 229 × 140
98610	3932	24.9	4.97	3221	343	3676	535	0.873	34.1	3.45	154	610 × 229 × 125
87320	3434	24.6	4.88	2874	301	3281	469	0.870	38.0	2.99	117	610 × 229 × 113
75780	2915	24.2	4.75	2515	256	2881	400	0.864	43.1	2.52	77.0	610 × 229 × 101
76040	3388	22.1	4.67	2793	320	3196	500	0.877	27.6	2.32	178	533 × 210 × 122
66820	2943	21.9	4.60	2477	279	2828	436	0.875	30.9	1.99	126	533 × 210 × 109
61520	2692	21.9	4.57	2292	256	2612	399	0.874	33.2	1.81	101	533 × 210 × 101
55230	2389	21.7	4.51	2072	228	2360	356	0.872	36.5	1.60	75.7	533 × 210 × 92
47540	2007	21.3	4.38	1800	192	2059	300	0.864	41.6	1.33	51.5	533 × 210 × 82

Table 11.3 (*Continued*)

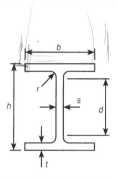

Designation	Area of section (cm²)	Mass per metre (kg/m)	Dimensions (mm)						Ratios for local buckling	
			h Section depth	b Section width	s Web thickness	t Flange thickness	r Root radius	d Depth between fillets	b/2t Flange	d/s Web
457 × 191 × 98	125	98.3	467.2	192.8	11.4	19.6	10.2	407.6	4.92	35.8
457 × 191 × 89	114	89.3	463.4	191.9	10.5	17.7	10.2	407.6	5.42	38.8
457 × 191 × 82	104	82.0	460.0	191.3	9.9	16.0	10.2	407.6	5.98	41.2
457 × 191 × 74	94.6	74.3	457.0	190.4	9.0	14.5	10.2	407.6	6.57	45.3
457 × 191 × 67	85.5	67.1	453.4	189.9	8.5	12.7	10.2	407.6	7.48	48.0
457 × 152 × 82	105	82.1	465.8	155.3	10.5	18.9	10.2	407.6	4.11	38.8
457 × 152 × 74	94.5	74.2	462.0	154.4	9.6	17.0	10.2	407.6	4.54	42.5
457 × 152 × 67	85.6	67.2	458.0	153.8	9.0	15.0	10.2	407.6	5.13	45.3
457 × 152 × 60	76.2	59.8	454.6	152.9	8.1	13.3	10.2	407.6	5.75	50.3
457 × 152 × 52	66.6	52.3	449.8	152.4	7.6	10.9	10.2	407.6	6.99	53.6
406 × 178 × 74	94.5	74.2	412.8	179.5	9.5	16.0	10.2	360.4	5.61	37.9
406 × 178 × 67	85.5	67.1	409.4	178.8	8.8	14.3	10.2	360.4	6.25	41.0
406 × 178 × 60	76.5	60.1	406.4	177.9	7.9	12.8	10.2	360.4	6.95	45.6
406 × 178 × 54	69.0	54.1	402.6	177.7	7.7	10.9	10.2	360.4	8.15	46.8
406 × 140 × 46	58.6	46.0	403.2	142.2	6.8	11.2	10.2	360.4	6.35	53.0
406 × 140 × 39	49.7	39.0	398.0	141.8	6.4	8.6	10.2	360.4	8.24	56.3
356 × 171 × 67	85.5	67.1	363.4	173.2	9.1	15.7	10.2	311.6	5.52	34.2
356 × 171 × 57	72.6	57.0	358.0	172.2	8.1	13.0	10.2	311.6	6.62	38.5
356 × 171 × 51	64.9	51.0	355.0	171.5	7.4	11.5	10.2	311.6	7.46	42.1
356 × 171 × 45	57.3	45.0	351.4	171.1	7.0	9.7	10.2	311.6	8.82	44.5
356 × 127 × 39	49.8	39.1	353.4	126.0	6.6	10.7	10.2	311.6	5.89	47.2
356 × 127 × 33	42.1	33.1	349.0	125.4	6.0	8.5	10.2	311.6	7.38	51.9
305 × 165 × 54	68.8	54.0	310.4	166.9	7.9	13.7	8.9	265.2	6.09	33.6
305 × 165 × 46	58.7	46.1	306.6	165.7	6.7	11.8	8.9	265.2	7.02	39.6
305 × 165 × 40	51.3	40.3	303.4	165.0	6.0	10.2	8.9	265.2	8.09	44.2
305 × 127 × 48	61.2	48.1	311.0	125.3	9.0	14.0	8.9	265.2	4.47	29.5
305 × 127 × 42	53.4	41.9	307.2	124.3	8.0	12.1	8.9	265.2	5.14	33.2
305 × 127 × 37	47.2	37.0	304.4	123.3	7.1	10.7	8.9	265.2	5.77	37.4
305 × 102 × 33	41.8	32.8	312.7	102.4	6.6	10.8	7.6	275.9	4.74	41.8
305 × 102 × 28	35.9	28.2	308.7	101.8	6.0	8.8	7.6	275.9	5.78	46.0
305 × 102 × 25	31.6	24.8	305.1	101.6	5.8	7.0	7.6	275.9	7.26	47.6
254 × 146 × 43	54.8	43.0	259.6	147.3	7.2	12.7	7.6	219.0	5.80	30.4
254 × 146 × 37	47.2	37.0	256.0	146.4	6.3	10.9	7.6	219.0	6.72	34.8
254 × 146 × 31	39.7	31.1	251.4	146.1	6.0	8.6	7.6	219.0	8.49	36.5
254 × 102 × 28	36.1	28.3	260.4	102.2	6.3	10.0	7.6	225.2	5.11	35.7
254 × 102 × 25	32.0	25.2	257.2	101.9	6.0	8.4	7.6	225.2	6.07	37.5
254 × 102 × 22	28.0	22.0	254.0	101.6	5.7	6.8	7.6	225.2	7.47	39.5
203 × 133 × 30	38.2	30.0	206.8	133.9	6.4	9.6	7.6	172.4	6.97	26.9
203 × 133 × 25	32.0	25.1	203.2	133.2	5.7	7.8	7.6	172.4	8.54	30.2
203 × 102 × 23	29.4	23.1	203.2	101.8	5.4	9.3	7.6	169.4	5.47	31.4
178 × 102 × 19	24.3	19.0	177.8	101.2	4.8	7.9	7.6	146.8	6.41	30.6
152 × 89 × 16	20.3	16.0	152.4	88.7	4.5	7.7	7.6	121.8	5.76	27.1
127 × 76 × 13	16.5	13.0	127.0	76.0	4.0	7.6	7.6	96.6	5.00	24.1

(*Source*: Structural sections to BS 4: Part 1 and BS 4848: Part 4 British Steel)

Table 11.3 (*Continued*)

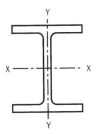

Second moment of area (cm⁴)		Radius of gyration (cm)		Elastic modulus (cm³)		Plastic modulus (cm³)		Parameters and constants				Designation
								u	*v*	*H*	*J*	
X–X axis	Y–Y axis	X–X axis	Y–Y axis	X–X axis	Y–Y axis	X–X axis	Y–Y axis	Buckling parameter	Torsional index	Warping constant (dm⁶)	Torsional constant (cm⁴)	
45730	2347	19.1	4.33	1957	243	2232	379	0.881	25.7	1.18	121	457 × 191 × 98
41020	2089	19.0	4.29	1770	218	2014	338	0.880	28.3	1.04	90.7	457 × 191 × 89
37050	1871	18.8	4.23	1611	196	1831	304	0.877	30.9	0.922	69.2	457 × 191 × 82
33320	1671	18.8	4.20	1458	176	1653	272	0.877	33.9	0.818	51.8	457 × 191 × 74
29380	1452	18.5	4.12	1296	153	1471	237	0.872	37.9	0.705	37.1	457 × 191 × 67
36590	1185	18.7	3.37	1571	153	1811	240	0.873	27.4	0.591	89.2	457 × 152 × 82
32670	1047	18.6	3.33	1414	136	1627	213	0.873	30.1	0.518	65.9	457 × 152 × 74
28930	913	18.4	3.27	1263	119	1453	187	0.869	33.6	0.448	47.7	457 × 152 × 67
25500	795	18.3	3.23	1122	104	1287	163	0.868	37.5	0.387	33.8	457 × 152 × 60
21370	645	17.9	3.11	950	84.6	1096	133	0.859	43.9	0.311	21.4	457 × 152 × 52
27310	1545	17.0	4.04	1323	172	1501	267	0.882	27.6	0.608	62.8	406 × 178 × 74
24330	1365	16.9	3.99	1189	153	1346	237	0.880	30.5	0.533	46.1	406 × 178 × 67
21600	1203	16.8	3.97	1063	135	1199	209	0.880	33.8	0.466	33.3	406 × 178 × 60
18720	1021	16.5	3.85	930	115	1055	178	0.871	38.3	0.392	23.1	406 × 178 × 54
15690	538	16.4	3.03	778	75.7	888	118	0.871	38.9	0.207	19.0	406 × 140 × 46
12510	410	15.9	2.87	629	57.8	724	90.8	0.858	47.5	0.155	10.7	406 × 140 × 39
19460	1362	15.1	3.99	1071	157	1211	243	0.886	24.4	0.412	55.7	356 × 171 × 67
16040	1108	14.9	3.91	896	129	1010	199	0.882	28.8	0.330	33.4	356 × 171 × 57
14140	968	14.8	3.86	796	113	896	174	0.881	32.1	0.286	23.8	356 × 171 × 51
12070	811	14.5	3.76	687	94.8	775	147	0.874	36.8	0.237	15.8	356 × 171 × 45
10170	358	14.3	2.68	576	56.8	659	89.1	0.871	35.2	0.105	15.1	356 × 127 × 39
8249	280	14.0	2.58	473	44.7	543	70.3	0.863	42.2	0.0812	8.79	356 × 127 × 33
11700	1063	13.0	3.93	754	127	846	196	0.889	23.6	0.234	34.8	305 × 165 × 54
9899	896	13.0	3.90	646	108	720	166	0.891	27.1	0.195	22.2	305 × 165 × 46
8503	764	12.9	3.86	560	92.6	623	142	0.889	31.0	0.164	14.7	305 × 165 × 40
9575	461	12.5	2.74	616	73.6	711	116	0.873	23.3	0.102	31.8	305 × 127 × 48
8196	389	12.4	2.70	534	62.6	614	98.4	0.872	26.5	0.0846	21.1	305 × 127 × 42
7171	336	12.3	2.67	471	54.5	539	85.4	0.872	29.7	0.0725	14.8	305 × 127 × 37
6501	194	12.5	2.15	416	37.9	481	60.0	0.866	31.6	0.0442	12.2	305 × 102 × 33
5366	155	12.2	2.08	348	30.5	403	48.5	0.859	37.4	0.0349	7.40	305 × 102 × 28
4455	123	11.9	1.97	292	24.2	342	38.8	0.846	43.4	0.0273	4.77	305 × 102 × 25
6544	677	10.9	3.52	504	92.0	566	141	0.891	21.2	0.103	23.9	254 × 146 × 43
5537	571	10.8	3.48	433	78.0	483	119	0.890	24.3	0.0857	15.3	254 × 146 × 37
4413	448	10.5	3.36	351	61.3	393	94.1	0.880	29.6	0.0660	8.55	254 × 146 × 31
4005	179	10.5	2.22	308	34.9	353	54.8	0.874	27.5	0.0280	9.57	254 × 102 × 28
3415	149	10.3	2.15	266	29.2	306	46.0	0.866	31.5	0.0230	6.42	254 × 102 × 25
2841	119	10.1	2.06	224	23.5	259	37.3	0.856	36.4	0.0182	4.15	254 × 102 × 22
2896	385	8.71	3.17	280	57.5	314	88.2	0.881	21.5	0.0374	10.3	203 × 133 × 30
2340	308	8.56	3.10	230	46.2	258	70.9	0.877	25.6	0.0294	5.96	203 × 133 × 25
2105	164	8.46	2.36	207	32.2	234	49.8	0.888	22.5	0.0154	7.02	203 × 102 × 23
1356	137	7.48	2.37	153	27.0	171	41.6	0.888	22.6	0.00987	4.41	178 × 102 × 19
834	89.8	6.41	2.10	109	20.2	123	31.2	0.890	19.6	0.00470	3.56	152 × 89 × 16
473	55.7	5.35	1.84	74.6	14.7	84.2	22.6	0.895	16.3	0.00199	2.85	127 × 76 × 13

between the wall and the beam, as shown in Fig. 11.Q10. (a) Calculate the safe inclusive floor load if the stress in the timber has a maximum value of 7.0 N/mm². (b) Choose a suitable section modulus for the main steel beam if the stress is not to exceed 165 N/mm². Ignore the weights of the joists and the beam.

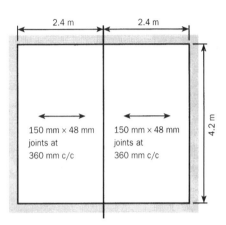

Fig. 11.Q10

11 (a) Calculate the dimension x to the centroid of the section shown in Fig. 11.Q11. (b) Determine I_{xx} and the two values of Z_{xx} for the section. (c) What safe inclusive uniformly distributed load can a beam of this section carry on a span of 3.6 m if the tension stress must not exceed 20 N/mm² and the compression stress 100 N/mm²?

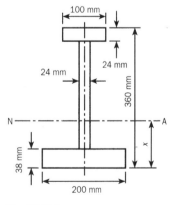

Fig. 11.Q11

12 How deep would a 150 mm wide timber beam need to be to carry the same load as the beam investigated in Question 11 if the maximum flexural stress equals 8 N/mm²?

13 Choose suitable section moduli and select appropriate UBs for the conditions shown in Fig. 11.Q13(a) to (f). $f = 165$ N/mm².

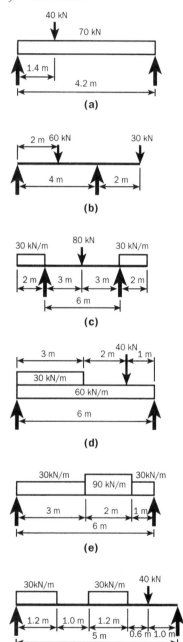

Fig. 11.Q13

14 A UB 620 mm deep has a section modulus (Z_{xx}) of 4.9×10^6 mm³. The beam spans 10 m carrying an inclusive uniform load of 100 kN and a central point load of 50 kN; (a) calculate the maximum stress due to bending; (b) what is the intensity of flexural stress at a point 150 mm above the neutral axis and 3.0 m from l.h. reaction?

15 A steel tank 1.8 m × 1.5 m × 1.2 m weighs 15 kN empty and is supported, as shown in Fig. 11.Q15, by two steel beams weighing 1 kN each. Choose a suitable section for the steel beams if the tank may be filled with water weighing 10 kN/m³. The permissible bending stress on the section is 165 N/mm².

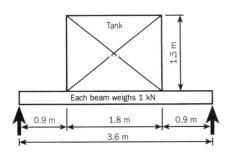

Fig. 11.Q15

16 Choose suitable timber joists and calculate sizes for UBs A and B for the floor shown in Fig. 11.Q16 if the inclusive floor loading is 8.4 kN/m².

Permissible timber stress = 7.2 N/mm²
Permissible steel stress = 165 N/mm²

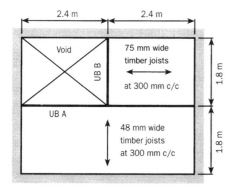

Fig. 11.Q16

17 The symmetrically loaded beam, shown in Fig. 11.Q17, carries three loads, and the internal span l has to be such that the negative bending moment at each support equals the positive bending moment at C. What is the span l? If each load W is 100 kN, choose a suitable UB (f = 165 N/mm²).

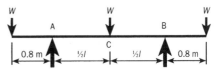

Fig. 11.Q17

18 A 610 × 305 UB149, 610 mm deep (I_{xx} = 1.25 × 10⁹ mm⁴) is simply supported on a span of l m. What will be the maximum permissible span if the stress in the beam under its own weight reaches 22 N/mm²?

19 The plan of a floor of a steel-framed building is as shown in Fig. 11.Q19. Reinforced concrete slabs spanning as indicated ↔ are supported by steel beams AD. Each beam AD carries a stanchion at C, and the point load from each stanchion is 90 kN. The total inclusive loading will be 9 kN/m². Select a suitable beam AB for the floor, using a safe bending stress of 160 N/mm².

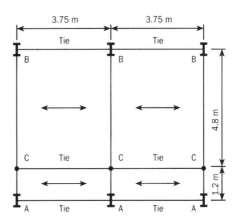

Fig. 11.Q19

20 A steel beam carries loads, as shown in Fig. 11.Q20. Calculate the position and amount of the maximum bending moment and draw the shear force and bending moment diagrams. Choose a suitable steel section for the beam, using a safe bending stress of 150 N/mm².

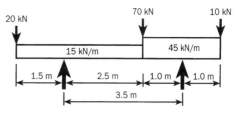

Fig. 11.Q20

21 A rectangular timber beam 300 mm deep and 250 mm
wide, freely supported on a span of 6 m, carries a uniform load of 3 kN and a triangular load of 12 kN as
shown in Fig. 11.Q21.

What is the greatest central point load that can be
added to this beam if the maximum bending stress is
8 N/mm²?

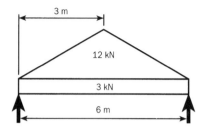

Fig. 11.Q21

Chapter twelve Beams of two materials

The basis of design of composite beams is explained by reference to flitch beams, where all the material of each constituent is taken into account in the calculations. The reinforced concrete beam is then dealt with, the elastic theory method receiving a fairly detailed treatment. The load factor method and the limit state design method are treated in their simplified form.

Flitch beams

The most common case of beams composed of two materials is the **reinforced concrete beam**, where steel and concrete combine with each other in taking stress. A special case and, therefore, not dealt with in this book is the use of steel and concrete in **composite construction**, where steel beams and concrete slabs are fastened together by means of **shear connectors**.

The simplest case, however, is that of the **flitch beam** which consists of timber and steel acting together, as shown in Fig. 12.1. A timber joist is strengthened by the addition of a steel plate on the top and bottom, the three members being securely bolted together at intervals.

The bolting together ensures that there is no slip between the steel and the timber at the sections A–A and B–B. Thus the steel and timber at these sections alter in length by the same amount. That is to say,

Strain in the steel at A–A = strain in timber at A–A

But, as shown in Chapter 7,

$$E = \frac{\text{stress}}{\text{strain}} \quad \text{and} \quad \text{stress} = \text{strain} \times E$$

Fig. 12.1 Flitch beam

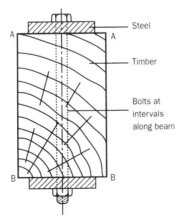

Steel

Timber

Bolts at
intervals
along beam

Fig. 12.2 Equivalent beams:
(a) composite beam of timber and
steel; (b) equivalent timber beam;
(c) equivalent steel beam

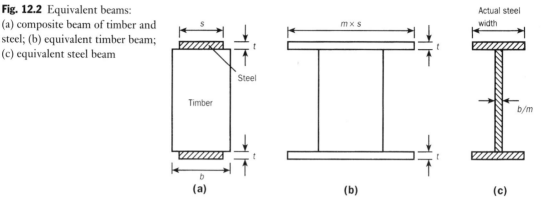

Strain in steel at A–A = E steel × strain
Strain in timber at A–A = E timber × (same strain in steel)

Thus

$$\frac{\text{Stress in steel}}{\text{Stress in timber}} = \frac{E\ \text{steel}}{E\ \text{timber}}$$

This ratio of the moduli of elasticity is called the **modular ratio** of the two
materials and is usually denoted by the letter m. Therefore,

Stress in steel = m × stress in timber

If, for example, E for steel is 210 000 N/mm^2 and E for timber is
7000 N/mm^2, then

$$m = \frac{210\,000\ \text{N/mm}^2}{7000\ \text{N/mm}^2} = 30$$

and, in effect, the steel plate can thus carry the same load as a timber member of equal thickness but m times the width of the steel.

The beam of the two materials can therefore be considered as an *all-timber
beam or equivalent timber beam* as shown in Fig. 12.2(b).

Similarly, an imaginary *equivalent steel beam* could be formed by substituting for the timber a steel plate having the same depth but only $1/m$ times
the thickness of the timber, as in Fig. 12.2(c).

It should be noted that, in forming these equivalent beams, the width only
has been altered. Any alteration in the vertical dimensions of the timber or
steel would affect the value of the strain and therefore only horizontal dimensions may be altered in forming the equivalent sections.

The strength of the *real* beam may now be calculated by determining the
strength of the equivalent timber beam or the equivalent steel beam in the
usual manner – treating the section as a normal homogeneous one, but making certain that neither of the maximum permissible stresses (timber and
steel) is exceeded.

Example 12.1

A composite beam consists of a 300 mm × 200 mm timber joist strengthened by the addition of two steel plates 180 mm × 12 mm, as in Fig. 12.3(a).
The safe stress in the timber is 5.5 N/mm^2, the safe stress in the steel is
165 N/mm^2 and $m = 30$. Calculate the moment of resistance in N mm.

Fig. 12.3 Example 12.1:
(a) composite beam;
(b) equivalent timber beam;
(c) equivalent steel beam

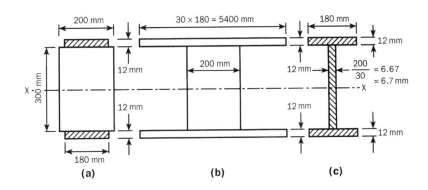

Solution The equivalent timber beam would be as shown in Fig. 12.3(b).

I_{xx} of equivalent timber beam is

$$\tfrac{1}{12} \times 200 \times 300^3 + 2(\tfrac{1}{12} \times 5400 \times 12^3 + 5400 \times 12 \times 156^2)$$

$$= (450.0 + 1.6 + 3153.9) \times 10^6 = 3605.5 \times 10^6 \text{ mm}^4$$

$$Z_{xx} = \frac{3605.5 \times 10^6}{150 + 12} = 22.3 \times 10^6 \text{ mm}^3$$

Stress in timber $= 5.5 \text{ N/mm}^2$ (given)

Therefore

$$M_T = f \times Z = 5.5 \times 22.3 \times 10^6 = 122.4 \times 10^6 \text{ N mm}$$

The equivalent steel beam would be as in Fig. 12.3(c).

I_{xx} of equivalent timber beam

$$= \tfrac{1}{12} \times 6.7 \times 300^3 + \tfrac{1}{12} \times 180(324^3 - 300^3)$$

$$= (15 + 105.2) \times 10^6 = 120.2 \times 10^6 \text{ mm}^4$$

$$Z_{xx} = \frac{120.2 \times 10^6}{162} = 0.74 \times 10^6 \text{ mm}^3$$

Stress in steel $= 165 \text{ N/mm}^2$ (given)

$$\text{or} = 30 \times \text{stress in timber} = 30 \times 5.5$$

$$= 165 \text{ N/mm}^2$$

Therefore

$$M_S = f \times Z = 165 \times 0.74 \times 10^6 = 122.4 \times 10^6 \text{ N mm}$$

Note: In this example, the ratio of the permissible stresses is equal to the modular ratio.

Example 12.2 A composite beam, shown in Fig. 12.4(a), consists of a 300 mm × 150 mm timber joist strengthened by the addition of two steel plates each 24 mm thick. The stress in the steel must not exceed 150 N/mm²; the stress in the timber must not exceed 8 N/mm²; the modular ratio for the materials is 30. Calculate the safe moment of resistance of the section in N mm and the safe uniform load in kN on a span of 4.0 m.

Fig. 12.4 Example 12.2:
(a) composite beam; (b) steel
stressed to 150 N/mm^2
(c) timber stressed to 8 N/mm^2;
(d) equivalent timber beam

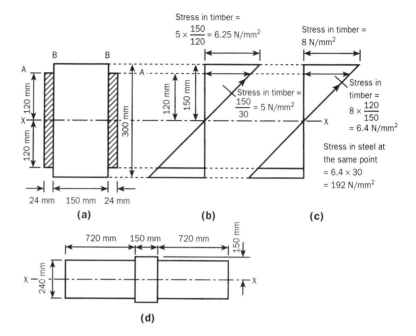

(a)   (b)   (c)

(d)

Solution

This case has one additional complication which was not present in Example 12.1. The stresses in steel and timber are not to exceed 150 and 8 N/mm^2 (Fig. 12.4(b) and (c)), respectively, but it may be that stressing the steel to 150 N/mm^2 would result in a stress of more than 8 N/mm^2 in the timber. Alternatively, a stress of 8 N/mm^2 in the timber may result in causing more than 150 N/mm^2 in the steel.

The critical case may be reasoned as follows. If the steel at A is stressed to 150 N/mm^2, then

Stress in timber at A = 150/30 = 5 N/mm^2

From the stress distribution triangles it follows that

Stress in timber at B = 5 × 150/120 = 6.25 N/mm^2

Both these stresses (150 and 6.25 N/mm^2) are permissible.

If, on the other hand, the timber at B is stressed to 8 N/mm^2, then (from the triangles of stress distribution)

Stress in timber at A = 8 × 120/150 = 6.4 N/mm^2
Stress in steel at A = 30 × 6.4 = 192 N/mm^2

This stress would be exceeding the safe allowable stress in the steel.

Thus it follows that, in calculating the strength of the composite beam, the stress in the steel may be kept at 150 N/mm^2, but the maximum permissible stress in the equivalent timber section must be reduced to 6.25 N/mm^2.

Figure 12.4(d) shows the equivalent timber section.

I_{xx} of equivalent section

$$= 2(\tfrac{1}{12} \times 720 \times 240^3) + \tfrac{1}{12} \times 150 \times 300^3$$

$$= (1658.9 + 337.5) \times 10^6$$

$$= 1996.4 \times 10^6 \text{ mm}^4$$

and

$$Z_{xx} = \frac{1996.4 \times 10^6}{150} = 13.3 \times 10^6 \text{ mm}^3$$

Therefore the safe moment of resistance is

$$M_T = fZ = 6.25 \times 13.3 = 83.13 \text{ kN m}$$

and the safe uniform load is given by

$$\tfrac{1}{8}W \times 4 = 83.13$$

$$W = \frac{83.13 \times 8}{4} = 166.4 \text{ kN}$$

Reinforced concrete beams

Concrete is a material strong in its resistance to compression, but very weak indeed in tension. A good concrete will safely take a stress upward of 7 N/mm^2 in compression, but the safe stress in tension is usually limited to no more than $\frac{1}{10}$th of its compressive stress. It will be remembered that in a homogeneous beam the stress distribution is as shown in Fig. 12.5(a), and in the case of a section symmetrical about the X–X axis, the actual stress in tension equals the actual stress in compression. If such a beam was of ordinary unreinforced concrete, then the stresses in tension and in compression would of necessity have to be limited to avoid overstressing in the tension face, whilst the compressive fibres would be taking only $\frac{1}{10}$th of their safe allowable stress. The result would be most uneconomical since the compression concrete would be understressed.

In order to increase the safe load-carrying capacity of the beam, and allow the compression concrete to use to the full its compressive resistance, steel bars are introduced in the tension zone of the beam to carry the whole of the tensile forces.

This may, at first, seem to be a little unreasonable – for the concrete is capable of carrying some tension, even though its safe tensile stress is very low. In fact it will be seen from Fig. 12.5(b) that the strain in the steel and in the concrete at A–A are the same. Therefore, if the stress in steel is

Fig. 12.5 Concrete beams: (a) homogeneous beam; (b) beam with steel bars

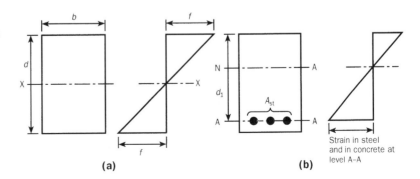

(a)   (b)

Strain in steel and in concrete at level A–A

140 N/mm^2, the concrete would be apparently stressed to $1/m$ times this and, since m, the ratio of the elastic moduli of steel and concrete, is usually taken as 15, the tensile stress in concrete would be in excess of 9 N/mm^2. The concrete would crack through failure in tension at a stress very much lower than this and thus its resistance to tension is disregarded.

Basis of design – elastic theory method

In the design of reinforced concrete beams the following assumptions are made:

- Plane sections through the beam before bending remain plane after bending.
- The concrete above the neutral axis carries *all* the compression.
- The tensile steel carries *all* the tension.
- Concrete is an elastic material, and therefore stress is proportional to strain.
- The concrete shrinks in and thus grips the steel bars firmly so that there is no slip between the steel and the surrounding concrete.

As the concrete below the neutral axis (NA) is ignored from the point of view of its tensile strength, the actual beam shown in Fig. 12.6(a) is, in fact, represented by an area of concrete above the neutral axis carrying compression, and a relatively small area of steel taking tension as in Fig. 12.6(b). Thus the position of the neutral axis will not normally be at $d_1/2$ from the top of the beam, but will vary in position according to the relation between the amount of concrete above the NA and the area of steel.

The distance from the top face of the beam to the NA which indicates the depth of concrete in compression in a beam is denoted by d_n.

Critical area of steel

In a beam of a given size, say 200 mm wide with a 300 mm effective depth, the area of steel could be varied – it could be 1000 mm^2 in area, or 2000 mm^2, etc. – and each addition of steel would make for a stronger beam, but not necessarily for an economical one.

If, for example, only a very small steel area is included, as in Fig. 12.7(a), and this small steel content is stressed up to its safe allowable value, then the concrete will probably be very much understressed, so that the concrete is not used to its full advantage. From this point of view, an uneconomical beam results.

Fig. 12.6 Elastic theory method: (a) beam cross-section; (b) representation of tensile strength

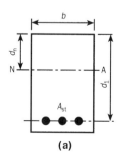
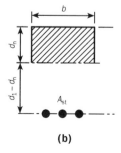

Fig. 12.7 Critical area of steel: (a) small steel area; (b) large steel area; (c) normal reinforced concrete beam; (d) normal strain diagram

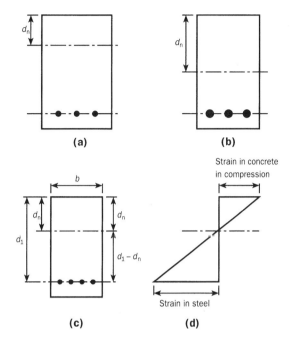

If, on the other hand, an overgenerous amount of steel is included, as in Fig. 12.7(b), then stressing this steel to its safe allowable value will probably result in overstressing the concrete. Thus, to keep the concrete stress to its safe amount, the stress in the steel will have to be reduced, and again the result is an uneconomical section.

Given that $m = 15$, there will be only one critical area of steel for a given size of beam which will allow the stress in the concrete to reach its permissible value at the same time as the stress in the steel reaches its maximum allowable value. In reinforced concrete beam calculations the designer is nearly always seeking to choose a depth of beam and an area of steel which will allow these stresses to be attained together thus producing what is called a 'balanced' or 'economical' section. The breadth of the beam is normally chosen from practical considerations.

If the two given safe stresses are reached together, then the required area of steel will always remain a constant fraction of the beam's area (bd_1) and the factor d_n/d_1 will also have a constant value.

Figure 12.7(c) shows a normal reinforced concrete (RC) beam and Fig. 12.7(d) the strain diagram for the beam.

From similar triangles it will be seen that

$$\frac{\text{strain in concrete}}{\text{strain in steel}} = \frac{d_n}{d_1 - d_n}$$

But stress/strain for any material is E (Young's modulus) for the material. Therefore strain = stress/E.

Let p_{cb} represent the permissible compressive stress for concrete in bending in N/mm^2 and let p_{st} represent the permissible tensile stress in the steel in N/mm^2.

Thus from equation (1)

$$\frac{d_n}{d_1 - d_n} = \frac{\text{stress in concrete}}{E_c} \div \frac{\text{stress in steel}}{E_s}$$

$$\frac{d_n}{d_1 - d_n} = \frac{p_{cb}}{E_c} \div \frac{p_{st}}{E_s} = \frac{p_{cb}}{E_c} \times \frac{E_s}{p_{st}}$$

But $E_s/E_c = m$, modular ratio, therefore

$$\frac{d_n}{d_1 - d_n} = \frac{mp_{cb}}{p_{st}} \tag{2}$$

Equation (2) shows that d_n and d_1 will always be proportional to each other when m, p_{cb}, and p_{st} are constant.

From equation (2)

$$d_1 - d_n = \frac{d_n p_{st}}{mp_{cb}}$$

$$d_1 = d_n\left(1 + \frac{p_{st}}{mp_{cb}}\right)$$

So

$$d_n = \frac{d_1}{1 + (p_{st}/mp_{cb})} \tag{3}$$

Note: d_n is the distance down to the neutral axis from the compression face of the beam, and is expressed in millimetres.

As d_n is, however, seen to be a constant fraction of d_1 for fixed values of m, p_{cb} and p_{st}, it is frequently useful to know the value of the constant d_n/d_1, and this constant is usually denoted by n, i.e.

$$d_n/d_1 = n$$

As d_n is equal to $\dfrac{d_1}{1 + (p_{st}/mp_{cb})}$ from equation (3), then

$$n = \frac{d_n}{d_1} = \frac{1}{1 + (p_{st}/mp_{cb})} \tag{4}$$

For the values $m = 15$, $p_{cb} = 7 \text{ N/mm}^2$, $p_{st} = 140 \text{ N/mm}^2$

$$n = \frac{1}{1 + \dfrac{140}{15 \times 7}} = 0.43$$

and $d_n = 0.43 d_1$

Figure 12.8 shows that the stress in the concrete from the top of the section to the NA follows the normal distribution for ordinary homogeneous sections, and the total compression force $C = \frac{1}{2}p_{cb}bd_n$, i.e. an area bd_n of concrete, taking stress at an average rate of $\frac{1}{2}p_{cb}$ N/mm².

The tension, however, will now all be taken by the steel and will be $T = A_{st} \times p_{st}$, i.e. an area of steel A_{st} taking stress at p_{st} N/mm². These two internal forces T and C are equal and opposite in their direction, and they form an internal couple acting at $d_1 - \frac{1}{3}d_n$ apart, where $d_1 - \frac{1}{3}d_n$ is the lever arm, denoted by l_2.

Fig. 12.8 Stress within concrete

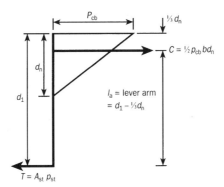

This internal couple or moment of resistance may thus be written as follows:

$$\text{In terms of concrete } \tfrac{1}{2}p_{cb}bd_n(d_1 - \tfrac{1}{3}d_n) = M_{rc} \left.\vphantom{\begin{array}{c}1\\1\end{array}}\right\} M_{max} \tag{5}$$

$$\text{In terms of steel } A_{st} \times p_{st} \times (d_1 - \tfrac{1}{3}d_n) = M_{rt} \tag{6}$$

But just as d_n/d_1 is a 'constant' n, similarly $(d_1 - \tfrac{1}{3}d_n)/d_1$ is also a constant, say j, and so the lever arm l_a may be expressed as jd_1.

For the values $m = 15$, $p_{cb} = 7\ \text{N/mm}^2$, $p_{st} = 140\ \text{N/mm}^2$,

$$j = \frac{d_1 - \tfrac{1}{3}d_n}{d_1} = \frac{1 - \tfrac{1}{3} \times 0.43}{1} = 0.86$$

and $l_a = 0.86d_1$

Substituting the constants n and j in (5),

$$M_{rc} = \tfrac{1}{2}p_{cb}bd_n(d_1 - \tfrac{1}{3}dn) = \tfrac{1}{2}p_{cb} \times b \times nd_1 \times jd_1$$

$$= \tfrac{1}{2}p_{cb} \times n \times jbd_1^2$$

As $\tfrac{1}{2}p_{cb}$, n and j are all constants, then $\tfrac{1}{2}p_{cb} \times n \times j$ has a constant value which may be called R, and therefore

$$M_{rc} = Rbd_1^2 \tag{7}$$

Using the values $m = 15$, $p_{cb} = 7\ \text{N/mm}^2$, $p_{st} = 140\ \text{N/mm}^2$

$$R = \tfrac{1}{2}p_{cb} \times n \times j = \tfrac{1}{2} \times 7 \times 0.43 \times 0.86 = 1.29$$

and

$$M_{rc} = 1.29\ bd_1^2$$

Hence

$$d_1^2 = \frac{M}{1.29b} \quad \text{and} \quad d_1 = \sqrt{\frac{M}{1.29b}} \tag{8}$$

As $M_{rt} = A_{st} \times p_{st} \times jd_1$ then the required area of steel is

$$A_{st} = \frac{M}{p_{st} \times jd_1} = \frac{M}{p_{st}l_a} \tag{9}$$

Note: As the area of steel is a constant fraction of the area $b \times d_1$ (i.e. $A_{st} = bd_1 \times$ constant), this relationship may be specified as the percentage area of steel, p, where

$$p = \frac{A_{st}}{bd_1} \times 100\% \quad \text{or} \quad A_{st} = bd_1 \times \frac{p}{100}$$

From equation (9)

$$A_{st} = \frac{M}{p_{st} j d_1}$$

but

$$M = Rbd_1^2 = \tfrac{1}{2} p_{cb} \times n \times j \times bd_1^2$$

therefore

$$A_{st} = \frac{p_{cb} \times n \times j \times bd_1^2}{2 \times p_{st} \times j d_1} = \frac{p_{cb} \times n \times bd_1}{2 \times p_{st}}$$

$$\frac{A_{st}}{bd_1} = \frac{p_{cb} n}{2 p_{st}}$$

So

$$p = \text{percentage of steel} = \frac{A_{st}}{bd_1} \times 100 = \frac{p_{cb} \times n \times 100}{2 p_{st}}$$

$$p = \frac{50 \times p_{cb} \times n}{p_{st}} \tag{10}$$

Using the values $p_{cb} = 7\,\text{N/mm}^2$, $p_{st} = 140\,\text{N/mm}^2$, $m = 15$,

$$p = \frac{50 \times 7 \times 0.43}{140} = 1.07\%$$

Hence

$$A_{st} = bd_1 \times (1.07/100) = 0.011\,bd_1$$

It must be remembered, however, that there is a wide range of concrete strengths, hence it is not advisable to memorize the numerical values of these design constants.

They may be derived, quite easily, from

$$n = \frac{1}{1 + (p_{st}/mp_{cb})} \qquad\qquad j = 1 - \tfrac{1}{3}n$$

$$R = \tfrac{1}{2} p_{cb} \times n \times j \qquad\qquad p = \frac{50 \times p_{cb} \times n}{p_{st}}$$

From all the above it will be seen that in designing simple reinforced concrete beams the procedure is as follows:

1 Calculate maximum bending moment, M_{max}, in N mm.
2 Choose a suitable breadth of beam, b.

3 Calculate effective depth to centre of steel

$$d_1 = \sqrt{\frac{M_{max}}{R \times b}}$$

4 Determine area of steel required from

$$A_{st} = \frac{M_{max}}{p_{st} \times j \times d_1} \quad \text{or} \quad A_{st} = \frac{pbd_1}{100}$$

Example 12.3

Design a simple reinforced concrete beam 150 mm wide to withstand a maximum bending moment of 20×10^6 N mm using the following permissible stresses: $p_{cb} = 7$ N/mm^2, $p_{st} = 140$ N/mm^2 and $m = 15$.

Solution

Solution by design constants:

$$n = \frac{1}{1 + (p_{st}/mp_{cb})} = \frac{7}{7 + 140/5} = 0.43$$

$$j = 1 - \tfrac{1}{3}n = 1 - \tfrac{1}{3} \times 0.43 = 0.86$$

$$R = \tfrac{1}{2}p_{cb} \times n \times j = \tfrac{1}{2} \times 7 \times 0.43 \times 0.86 = 1.29$$

$$p = \frac{50 \times p_{cb} \times n}{p_{st}} = \frac{50 \times 0.43 \times 7}{140} = 1.07\%$$

$$d_1 = \sqrt{\frac{20\,000\,000}{1.29 \times 150}} = \sqrt{103\,352} = 321 \text{ mm}$$

$$A_{st} = 0.0107 \times 150 \times 321 = 516 \text{ mm}^2$$

Alternative solution:
From equation (3)

$$d_n = \frac{d_1}{1 + \dfrac{140}{15 \times 7}} = \tfrac{3}{7}d_1 = 0.43d_1$$

From equation (5) and substituting $0.43d_1$ for d_n

$$M_{rc} = \tfrac{1}{2}p_{cb} \times b \times 0.43d_1 \times (d_1 - \tfrac{1}{3} \times 0.43d_1)$$

$$20 \times 10^6 = \tfrac{1}{2} \times 7 \times 150 \times 0.43 \times 0.86 \times d_1^2$$

$$d_1 = \sqrt{\frac{20\,000\,000 \times 2}{7 \times 150 \times 0.43 \times 0.86}} = 321 \text{ mm}$$

From equation (6)

$$A_{st} = \frac{M}{p_{st} \times 0.86 \times d_1} = \frac{20\,000\,000}{140 \times 0.86 \times 321} = 516 \text{ mm}^2$$

Use five 12 mm diameter bars (area $= 5 \times 113.1 = 566$ mm^2). The overall depth d is determined as follows.

Fig. 12.9 Example 12.3

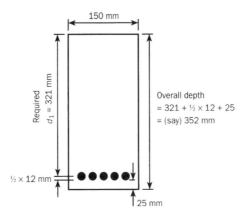

The effective depth d_1 is measured from the top face to the centre of the steel bars. The bars themselves need an effective cover of at least 25 mm, so to determine the overall depth d it is necessary to add (half the bar diameter + 25 mm) to the effective depth d_1 (see Fig. 12.9).

In this case overall depth $d = 321 + 6 + 25 = 352$ mm. Say 150 mm $\times$ 355 mm overall with five 12 mm diameter bars.

BS 5328: 1991, in its 'Guide to specifying concrete', lists four types of concrete mix (see Part 1, Clause 7.2) and gives detailed methods for specifying concrete mixes in Part 2.

For the purpose of exercises in this chapter the permissible stresses in compression due to bending, p_{cb}, may be taken as follows:

$$10.0 \text{ N/mm}^2 \text{ for } 1{:}1{:}2 \text{ mix (or C30 grade)}$$

$$8.5 \text{ N/mm}^2 \text{ for } 1{:}1\tfrac{1}{2}{:}3 \text{ mix (or C25 grade)}$$

$$7.0 \text{ N/mm}^2 \text{ for } 1{:}2{:}4 \text{ mix (or C20 grade)}$$

The permissible tensile stress, p_{st}, for steel reinforcement may be taken as $0.55f_y$, which for hot rolled mild steel bars is generally accepted to be 140 N/mm².

Example 12.4

A simply supported reinforced concrete beam is to span 5 m carrying a total uniform load of 60 kN inclusive of self-weight, and a point load of 90 kN from a secondary beam as shown in Fig. 12.10. The beam is to be 200 mm wide. Choose a suitable overall depth and area of tensile steel reinforcement for the maximum bending moment. Assume: $p_{cb} = 8.5$ N/mm², $p_{st} = 140$ N/mm², $m = 15$.

Fig. 12.10 Example 12.4

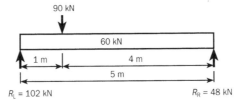

Solution

$$R_R = \tfrac{1}{2} \times 60 + \tfrac{1}{5} \times 90 \times 1 = 30 + 18 = 48 \text{ kN}$$

$$R_1 = 60 + 90 - 48 = 102 \text{ kN}$$

M_{max} occurs at a distance x, where

$$x = \frac{(102 - 90) \times 5}{60} = 1 \text{ m from } R_R, \text{ i.e. at the point load}$$

$$M_{max} = 102 \times 1 - \tfrac{1}{5} \times 60 \times \tfrac{1}{2} = 96 \times 10^6 \text{ N mm}$$

$$d_n = \frac{8.5}{8.5 + (140/15)} = 0.48d_1 \quad \text{and} \quad l_2 = 0.84d_1$$

$$d_1 = \sqrt{\frac{2 \times 96\,000\,000}{8.5 \times 200 \times 0.48 \times 0.84}} = 530 \text{ mm}$$

$$A_{st} = \frac{96\,000\,000}{140 \times 0.84 \times 530} = 1540 \text{ mm}^2$$

Use five 20 mm diameter bars (1570 mm²).

Then $d = 530 + \tfrac{1}{2} \times 20 + 25 = 565$ mm.

Example 12.5

A simply supported beam is to span 3.6 m carrying a uniform load of 60 kN inclusive of self-weight. The beam is to be 150 mm wide, and the stresses in steel and concrete respectively are to be 140 N/mm² and 10 N/mm² ($m = 15$).

Determine the constants n, j, R and p, and choose a suitable effective depth, overall depth and 676 of steel for the beam.

Solution

$$n = \frac{1}{1 + (p_{st}/mp_{cb})} = \frac{1}{1 + (140/15 \times 10)} = 0.517$$

$$j = 1 - \tfrac{1}{3}n = 1 - \tfrac{1}{3} \times 0.517 = 0.828$$

$$R = \tfrac{1}{2}p_{cb} \times n \times j = \tfrac{1}{2} \times 10 \times 0.517 \times 0.828 = 2.140$$

$$p = \frac{50 \times p_{cb} \times n}{p_{st}} = \frac{50 \times 10 \times 0.517}{140} = 1.846\%$$

$$M_{max} = \tfrac{1}{8}Wl = \tfrac{1}{8} \times 60 \times 3.6 \times 10^6 = 27 \times 10^6 \text{ N mm}$$

$$d_1 = \sqrt{\frac{27\,000\,000}{2.140 \times 150}} = 290 \text{ mm}$$

$$A_{st} = 0.01846 \times 150 \times 290 = 803 \text{ mm}^2$$

Use four 16 mm diameter bars (804 mm²).

Then $d = 290 + \tfrac{1}{2} \times 16 + 25 = 323$, say 325 mm.

Basis of design – load factor method

The introduction of this method alongside the elastic theory method to CP114 in 1957 was the result of tests to destruction. They have shown that, at failure, the compressive stresses adjust themselves to give a compressive resistance greater than that obtained by the elastic theory method.

Fig. 12.11 Load factor method

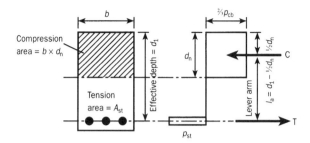

The load factor method, therefore, does not use the modular ratio nor does it assume a proportionality between stress and strain in concrete. It requires the knowledge of plastic behaviour, which is outside the scope of this textbook. CP114, however, introduced simplified formulae for rectangular beams and slabs which give acceptable results. They are based on the assumption that the compressive stress in the concrete is two-thirds of the permissible compressive stress in the concrete in bending (i.e. $\frac{2}{3}p_{cb}$). The stress is considered to be uniform over a depth d_n, not exceeding one-half of the effective depth ($\frac{1}{2}d_1$).

Figure 12.11 shows the stress distribution diagram based on the above assumptions.

$$\text{Total compressive force } C = \tfrac{2}{3}p_{cb} \times b \times d_n$$

$$\text{Total tensile force } T = p_{st} \times A_{st}$$

For equilibrium $C = T$ and therefore

$$d_n = \tfrac{3}{2} \times \frac{p_{st} \times A_{st}}{p_{cb} \times b}$$

Since the compressive stress is uniform

$$\text{Lever arm } l_a = d_1 - \tfrac{1}{2}d_n$$

$$= d_1 - \frac{3 \times p_{st} \times A_{st}}{4 \times p_{cb} \times b}$$

but, for an 'economical' or balanced section, the code recommends $d_n = \tfrac{1}{2}d_1$

$$l_a = d_1 - \tfrac{1}{2} \times \tfrac{1}{2}d_1 = \tfrac{3}{4}d_1$$

Hence, the moment of resistance, based on concrete in compression, is

$$M_{rc} = \tfrac{2}{3}p_{cb} \times b \times \tfrac{1}{2}d_1 \times \tfrac{3}{4}d_1$$

$$= \tfrac{1}{4}p_{cb} \times b \times (d_1)^2$$

and the moment of resistance, based on tensile reinforcement, is

$$M_{rt} = A_{st} \times p_{st} \times l_a$$

Example 12.6

Consider the beam in Example 12.5 and, using the same concrete mix, design the beam by the load factor method.

Solution

$$M_{max} = 27 \times 10^6 \, \text{N mm}$$

$$= M_{rc} = \tfrac{1}{4} \times 10 \times 150 \times (d_1)^2$$

$$d_1 = \sqrt{\frac{27\,000\,000 \times 4}{10 \times 150}} = 268 \, \text{mm}$$

and

$$A_{st} = \frac{27\,000\,000 \times 4}{140 \times 3 \times 268} = 960 \, \text{mm}^2$$

It may be correctly deduced from the above that the load factor method of design results in shallower beams with more reinforcement.

The load factor method gained fairly general acceptance in a relatively short time, so that, when in 1972 CP110, the code of practice for the structural use of concrete, appeared, it did not contain any explicit reference to the modular ratio or elastic theory. This code introduced yet another approach to structural design in what it called 'strict conformity with the theory of limit states'.

CP110 has been superseded by BS 8110: 1997, with 'no major changes in principle'. The new code is in three parts: Part 1 deals with design and construction, Part 2 is 'for special circumstances' and Part 3 contains design charts for beams and columns.

Basis of design – limit state design

A brief introduction to limit states is given in Chapter 11.

Consider the design of a reinforced concrete beam for the ultimate limit state.

Loads

The design load is obtained by multiplying the characteristic load by an appropriate partial safety factor, γ_f.

The characteristic loads (G_k = dead load, Q_k = imposed load, and W_k = wind load) are based on values given in BS 6399: Part 1: 1996 for dead and imposed (live) loads, and BS 6399–2: 1997 for wind loads.

The partial safety factors γ_f depend on the combination of loads. For example, for dead + live loads, the design load would be

$$(1.4 \times G_k + 1.6 \times Q_k)$$

and for dead + live + wind loads, the design load would be

$$1.2 \times (G_k + Q_k + W_k)$$

Strength of materials

The design strength is obtained by dividing the characteristic strength, f_k, by the appropriate partial safety factor γ_m.

Characteristic strength is defined as that value of the cube strength of concrete (f_{cu}) or the yield (or proof) stress of reinforcement (f_y) below which not more than 5 per cent of the test results will fall.

The partial safety factors γ_m for ultimate limit state are 1.5 for concrete and 1.05 for the reinforcement.

Ultimate moment of resistance

In their simplest form, as given in Clause 3.4.4.4 of BS 8110: Part 1: 1997, the calculations are based on assumptions similar to the load factor method, i.e. rectangular stress block with the stress value of $0.477f_{cu}$ over $0.9 \times$ distance to NA.

Therefore

$$M_{uc} = 0.477f_{cu} \times b \times 0.446d \times 0.777d$$

$$= 0.156f_{cu} \times b \times d^2$$

and

$$M_{ut} = 0.95f_y \times A_s \times z$$

where
M_{uc} = ultimate resistance moment (compression)
M_{ut} = ultimate resistance moment (tension)
b = width of beam section
d = effective depth to tension reinforcement
A_s = area of tension reinforcement
z = lever arm (in this case $0.777d$)

Example 12.7

Take the beam in Example 12.5 and adapt the data to limit state design method for the ultimate limit state.

Solution

Assume the 60 kN load to consist of 60 per cent dead and 40 per cent live loads and

$$f_y = 250 \text{ N/mm}^2 \text{ (mild steel)} \quad f_y = 460 \text{ N/mm}^2 \text{ (high yield steel)}$$

$$f_{cu} = 30 \text{ N/mm}^2$$

Therefore,

$$\text{Design load} = (1.4 \times 36 + 1.6 \times 24) = 88.8 \text{ kN}$$

and

$$M_{max} = \tfrac{1}{8} \times 88.8 \times 3.6 \times 10^6 = 40 \times 10^6 \text{ N mm}$$

Hence

$$M_{uc} = 0.156 \times 30 \times 150 \times d^2 = 40 \times 10^6 \text{ N mm}$$

$$d = \sqrt{\frac{40\,000\,000}{0.156 \times 30 \times 150}} = 238 \text{ mm}$$

and

$$M_{ut} = 0.95 \times 250 \times A_s \times 0.777 \times 238 = M_{max}$$

$$A_s = \frac{40\ 000\ 000}{0.95 \times 259 \times 0.777 \times 238} = 879.10\ \text{mm}^2$$

It must be emphasized that the above exposition of the reinforced concrete beam design methods is very much simplified. It is essential to consult textbooks on concrete design for more detailed discussion of both principles and their application.

Summary

Timber and steel composite (flitch) beams   Replace the steel by its equivalent area of timber by multiplying the width of the steel (parallel to the axis of bending) by the modular ratio E_s/E_t. Calculate the value of Z for the equivalent timber section. Determine the maximum permissible stress f in the extreme fibres of the beam such that neither the maximum permissible steel stress nor the maximum permissible timber stress is exceeded. Then

$$M_r = fZ$$

Reinforced concrete beams
- Elastic theory method

$$d_n = \frac{1}{1 + p_{st}/mp_{cb}} \times d_1$$

$$M_{rc} = \tfrac{1}{2}p_{cb} \times b \times d_n \times l_a$$

$$M_{rt} = p_{st} \times A_{st} \times l_a$$

where $l_a = d_1 - \tfrac{1}{3}d_n$.

- Load factor method

$$M_{rc} = \tfrac{1}{4}p_{cb} \times b \times (d_1)^2$$

$$M_{rt} = p_{st} \times A_{st} \times \tfrac{3}{4}d_1$$

- Limit state design method

$$M_{uc} = 0.156f_{cu} \times b \times d^2$$

$$M_{ut} = 0.95f_y \times A_{st} \times Z$$

where $Z = 0.777d$.

Exercises

1 A composite beam is formed using a 400 mm × 180 mm timber beam with a 300 mm × 12 mm steel plate securely fixed to each side as shown in Fig. 12.Q1. The maximum stresses in the steel and timber respectively must not exceed 140 and 8 N/mm^2, and the modular ratio is 20. (a) What will be the actual stresses used for the steel and for the timber? (b) What is the safe moment of resistance in N mm for the beam section?

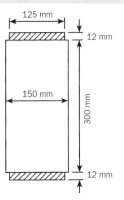

Fig. 12.Q3

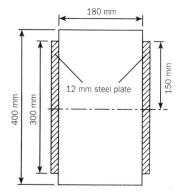

Fig. 12.Q1

2 A timber flitch beam is composed of two 300 mm × 150 mm timber beams and one 250 mm × 20 mm steel plate placed between the timbers so that, when properly bolted together, the centre lines of all three members coincide. Calculate the maximum safe uniformly distributed load in kilonewtons that this beam could carry over a span of 4.5 m if the stress in the steel is not to exceed 125 N/mm^2 and that in the timber 7 N/mm^2, and given that the modular ratio $E_s/E_t = 20$.

3 A timber beam 150 mm × 300 mm deep has two steel plates, each 125 mm × 12 mm, bolted to it as shown in Fig. 12.Q3. Assuming the safe steel stress is 140 N/mm^2, the safe timber stress is 7 N/mm^2, E for steel is 205 000 N/mm^2 and E for timber is 8200 N/mm^2, calculate the moment of resistance of the beam. (Ignore bolt holes.)

4 A timber beam in an existing building is 200 mm wide and 380 mm deep and is simply supported at the ends of a 6 m span. (a) Calculate the maximum safe uniformly distributed load for the timber alone if the bending stress must not exceed 7 N/mm^2. (b) It is proposed to strengthen the beam to enable it to carry a uniformly distributed load of 150 kN by bolting two steel plates to the beam as indicated in Fig. 12.Q4. Calculate the required thickness t of the plates if the maximum permissible stress for the steel is 140 N/mm^2 and the modular ratio is 24.

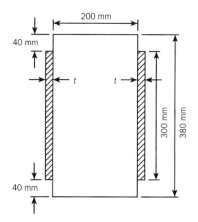

Fig. 12.Q4

5 A short concrete beam is to be constructed without any steel reinforcement to span 2.8 m, carrying a total inclusive uniform load of 20 kN. If the concrete has a safe tensile stress of only 0.6 N/mm^2, state what depth would be needed for a suitable beam 200 mm wide.

6 Design a reinforced concrete section for the loading conditions as in Question 5 if the beam remains 200 mm wide, using the following stresses: $p_{cb} = 10 \text{ N/mm}^2$, $p_{st} = 140 \text{ N/mm}^2$, $m = 15$.

7 A simply supported reinforced concrete beam, 240 mm wide, carries inclusive loads as shown in Fig. 12.Q7. Determine (a) the effective depth d_1 in mm, (b) the required steel area if $p_{cb} = 7 \text{ N/mm}^2$, $p_{st} = 140 \text{ N/mm}^2$, $m = 15$. The weight of the beam may be assumed to be included in the given loads.

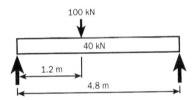

Fig. 12.Q7

8 Referring to Example 11.5 design a reinforced concrete beam as an alternative to the 400 mm × 150 mm timber beam. Assume that the floor load is 7 kN/m² and allow 12 kN for the weight of the reinforced concrete beam. Take the breadth of the beam as 250 mm and $p_{cb} = 7 \text{ N/mm}^2$, $p_{st} = 140 \text{ N/mm}^2$, $m = 15$.

9 Referring to Example 11.7 for loading conditions but substituting 18 kN for the weight of the beam, design a reinforced concrete beam 225 mm wide. When the design is complete, check the assumed weight of beam, taking the weight of reinforced concrete as 24 kN/m³. Use $1:1\frac{1}{2}:3$ mix of concrete.

10 A small floor is to be supported as shown in Fig. 12.Q10. The total floor load is 5 kN/m², and 12 kN can be assumed as the weight of the beam. Design the beam assuming a breadth of 250 mm and a 1:2:4 mix of concrete.

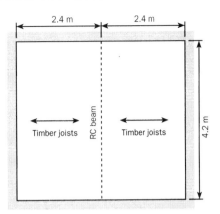

Fig. 12.Q10

11 Referring to Exercise 15 in Chapter 11, but taking 8 kN as the weight of each beam instead of the value given, design reinforced concrete beams assuming a breadth of 225 mm and a 1:1:2 mix of concrete.

12 Referring to Exercise 16 in Chapter 11 design beams A and B in reinforced concrete (1:2:4 mix). Take the weight of beam B as 1.8 kN and the weight of beam A as 16.0 kN (in addition to the floor load given).

13 A reinforced concrete beam 300 mm wide simply supported on a span of 6 m carries a triangular load of 80 kN in addition to its own weight, which may be assumed to be 20 kN (Fig. 12.Q13). Design the beam in $1:1\frac{1}{2}:3$ mix concrete.

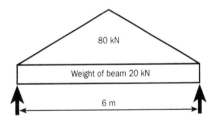

Fig. 12.Q13

Chapter thirteen Deflection of beams

So far in this book the beam has been considered from the point of view of its safety and strength in its resistance to bending. This chapter investigates the deformation of beams as the direct effect of that bending tendency, which affects their serviceability and stability, and does so in terms of their deflection.

A beam may be strong enough to resist safely the bending moments due to the applied loading and yet not be suitable because its deflection is too great. Excessive deflection might not only impair the strength and stability of the structure but also give rise to minor troubles such as cracking of plaster ceilings, partitions and other finishes, as well as adversely affecting the functional needs and aesthetic requirements or simply being unsightly.

The relevant BS specifications and codes of practice stipulate that the deflection of a beam shall be restricted within limits appropriate to the type of structure. In the case of structural steelwork the maximum deflection due to unfactored imposed loads for beams carrying plaster or other brittle finish must not exceed 1/360 of the span, but for all other beams it may be span/200 (Clause 2.5.1 of BS 5950: Part 1: 2000). For timber beams, on the other hand, the figure is 0.003 of the span when the supporting member is fully loaded (Clause 2.10.7 of BS 5268: Part 2: 1996). In reinforced concrete the deflection is generally governed by the span/depth ratio (Clause 3.4.6.3 of BS 8110: Part 1: 1997).

Factors affecting deflection

For many beams in most types of buildings, e.g. flats, offices, warehouses, it will usually be found that, if the beams are made big enough to resist the bending stresses, the deflections will not exceed the permitted values. In beams of long spans, however, it may be necessary to calculate deflections to ensure that they are not excessive. The derivation of formulae for calculating deflections usually involves calculus. In this chapter, therefore, only a general treatment will be attempted and deflection formulae for a few common cases of beam loadings will be given without proof. General methods of calculation of deflections are given in standard books on theory of structures or strength of materials.

Load

AB (Fig. 13.1) represents a beam of span l metres supported simply at its ends and carrying a point load of W kN at midspan. Let us assume that the

Fig. 13.1 Deflection of a beam under loading

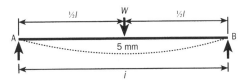

deflection due to the load is 5 mm. It is obvious that, if the load is increased, the deflection will increase. It can be proved that the deflection is directly proportional to the load, i.e. a load of $2W$ will cause a deflection of 10 mm, $3W$ will produce a deflection of 15 mm and so on. W must therefore be a term in any formula for calculating deflection.

Span

In Fig. 13.2(a) and (b) the loads W are equal and the weights of the beams, which are assumed to be equal in cross-section, are ignored for the purposes of this discussion. The span of beam (b) is twice that of beam (a). It is obvious that the deflection of beam (b) will be greater than that of beam (a), but the interesting fact (which can be demonstrated experimentally or proved by mathematics) is that, instead of the deflection of (b) being twice that of (a), it is 8 times (e.g. 40 mm in this example). If the span of beam (b) were $3l$, its deflection would be 27 times that of beam (a). In other words, the deflection of a beam is proportional to the cube of the span, therefore l^3 is a term in the deflection formula.

Fig. 13.2 Effect of span upon deflection: the span of (a) is twice that of (b)

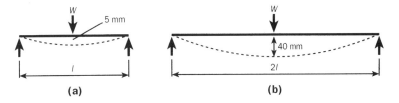

(a)   (b)

Size and shape of beam

Figure 13.3(a) and (b) represents two beams (their weights being ignored) of equal spans and loading but the moment of inertia of beam (b) is twice that of beam (a). Obviously, the greater the size of the beam, the smaller the deflection (other conditions being equal). It can be proved that the deflection is inversely proportional to the moment of inertia, e.g. the deflection of beam (b) will be one-half that of beam (a). Moment of inertia I is therefore a term in the denominator of the deflection formula. (It may be noted that, since the moment of inertia of a rectangular cross-section beam is $bd^3/12$, doubling the breadth of a rectangular beam decreases the deflection by one-half,

Fig. 13.3 Effect of size and shape upon deflection: (a) moment of inertia of beam = 1 unit; (b) moment of inertia of beam = 2 units

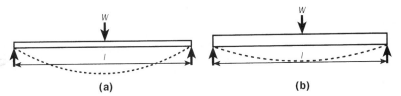

(a)   (b)

whereas doubling the depth of a beam decreases the deflection to one-eighth of the previous value.)

'Stiffness' of material

The stiffer the material of a beam, i.e. the greater its resistance to bending, the smaller will be the deflection, other conditions such as span, load, etc., remaining constant. The measure of the 'stiffness' of a material is its modulus of elasticity E and deflection is inversely proportional to the value of E.

Derivation of deflection formulae

A formula for calculating deflection must therefore contain the load W, the cube of the span l^3, the moment of inertia I, and the modulus of elasticity E. For standard cases of loading, the deflection formula can be expressed in the form cWl^3/EI, where c is a numerical coefficient depending on the disposition of the load and also on the manner in which the beam is supported, that is, whether the ends of the beam are simply supported or fixed, etc. For Fig. 13.4 the values of c are respectively 1/48 and 5/384. W and l^3 are in the numerator of the formula because an increase in their values means an increase of deflection, whereas E and I are in the denominator because an increase in their values means a decrease of deflection.

Referring to Fig. 13.4 it should be obvious (neglecting the weights of the beams) that although the beams are equally loaded, the deflection of beam (b) will be less than that of beam (a). In fact, the maximum deflection of beam (a) is

$$\frac{1}{48}\frac{Wl^3}{EI} = \frac{8}{384}\frac{Wl^3}{EI}$$

and the maximum deflection of beam (b) is

$$\frac{5}{384}\frac{Wl^3}{EI}$$

Table 13.1 gives the values of c for some common types of loading, etc. When the load system is complicated, e.g. several point loads of different magnitudes, or various combinations of point loads and uniformly distributed loads, the deflections must be calculated from first principles.

In certain simple cases it is possible to derive deflection formulae mathematically without using the calculus, and the following example is given for the more mathematically minded student. Neglecting the weight of the beam, shear force and bending moment diagrams are given in Fig. 13.5(b) and (c) for the beam loaded as shown. The maximum bending moment is $Wa/2$ and this moment is constant along the length AB, the shear force being zero. Since the

Fig. 13.4 Deflection formulae

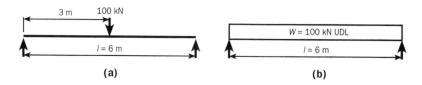

(a)

(b)

Table 13.1 Values of coefficient c for deflection formula $\delta = cWl^3/EI$

Condition of loading	Value of c (δ_{max} at A)
	$\dfrac{1}{48} = 0.02083$
	$\dfrac{23}{1296} = 0.01775$
	$\dfrac{11}{768} = 0.01432$
	$\dfrac{5}{384} = 0.01302$
	$\dfrac{1}{192} = 0.00521$
	$\dfrac{1}{384} = 0.00260$
	$\dfrac{1}{3} = 0.33333$
	$\dfrac{1}{8} = 0.12500$

moment is constant, the portion AB of the beam bends into the arc of a circle with a radius of curvature R. In triangle OBC (Fig. 13.6),

$$R^2 = (R - \delta)^2 + (\tfrac{1}{2}l)^2$$

$$= R^2 - 2R\delta + \delta^2 + \tfrac{1}{4}l^2$$

Fig. 13.5 Derivation of the deflection formulae: (a) loading diagram; (b) shear force diagram; (c) bending moment diagram

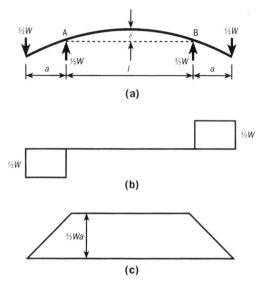

Fig. 13.6 Beam bending into an arc of a circle

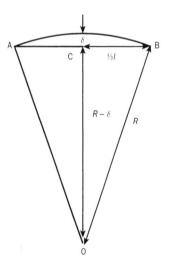

$$R^2 - R^2 + 2R\delta - \delta^2 = \tfrac{1}{4}l^2$$

$$2R\delta - \delta^2 = \tfrac{1}{4}l^2 \tag{1}$$

δ^2 is the square of a small quantity and can be ignored. Therefore

$$2R\delta = \tfrac{1}{4}l^2 \tag{2}$$

For example, if $l = 3000$ mm and $\delta = 30$ mm, which is a big deflection for such a small span, then from equation (1)

Fig. 13.7 Effect of beam overhang upon deflection

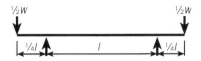

$$2R \times 30 - 900 = \tfrac{1}{4} \times 9 \times 10^6$$

$$60R - 900 = 2.25 \times 10^6$$

$$60R = 2\,249\,100$$

$$R = 37\,485 \text{ mm}$$

If we ignore the term δ^2, $R = 37\,500$ mm, so the inaccuracy is very small, even with such a comparatively large deflection.

It was shown in Chapter 11, page 206, that

$$\frac{M}{I} = \frac{E}{R} \quad \text{or} \quad R = \frac{EI}{M}$$

Substituting this value of R in equation (2):

$$2 \times \frac{EI}{M} \times \delta = \tfrac{1}{4}l^2$$

$$\delta = \frac{Ml^2}{8EI}$$

M is the bending moment and equals $\tfrac{1}{2}Wa$. Therefore

$$\delta = \frac{Wal^2}{16EI}$$

a can be expressed in terms of the span l. If, for example, the overhanging portion a is a quarter $(\tfrac{1}{4}l)$ of the interior span (Fig. 13.7) then, substituting in equation (3)

$$\delta = \frac{W \times \tfrac{1}{4}l \times l^2}{16EI} = \frac{1}{64}\frac{Wl^3}{EI}$$

Example 13.1

A 406×178 UB54 simply supported at the ends of a span of 5 m carries a uniformly distributed load of 60 kN/m. Calculate the maximum deflection ($E = 205\,000$ N/mm^2).

Solution

The formula for the maximum deflection is (from Table 13.1)

$$\delta = \frac{5}{384}\frac{Wl^3}{EI}$$

where

$$W = 60 \times 5 = 300 \text{ kN} = 0.3 \times 10^6 \text{ N}$$

$$l = 5000 \text{ mm}$$

$$E = 205\,000 \text{ N/mm}^2$$

$$I = 187.2 \times 10^6 \text{mm}^4$$

Therefore

$$\delta = \frac{5}{384} \times \frac{300\,000 \times 125 \times 10^9}{205\,000 \times 187.2 \times 10^6}$$

$$= 13 \text{ mm}$$

Example 13.2

Calculate the safe inclusive uniformly distributed load for a 457×152 UB52 simply supported at its ends if the span is 6 m and if the span is 12 m. The maximum permissible bending stress is 165 N/mm² and the maximum permissible deflection is 1/360 of the span. E is 205 000 N/mm².

Solution

$$Z = 950\,000 \text{ mm}^3 \text{ (from Table 11.3)}$$

$$M_r = f \times Z = 165 \times 950\,000 = 156.75 \times 10^6 \text{ N mm}$$

$$M_{max} = \tfrac{1}{8}Wl = \tfrac{1}{8}W \times 6000 = 156.75 \times 10^6 \text{ N mm}$$

Therefore

$$W = \frac{156.75 \times 10^6 \times 8}{6000} = 209 \text{ kN}$$

$$\text{Maximum deflection } \delta = \frac{5}{384} \frac{Wl^3}{EL}$$

where

$$W = 209\,000 \text{ N}$$

$$l = 6000 \text{ mm}$$

$$E = 205\,000 \text{ N/mm}^2$$

$$I = 213.7 \times 10^6 \text{ mm}^4$$

Therefore

$$\delta = \frac{5}{384} \times \frac{209 \times 6^3 \times 10^{12}}{0.205 \times 213.7 \times 10^{12}} = 13.4 \text{ mm}$$

$$\text{Maximum permissible deflection} = \frac{6000}{360} = 17 \text{ mm}$$

Therefore

$$\text{Safe load} = 209.0 \text{ kN}$$

For a 12 m span $M_r = 156.75 \times 10^6 \text{ N mm}$ as before.

Therefore

$$W = \frac{156.75 \times 10^6 \times 8}{12\,000} = 104.5 \text{ kN}$$

Maximum actual deflection due to this load

$$\delta = \frac{5}{384} \times \frac{104.5 \times 12^3 \times 10^{12}}{0.205 \times 213.7 \times 10^{12}} = 53.7 \text{ mm}$$

$$\text{Maximum permissible deflection} = \frac{12\,000}{360} = 33 \text{ mm}$$

This means that, although the beam is quite satisfactory from the strength point of view, the deflection is too great, therefore the load must be reduced

$$\text{Now } \delta = \frac{5}{384} \frac{Wl^3}{EI}$$

Therefore

$$33 = \frac{5}{384} \times \frac{W \times (12\,000)^3}{205\,000 \times 213.7 \times 10^6}$$

giving $W = 64.2$ kN and this is the maximum permitted load for the beam. (Instead of being obtained from the deflection formula, W can also be obtained from $W = (33/53.7) \times 104.5 = 64.2$ kN.)

Example 13.3

Calculate the safe inclusive uniformly distributed load for a 200 mm × 75 mm timber joist, simply supported at its ends, if the span is 4 m and if the span is 8 m. The maximum permissible bending stress is 6 N/mm² and the maximum permissible deflection is 0.003 of the span. E is 9500 N/mm².

Solution

From $M_r = \frac{1}{6}fbd^2 = M_{max} = \frac{1}{8}Wl$

$$\tfrac{1}{6} \times 6 \times 75 \times 200^2 = \tfrac{1}{8} \times W \times 4000$$

and so $W = 6.0$ kN.

$$\text{Maximum actual deflection} = \frac{5}{384} \frac{Wl^3}{EI}$$

where $W = 6000$ N

$l = 4000$ mm

$E = 9500$ N/mm²

$I = \tfrac{1}{12} \times 75 \times 200^3 = 50 \times 10^6$ mm⁴

$$\delta = \frac{5}{384} \times \frac{6 \times 4^3 \times 10^{12}}{9.5 \times 50 \times 10^9} = 11 \text{ mm}$$

Maximum permissible deflection $= 0.003 \times 4000$

$$= 12 \text{ mm}$$

The safe UDL for the 4 mm span $= 6.0$ kN

When the span is doubled only half the previous load will be applied: $W = 3.0$ kN.

Maximum actual deflection due to this load

$$= \frac{5}{384} \times \frac{3 \times 8^3 \times 10^2}{9.5 \times 50 \times 10^9} = 42 \text{ mm}$$

Maximum permissible deflection $= 0.003 \times 8000 = 24$ mm

The load must be reduced so that $\dfrac{5}{384} \times \dfrac{Wl^3}{EI} = 24$ mm

Or the answer can be obtained by multiplying the load of 3 kN by 24/42 which gives 1.7 kN. Therefore, for this span, the safe load is 1.7 kN.

Span/depth ratios

It appears from the above examples that, before a beam can be passed as suitable, deflection calculations must be made in addition to bending calculations. Fortunately, it is possible to derive simple rules which replace deflection calculations in many cases. For example, if having designed a UB simply supported at its ends and carrying a UDL over its full length it is found that the span of the beam does not exceed 17 times its depth, the beam will be suitable from the deflection point of view. This rule is derived as follows.

The maximum deflection for a simply supported beam with a UDL is

$$\delta = \frac{5}{384} \times \frac{Wl^3}{EI}$$

This formula can be rearranged as follows:

$$\delta = \frac{5}{48} \times \frac{Wl}{8} \times \frac{l^2}{EI}$$

Now $\frac{1}{8}Wl$ is the maximum bending moment, which is M. Thus

$$\delta = \frac{5}{48} \times \frac{M}{I} \times \frac{l^2}{E}$$

But $M/I = f/y$ (see page 206)

$$\delta = \frac{5}{48} \times \frac{f}{y} \times \frac{l^2}{E}$$

where $y = \frac{1}{2}d$ (see page 207)

$$f = 0.6 \times p_y = 165 \text{ N/mm}^2$$

$$E = 205\,000 \text{ N/mm}^2$$

Therefore

$$\delta = \frac{5}{48} \times \frac{165 \times 2}{d} \times \frac{l^2}{205\,000} = \frac{1650l^2}{9.84 \times 10^6\,d} = \frac{l^2}{5964\,d}$$

but δ must not exceed $l/360$, hence

$$\frac{l}{360} = \frac{l^2}{5964\,d}$$

$$360l = 5964\,d$$

$$\frac{l}{d} = \frac{5964}{360} = 17 \quad \text{or} \quad l = 17\,d$$

Similar rules can be derived in the same manner for rectangular timber sections. Table 13.2 gives values for the stresses and values of E taken from BS 5268: Part 2: 1996. These rules are only applicable to beams simply supported at each end and carrying a UDL over their full length.

Note that for each of the grades of timber mentioned in Table 13.2 two values of the elastic modulus are given. Clause 2.10.7 of BS 5268: Part 2: 1996 states: 'The deflections of solid timber members acting alone should be calculated using the minimum modulus of elasticity for the strength class or species and grade.' For load-sharing systems (floor and ceiling joists, rafters, etc.) the mean value should be used subject to limitations given in the code.

Table 13.2 Span/depth ratios for timber beams with UDL over full span and $\delta \not> 0.003 \times$ span (based on Table 7, BS 5268: Part 2: 1996)

Strength class	Bending stress parallel to grain (N/mm²)	E (N/mm²) Minimum	Mean	L_e/h
C14	4.1	4600		17.6
			6800	28.1
C16	5.3	5800		15.8
			8800	23.9
C18	5.8	6000		12.7
			9100	19.0
C22	6.8	6500		10.2
			9700	15.4

effective length [handwritten annotation]

Deflection of reinforced concrete beams

The composite nature of reinforced concrete beams complicates the determination of their deflections and, although these may be calculated, the process is tedious and time-consuming. Clause 3.4.6.3 of BS 8110: Part 1: 1997 states that normally the beam will not deflect excessively provided the span/effective depth ratios are kept within the following limits:

Cantilever	7
Simply supported	20
Continuous	26

Summary

For beams which behave elastically (steel, timber) and for standard types of loading, the *actual deflection* is

$$\delta = c \times \frac{W}{E} \times \frac{l^3}{I}$$

where　c = a numeral coefficient taking into account the load system (UDL, point load etc.) and the manner in which the beam is supported (fixed or simple supports); see Table 13.1

W = the total (unfactored) load on the beam

l = the span

E = modulus of clasticity of the material

I = moment of inertia (second moment of area) of the section

Limitations are imposed on the maximum amount of deflection by the British Standard specifications and codes of practice. In general:

For steel the limit is $(1/360) \times$ span

For timber the limit is $0.003 \times$ span

Reinforced concrete beams (and slabs) are deemed to satisfy the limitation if their specified span/depth ratio is not exceeded.

Exercises

Note: The value of E for steel is 205 000 N/mm^2. The values of E for timber are given in Table 13.2.

1　A 457 × 191 UB98 is simply supported at the ends of a span of 7.2 m. The beam carries an inclusive UDL of 350 kN. Calculate the maximum deflection.

2　A 406 × 178 UB60 is simply supported at the ends of a span of 6.0 m. The beam carries a point load of 140 kN at midspan. Calculate the deflection due to this load, ignoring the weight of the beam.

3　A 356 × 171 UB45 is simply supported at the ends of a span of 5.5 m. It carries an inclusive UDL of 12

kN/m and a central point load of 75 kN. Calculate the maximum deflection.

4　Calculate the safe, inclusive UDL for a 356 × 171 UB67 simply supported at the ends of a span of 9 m. The permissible bending stress is 165 N/mm^2 and the deflection must not exceed $(1/360) \times$ span.

5　Calculate the maximum deflection of a 305 × 165 UB40 cantilevered 3 m beyond its fixed support and carrying an inclusive UDL of 30 kN.

6　A 203 × 133 UB30 is fixed at one end and cantilevered for a distance of 1.2 m. The beam supports a point load of 15 kN at its free end. Calculate the maximum deflection ignoring the weight of the beam.

7 A 254 × 146 UB43 is fixed at one end and cantilevered for a distance of 1.5 m. The beam carries a UDL of 12 kN/m and a point load of 10 kN at the free end. Calculate the maximum deflection.

8 A 75 mm wide and 150 mm deep beam in C18 class timber carries an inclusive UDL over a simply supported span of 1.8 m. Calculate the maximum deflection for UDL of 10 kN.

9 A timber (C14) beam 75 mm wide and 240 mm deep carries a point load of 5.5 kN at the centre of its simply supported span of 3 m. Calculate the maximum deflection due to this load.

10 A timber (C16) beam, 100 mm × 300 mm, spans 4 m and carries a central point load of 5 kN in addition to an inclusive UDL of 10 kN. Calculate the maximum deflection.

11 A 75 mm × 225 mm C18 timber beam spans 5 m on simple supports. Calculate the value of the safe UDL for the following conditions: (a) permissible bending stress of 7.5 N/mm^2 and deflection is not important; (b) the deflection must not exceed 0.003 × span.

12 A beam in C16 timber carries a UDL of 7.5 kN on a simply supported span of 5 m. Assuming the beam to be 89 mm wide calculate its depth when (a) the permissible bending stress is 5.3 N/mm^2 and deflection is not important; (b) maximum deflection is limited to 0.003 × span.

13 Calculate the maximum deflection of a cantilever beam in C18 timber 75 mm wide and 240 mm deep. The beam carries an inclusive UDL of 5 kN over a span of 2 m.

14 A timber (C14) cantilever beam is 1.2 m long, 75 mm wide and 150 mm deep. It carries a UDL of 1.0 kN and a point load of 0.9 kN at its free end. Calculate the maximum deflection.

15 Floor joists in C14 timber spaced at 360 mm c/c are to carry an inclusive load of 2.0 kN/m^2 over a simply supported span of 4.0 m. Determine a suit-

able size for the joists if the bending stress is limited to 4.5 N/mm^2 and the deflection must not exceed 0.003 × span. (Use the mean value of E (see Table 13.2, page 259).

16 Calculate the maximum span/depth ratio for a steel cantilever beam supporting a UDL so that deflection does not exceed (1/360) × span. The permissible bending stress is 155 N/mm^2.

17 Calculate the minimum depth for a steel beam simply supported at the ends of a 7.5 m span carrying a point load at midspan. The weight of the beam may be ignored. The deflection must be limited to the usual (1/360) × span and the permissible bending stress is 165 N/mm^2.

18 Determine the deflection of a 150 mm × 400 mm C14 timber beam which is simply supported at the ends of its 5 m length. The beam is subjected to a maximum bending moment of 21×10^6 N mm due to an inclusive UDL.

19 A solid rectangular beam is simply supported at the ends of a 4.8 m span. The beam is subjected to a maximum bending moment of 90×10^6 N mm due to an inclusive UDL. Determine a suitable section for the beam given:

Modulus of elasticity of the material = 4000N/mm^2

Maximum permissible bending stress = 7.5 N/mm^2

Deflection must be limited to 15 mm

20 Calculate the minimum depth for a simply supported C18 timber beam, which is to carry a UDL over a span of 4.0 m, if the deflection must not exceed 0.003 × span and the permissible stress in bending is 7.5 N/mm^2. Assuming the beam to be 150 mm wide, determine the value of the UDL.

21 A timber beam 75 mm × 150 mm simply supported at the ends of a 2.0 m span deflects 5 mm under a 10 kN UDL. Without calculating the modulus of elasticity, determine the maximum deflection of a beam of similar timber 150 mm × 300 mm due to a UDL of 40 kN on a span of 4.0 m.

In this chapter the factors affecting the column's load-carrying capacity are investigated. The connection between the slenderness of the column and its tendency to buckle is discussed. The influence of the 'fixity' of the ends of the column, and the shape of its section on that slenderness, is considered in relation to timber, steel and reinforced concrete columns.

When the line of action of the resultant load is coincident with the centre of gravity axis of the column (Fig. 14.1(a)), the column is said to be **axially loaded** and the stress produced in the material is said to be a *direct compressive stress*. This stress is uniform over the cross-section of the column. The term *concentric loading* is sometimes used instead of *axial loading*.

When the load is not axial, it is said to be eccentric (i.e. off-centre) and bending stress is induced in the column as well as a direct compressive stress (Fig. 14.1(b)). Eccentric loading is dealt with in Chapter 16.

Other words used to describe members which are subjected to compressive stress are *pillar*, *post*, *stanchion*, *strut*. There are no definite rules as to when any one of these words should be used, but the following convention is fairly general.

Column and pillar can usually be applied to any material, e.g. timber, stone, concrete, reinforced concrete, steel. Post is usually confined to timber. Stanchion is often used for rolled steel I-sections and channel sections. Strut has a more general significance than stanchion or post but normally it is not applied to a main supporting member of a building. The word is often used for compression members of roof trusses whether the material is timber or steel.

Fig. 14.1 Loading of columns: (a) axial loading; (b) eccentric loading

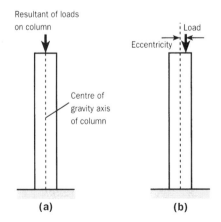

Design factors

The maximum axial load a column can be allowed to support depends on:

- the material of which the column is made
- the slenderness of the column

The slenderness involves not only the height or length of the column, but also the size and shape of its cross-section and the manner in which the two ends of the column are supported or fixed.

The majority of columns are designed by reference to tables of permissible stresses contained in British Standard specifications and codes of practice. These tables of permissible stresses (which are reproduced on pages 264 and 270) have been constructed from complex formulae which have been derived as the result of a great deal of research, mathematical and experimental, into the behaviour of columns under load. It is not possible in this book to deal with the mathematical theories of column design but an attempt will be made to give an explanation of the general principles.

A very short column will fail due to crushing of the material, but long columns are likely to fail by 'buckling', the failing load being much less than that which would cause failure in a short column of identical cross-sectional dimensions.

Consider Fig. 14.2(a), which represents a strip of pliable wood, say 6 mm × 54 mm in cross-section and 600 mm long. A small vertical force applied as shown will cause buckling. It should be obvious (but can be confirmed by experiment if necessary) that a larger force would be required to cause failure if the member were only 300 mm high. In other words, the 600 mm high member is more slender than the 300 mm high member of equal cross-sectional dimensions.

Now consider Fig. 14.2(b). All three members are of equal cross-sectional area (324 mm²) and of equal height, yet member (ii) will require more load to cause buckling than member (i) and member (iii) is the strongest column.

Fig. 14.2 Design factors:
(a) buckling; (b) three members of equal cross-sectional area

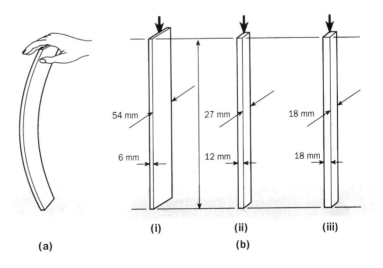

Slenderness ratio

By reference to Fig. 14.2(b) it can be seen that the smaller cross-sectional dimension of the column is very important from the point of view of buckling and it appears that the **slenderness ratio** of a column can be defined as

$$\frac{\text{Effective length of column (in millimetres)}}{\text{Least width of column (in millimetres)}} = \left(\frac{l}{b}\right); \left(\frac{L_e}{b}\right)$$

e.g. slenderness ratio of

$$\text{Column (i)} = 600/6 = 100$$

$$\text{Column (ii)} = 600/12 = 50$$

$$\text{Column (iii)} = 600/18 = 33 \text{ approx.}$$

Since most timber posts and struts are rectangular in cross-section, it is reasonable to express the slenderness ratio in terms of the length and least width (i.e. least lateral dimension). This method is, in fact, adopted by BS 5268 (see Table 14.1) but an alternative slenderness ratio is also given, i.e.

$$\frac{\text{Effective length}}{\text{Least radius of gyration}} = \left(\frac{l}{r}\right) \text{ or } \left(\frac{L_e}{i}\right)$$

and it is this slenderness ratio which must be used when the post is not of solid rectangular cross-section.

For explanation of the terms *least radius of gyration* and *effective length* see pages 268 and 269, respectively.

A Swiss mathematician named Leonhard Euler (1707–1783) showed that a long thin homogeneous column, axially loaded, suffers no deflection as the load is gradually applied until a critical load (the collapsing or buckling load P) is reached. At this load, instability occurs and the column buckles into a curve.

The curve of Fig. 14.3 is not the arc of a circle, and Euler found (with the aid of calculus) that the buckling load P gets smaller as the slenderness of the column increases.

Table 14.1 Permissible compression stresses for timber struts (compression members)

L_c/i (N/mm^2)	L_c/b (N/mm^2)	Strength class for service class 1 and 2		L_c/i (N/mm^2)	L_c/b (N/mm^2)	Strength class for service class 1 and 2	
		C14	C24			C14	C24
<5	1.4	5.2	7.90	90	26.0	2.30	3.6
5	1.4	5.07	7.70	100	28.9	2.00	3.14
10	2.9	4.95	7.52	120	34.7	1.52	2.4
20	5.8	4.70	7.13	140	40.5	1.18	1.87
30	8.7	4.43	6.73	160	46.2	0.94	1.48
40	11.6	4.13	6.27	180	52.0	0.76	1.2
50	14.5	3.79	5.78	200	57.8	0.63	1.0
60	17.3	3.42	5.25	220	63.6	0.53	0.84
70	20.2	3.02	4.68	240	69.4	0.45	0.72
80	23.1	2.65	4.12	250	72.3	0.42	0.66

(*Source*: Adapted from Tables 7 and 19, BS 95268: Part 2: 1996)

Fig. 14.3 Euler's formula

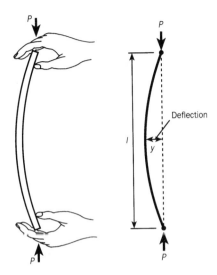

Euler's formula is not used for design, since (except for very long columns) it gives a value of the collapsing load which is much higher than the actual collapsing load of practical columns, but it still forms part of modern column formulae.

The values of permissible compressive stresses for timber struts are the product of the *grade stress* and *modification factors* appropriate to given conditions of services. Those given in Table 14.1 have been compiled for use in the examples and exercises in this chapter only.

Example 14.1

Calculate the permissible axial load for the following timber posts of C14 class timber, all the posts having an effective length of 3.47 m:

- 150 mm × 150 mm
- 225 mm × 100 mm
- 300 mm × 75 mm

Note: All posts have a cross-sectional area of 22 500 mm².

Solution

150 mm × 150 mm

Slenderness ratio = 3470/150 = 23.1

From Table 14.1,

Permissible stress = 2.65 N/mm²

Permissible axial load = 2.65 × 22 500 = 59.625 kN

225 mm × 100 mm

Slenderness ratio = 347/100 = 34.7

By interpolation,

Permissible stress = 1.52 N/mm²

Permissible axial load = 1.52 × 22 500 = 34.20 kN

300 mm $\times$ 75 mm

$$\text{Slenderness ratio} = 3470/75 = 46$$

Again, by interpolation,

$$\text{Permissible stress} = 0.94 \text{ N/mm}^2$$

$$\text{Permissible axial load} = 0.94 \times 22\,500 = 21.15 \text{ kN}$$

It is interesting to note that the 150 mm $\times$ 150 mm post can carry almost three times the load permitted for the 300 mm $\times$ 75 mm post.

Example 14.2

A post made from C14 class timber of 3.45 m effective length is required to support an axial load of 35 kN. Determine suitable dimensions for the cross-section of the post.

Solution

Dimensions must be assumed because it is not possible to determine the permissible stress until the slenderness ratio is known. For example, let the first trial be a post 150 mm square (area of cross-section = 22 500 mm^2).

$$\text{Actual stress due to the load } = 35\,000/22\,500$$

$$= 1.56 \text{ N/mm}^2$$

$$\text{Slenderness ratio} = l/b = 3450/150 = 23$$

The permissible stress for a slenderness ratio of 23 is 2.62 N/mm^2 therefore the assumed size of 150 mm square is too large. As a second trial, assume a post 125 mm square.

$$\text{Actual stress} = 35\,000/15\,625 = 2.24 \text{ N/mm}^2$$

$$\text{Slenderness ratio} = l/b = 3450/125 = 27.6$$

From Table 14.1

$$\text{Permissible stress for } l/b \text{ of } 26 \quad = 2.30 \text{ N/mm}^2$$

$$\text{for } l/b \text{ of } 28.9 = 2.0 \text{ N/mm}^2$$

$$\text{Difference in stress for } 2.9 = (2.3 - 2.0) = 0.30$$

$$\text{Difference in stress for } 1.3 = \frac{1.3}{2.9} \times 0.30 = 0.134$$

$$\text{Permissible stress for } l/b \text{ of } 27.6 = 2.0 + 0.134$$

$$= 2.134 \text{ N/mm}^2$$

Since the permissible stress is slightly more than the actual stress due to the load, a 125 mm square post is suitable.

Radius of gyration

It has already been stated that the shape of the cross-section has an important influence on the load-carrying capacity of a column. A square timber

Fig. 14.4 Typical steel column cross-sections

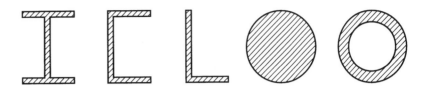

post can support more load than a post of rectangular cross-section of equal area and equal height. It has been shown also that in the case of rectangular cross-section columns it is reasonable to base permissible stresses on a slenderness ratio obtained by dividing the length of the column by its least width. It is not possible, however, to use the dimension of least width when designing columns which have other than solid rectangular sections. Steel columns, for example, are made in various shapes, some of which are shown in Fig 14 4

For such columns it is necessary to use some method of calculating slenderness ratios which can be applied to any shape of cross-section. A property which takes into account not only the size of the section (i.e. area) but also its shape (i.e. the arrangement of the material in the cross-section) is the **radius of gyration**. It is obtained by dividing the moment of inertia I of the section by its area A and then extracting the square root. The symbol r is commonly used to denote radius of gyration but g, i and k are sometimes used.

$$\text{Radius of gyration } r = \sqrt{\left(\frac{I}{A}\right)}$$

The use of the word *gyration* when applied to stationary columns in buildings may appear strange until it is realized that the word is also used in dynamics, the branch of mechanics dealing with bodies in motion. For example, consider Fig. 14.5(a) which represents a disc (such as a flywheel) rotating about its centre C. In dynamics, it is usually the mass (weight) of the wheel which enters into calculations but in this instance the area A of the disc will be considered. Different particles of the disc travel at different velocities. For example, in one revolution of the wheel, particle 1 travels a greater distance than particle 2. In estimating the total energy of the disc, the term $\sum ay^2$ enters into the calculations. $\sum ay^2$ is the sum of the second moments of all the particles of area about the centre C of the disc, i.e. the moment of inertia I.

Fig. 14.5 Radius of gyration: (a) flywheel; (b) structural beam

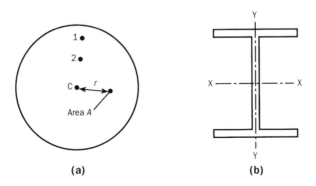

Imagine that all the area A of the disc is concentrated into *one* imaginary particle at a distance r from the centre of the disc. The moment of inertia about C of this particle is Ar^2 and, if the total energy of the disc is to remain unaltered, Ar^2 must equal the total moment of inertia I of the disc. Therefore

$$Ar^2 = I$$

and

$$r = \sqrt{\left(\frac{I}{A}\right)}$$

The radius of gyration is, in this connection, the distance from the centre C of the disc to the point at which the whole area of the wheel can be assumed to be concentrated so that the total energy remains unaltered.

In structural work, it is convenient to use the property $\sqrt{(I/A)}$ in conjunction with the length of the column for estimating slenderness ratios.

Least radius of gyration

The structural engineer is not concerned with moment of inertia about a point as in the disc discussed above, but with moment of inertia with reference to a given axis. If a column of I-section buckles under its load, the bending will be about the weaker axis (axis Y–Y), as indicated in Fig. 14.5(b). Therefore, the radius of gyration must be calculated from I_{yy} which is the least moment of inertia.

$$\text{Least radius of gyration} = \sqrt{\left(\frac{\text{least moment of inertia}}{\text{area of cross-section}}\right)}$$

$$\text{i.e. } r = \sqrt{\left(\frac{I_{yy}}{A}\right)}$$

$$\text{Slenderness ratio} = \frac{\text{effective length of column (mm)}}{\text{least radius of gyration (mm)}}$$

$$= \left(\frac{l}{r}\right) \text{ or } \left(\frac{L_e}{i}\right)$$

Example 14.3

A timber post is 150 mm $\times$ 100 mm in cross-section and has an effective length of 2.6 m. Calculate its least radius of gyration and the slenderness ratio.

Solution

From Fig. 14.6,

$$\text{least } I = I_{yy} = \frac{db^3}{12} \text{ and the area } A = db$$

Therefore

$$\text{least } r = \sqrt{\left(\frac{I}{A}\right)} = \sqrt{\left(\frac{db^3}{12} \times \frac{1}{db}\right)} = \sqrt{\left(\frac{b^2}{12}\right)} = 0.289b$$

Fig. 14.6 Example 14.3

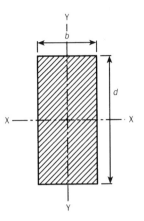

$$r_{yy} = 0.289b$$

(Note that $r_{xx} = 0.289d$)

Least $r = 0.289b = 0.289 \times 100 = 28.9$ mm

Effective length $= 2600$ mm

Slenderness ratio $= l/r = 2600/28.9 = 90$

By reference to Table 14.1 it will be seen that $l/r = 90$ corresponds to $l/b = 26.0$; and in this case $l/b = 2600/100 = 26.0$. Thus, when designing timber posts of solid rectangular cross-section by reference to Table 14.1, it is immaterial whether the slenderness ratio is taken as effective height divided by least radius of gyration (l/r) or as effective height divided by least width (l/b).

No such dilemmas arise, however, in the design of steel columns. Because of their varied shapes the allowable compressive strength, p_c, shown in Table 14.2, is based on the l/r slenderness ratio.

Effective length of columns

It should be noted that in discussing the slenderness ratio the length of the column was qualified by the term 'effective'. This is in accordance with the provisions of the relevant BS codes of practice which state that, for the purpose of calculating the slenderness ratio of columns, an effective length should be assumed. This **effective length** can be defined as that length of the column which is subject to buckling.

The relevant codes of practice give guidance on the relationship of the actual length of the column between lateral supports, L, to its effective length. This is summarized in Table 14.3.

The reason why the effective length of a column may be less than or greater than its actual length in a building or structure is as follows. The safe compressive stress for a column depends not only on the actual length and cross-sectional dimensions of the column but also on the manner in which the ends of the column are restrained or fixed. Tables 14.1 and 14.2 have been derived for one condition of end-fixing (both ends pinned or hinged). To make allowance for other conditions of end-fixing, instead of constructing further tables of permissible stresses, adjustment is made by using a different length of column when calculating the slenderness ratio.

Table 14.2 Compressive strength p_c for rolled section structs with flange maximum thickness ≤ 40 mm

λ^*	Compressive strength p_c (N/mm²)							
	P_y axis of buckling X–X (a) (rolled I sections)				P_y axis of buckling Y–Y (b) (for rolled I and H sections)			
	S275		S355		S275		S355	
	265	275	315	355	265	275	315	355
15	265	275	315	355	265	275	315	355
25	261	270	309	347	258	267	304	342
35	254	264	301	338	247	256	219	327
50	242	251	285	318	229	237	267	298
60	232	239	269	298	214	221	247	272
70	217	224	248	270	196	202	223	242
80	198	203	221	235	177	181	197	211
90	177	180	192	201	157	161	172	181
100	155	157	165	171	138	141	149	155
110	135	137	142	145	121	123	129	134
120	118	119	122	125	107	108	112	116
130	103	103	106	108	94	95	98	101
135	96	97	99	101	88	89	92	94
140	90	91	93	94	83	84	86	88
145	85	85	87	88	78	79	81	83
150	80	80	82	83	74	74	76	78
155	75	75	77	78	70	70	72	73
160	71	71	72	73	66	66	68	69
165	67	67	68	69	62	63	64	66
170	63	64	64	65	59	60	61	62
175	60	60	61	62	56	57	58	59
180	57	57	58	58	54	54	55	56
250	30	30	31	31	29	29	30	30
350	16	16	16	16	15	15	16	16

*λ = (effective length of strut)/(radius of gyration about relevant axis)
(*Source*: Adapted from Tables 24(a) and (b), BS 5950: Part 1: 2000)

Before considering the end-fixing of columns, it may be instructive to study the behaviour of beams as indicated in Fig. 14.7(a).

In Fig. 14.7(a)(i) both ends of the beam are free to bend upwards when the load is applied; in other words, the ends of the beam are not restrained in direction, although they are held in position.

In Fig. 14.7(a)(ii) one end of the beam is firmly fixed so that the end is restrained in direction, and in Fig. 14.7(a)(iii) both ends of the beam are restrained in direction.

The load-carrying capacities of these three beams will be different (other conditions being equal) because of the manner in which the ends of the beams are held or supported.

Table 14.3 Effective length of columns

Type of 'fixity'	Effective length of column	
	BS 5950, BS 5268	*BS 8110*
1 Effectively held in position and restrained in direction at both ends.	0.7L	0.75L
2 Effectively held in position at both ends and:		
(a) restrained in direction at one end,	0.85L	0.75L–L
(b) partially restrained in direction at both ends.	0.85L	
3 Effectively held in position at both ends, restraint in the plane under consideration by other parts of the structure.	L	0.75L–L
4 Effectively held in position and restrained in direction at one end, and at the other partially restrained in direction but not held in position.	1.5L	L–2.0L
5 Effectively held in position and restrained in direction at one end, but not held in position nor restrained in direction at the other end.	2.0L	L–2.0L

(*Source*: Adapted from Table 18, BS 5268: 1996, Table 22, BS 5950: 2000 and Tables 3.21 and 3.22, BS 8110)

In a comparable manner, columns will buckle into differently shaped curves according to the way in which the ends of the columns are held.

The columns of Fig. 14.7(b) are all held in position, i.e. the ends of the columns are not free to move sideways or backwards or forwards. A greater load is required to cause column (ii) to fail than column (i), and column (iii) is the strongest. The length of curve marked AB in columns (ii) and (iii) is

Fig. 14.7 Behaviour of beams and columns: (a)(i) both ends of beam freely supported (bending exaggerated), (ii) one end fixed, (iii) both ends fixed; (b) all column ends are held in position with (i) top and bottom not restrained in direction, (ii) top not restrained, (iii) both ends restrained

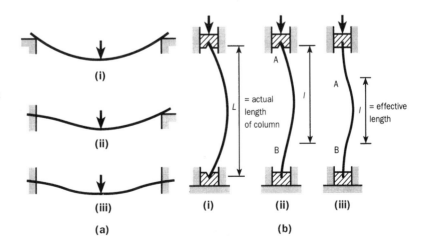

similar to the whole length of curve of column (i), and this length is called the effective length of the column.

Column (i) is similar to case 3 of Table 14.3, column (ii) is similar to case 2a, and column (iii) is similar to case 1.

Case 5 in Table 14.3 can be explained by reference to Fig. 14.8(a). The tops of the columns are not restrained in either position or direction and will tend to buckle as shown. The effective length is therefore taken as twice the actual length of the column.

Figure 14.8(b) represents columns which are held in position and restrained in direction at the bottom and only restrained in direction at the top. This case is a little better than case 4 of Table 14.3.

Fig. 14.8 Representation of 'fixity' types: (a) case 5 from Table 14.3; (b) restrained at top and bottom, but only held in position at the bottom – similar to case 4 from Table 14.3

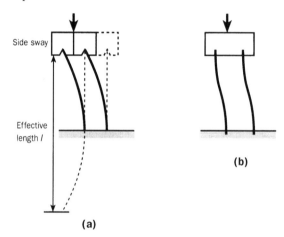

Practical interpretation

Appendix D to BS 5950: Part 1: 1990 gives typical examples of stanchions for single-storey buildings and the effective lengths which may be used in their design. Attention is also drawn to the fact that, although Tables 14.1 and 14.2 give values of permissible stresses for slenderness ratios up to 250 and 350, respectively, it is only in special cases that the slenderness ratio is allowed to exceed 180. Clauses 15.4 of BS 5268: Part 2: 1996 and 4.3.5 of BS 5950: Part 1: 2000 should be consulted for details.

Example 14.4

Calculate the compression resistance of a 203 × 203 UC60 stanchion which is 3.6 m high (between lateral supports). Both ends of the stanchion are held effectively in position but only one is also restrained in direction. (Grade S275 steel.)

Solution

The end fixity of the stanchion corresponds to case 2a of Table 14.3, so

$$\text{Effective length of the column, } l = 0.85L = 3060 \text{ mm}$$

From Table 14.4 on page 274

$$r_{yy} = 52.0 \text{ mm} \qquad A_g = 7640 \text{ mm}^2 \qquad t = 14.2 < 16 \text{ m}$$

Therefore

$$l/r_{yy} = 3060/52.0 = 59$$

From Table 14.2 (by interpolation)

$$P_c \text{ for } l/r_{yy} \text{ of } 59 = 222 \text{ N/mm}^2$$

Hence, the compression resistance,

$$P_c = p_c \times A_g$$
$$= 222 \times 7640 = 1696 \text{ kN}$$

Example 14.5

A stanchion of I-section is required to support a factored axial load of 2000 kN. The effective length of the stanchion is 4 m. Choose a suitable section in Grade S275 steel.

Solution

It is necessary to make a guess at the size of the stanchion. For example, assume a 203 × 203 UC60 stanchion then, working from first principles:

From Table 14.4

$$r_{yy} = 52.0 \text{ mm} \qquad A_g = 7640 \text{ mm}^2 \qquad t = 14.2 \text{ mm} < 16 \text{ m}$$

Therefore

$$l/r_{yy} = 4000/52.0 = 77$$

From Table 14.2, $p_c = 188 \text{ N/mm}^2$. The compression resistance, P_c, of the stanchion is

$$P_c = p_c \times A_g$$
$$= 188 \times 7640 = 1436 \text{ kN} < \text{applied factored load}$$

i.e. the assumed section is not adequate.

By trying a 254 × 254 UC73 in the same way it will be found that its compression resistance, P_c, is 2020 kN, which is satisfactory.

Reinforced concrete columns

The design of reinforced concrete columns by the elastic method was discontinued on the introduction of the 1957 edition of CP114, after test results had shown that steel and concrete do not behave elastically at failure.

The calculations for the permissible axial load, P_0, on a short braced reinforced concrete column, i.e. one where the ratio of effective height to the least lateral dimension does not exceed 15, were based on

$$P_0 = p_{cc}A_c + p_{sc}A_{sc}$$

where

p_{cc} = the permissible stress for the concrete in direct compression

A_c = the cross-sectional area of concrete excluding any finishing material and reinforcing steel

p_{sc} = the permissible compression stress for column bars

A_{sc} = the cross-sectional area of the longitudinal steel

Table 14.4 Dimensions and properties of universal columns

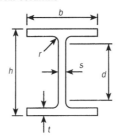

Designation	Area of section (cm²)	Mass per metre (kg/m)	Dimensions (mm)						Ratios for local buckling	
			h Section depth	b Section width	s Web thickness	t Flange thickness	r Root radius	d Depth between fillets	b/2t Flange	d/s Web
356 × 406 × 634	808	633.9	474.6	424.0	47.6	77.0	15.2	290.2	2.75	6.10
356 × 406 × 551	702	551.0	455.6	418.5	42.1	67.5	15.2	290.2	3.10	6.89
356 × 406 × 467	595	467.0	436.6	412.2	35.8	58.0	15.2	290.2	3.55	8.11
356 × 406 × 393	501	393.0	419.0	407.0	30.6	49.2	15.2	290.2	4.14	9.48
356 × 406 × 340	433	339.9	406.4	403.0	26.6	42.9	15.2	290.2	4.70	10.9
356 × 406 × 287	366	287.1	393.6	399.0	22.6	36.5	15.2	290.2	5.47	12.8
356 × 406 × 235	299	235.1	381.0	394.8	18.4	30.2	15.2	290.2	6.54	15.8
356 × 368 × 202	257	201.9	374.6	374.7	16.5	27.0	15.2	290.2	6.94	17.6
356 × 368 × 177	226	177.0	368.2	372.6	14.4	23.8	15.2	290.2	7.83	20.2
356 × 368 × 153	195	152.9	362.0	370.5	12.3	20.7	15.2	290.2	8.95	23.6
356 × 368 × 129	164	129.0	355.6	368.6	10.4	17.5	15.2	290.2	10.5	27.9
305 × 305 × 283	360	282.9	365.3	322.2	26.8	44.1	15.2	246.7	3.65	9.21
305 × 305 × 240	306	240.0	352.5	318.4	23.0	37.7	15.2	246.7	4.22	10.7
305 × 305 × 198	252	198.1	339.9	314.5	19.1	31.4	15.2	246.7	5.01	12.9
305 × 305 × 158	201	158.1	327.1	311.2	15.8	25.0	15.2	246.7	6.22	15.6
305 × 305 × 137	174	136.9	320.5	309.2	13.8	21.7	15.2	246.7	7.12	17.9
305 × 305 × 118	150	117.9	314.5	307.4	12.0	18.7	15.2	246.7	8.22	20.6
305 × 305 × 97	123	96.9	307.9	305.3	9.9	15.4	15.2	246.7	9.91	24.9
254 × 254 × 167	213	167.1	289.1	265.2	19.2	31.7	12.7	200.3	4.18	10.4
254 × 254 × 132	168	132.0	276.3	261.3	15.3	25.3	12.7	200.3	5.16	13.1
254 × 254 × 107	136	107.1	266.7	258.8	12.8	20.5	12.7	200.3	6.31	15.6
254 × 254 × 89	113	88.9	260.3	256.3	10.3	17.3	12.7	200.3	7.41	19.4
254 × 254 × 73	93.1	73.1	254.1	254.6	8.6	14.2	12.7	200.3	8.96	23.3
203 × 203 × 86	110	86.1	222.2	209.1	12.7	20.5	10.2	160.8	5.10	12.7
203 × 203 × 71	90.4	71.0	215.8	206.4	10.0	17.3	10.2	160.8	5.97	16.1
203 × 203 × 60	76.4	60.0	209.6	205.8	9.4	14.2	10.2	160.8	7.25	17.1
203 × 203 × 52	66.3	52.0	206.2	204.3	7.9	12.5	10.2	160.8	8.17	20.4
203 × 203 × 46	58.7	46.1	203.2	203.6	7.2	11.0	10.2	160.8	9.25	22.3
152 × 152 × 37	47.1	37.0	161.8	154.4	8.0	11.5	7.6	123.6	6.71	15.4
152 × 152 × 30	38.3	30.0	157.6	152.9	6.5	9.4	7.6	123.6	8.13	19.0
152 × 152 × 23	29.2	23.0	152.4	152.2	5.8	6.8	7.6	123.6	11.2	21.3

(*Source*: Structural sections to BS 4: Part 1 and BS 4848: Part 4 British Steel)

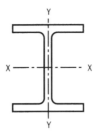

Second moment of area (cm^4)		Radius of gyration (cm)		Elastic modulus (cm^3)		Plastic modulus (cm^3)		Parameters and constants				Designation
								u	x	H	J	
$X-X$ axis	$Y-Y$ axis	$X-X$ axis	$Y-Y$ axis	$X-X$ axis	$Y-Y$ axis	$X-X$ axis	$Y-Y$ axis	Buckling parameter	Torsional index	Warping constant (dm^6)	Torsional constant (cm^4)	
274800	98130	18.4	11.0	11580	4629	14240	7108	0.843	5.46	38.8	13720	356 × 406 × 634
226900	82670	18.0	10.9	9962	3951	12080	6058	0.841	6.05	31.1	9240	356 × 406 × 551
183000	67830	17.5	10.7	8383	3291	10000	5034	0.839	6.86	24.3	5809	356 × 406 × 467
146600	55370	17.1	10.5	6998	2721	8222	4154	0.837	7.86	18.9	3545	356 × 406 × 393
122500	46850	16.8	10.4	6031	2325	6999	3544	0.836	8.85	15.5	2343	356 × 406 × 340
99880	38680	16.5	10.3	5075	1939	5812	2949	0.835	10.2	12.3	1441	356 × 406 × 287
79080	30990	16.3	10.2	4151	1570	4687	2383	0.834	12.1	9.54	812	356 × 406 × 235
66260	23690	16.1	9.60	3538	1264	3972	1920	0.844	13.4	7.16	558	356 × 368 × 202
57120	20530	15.9	9.54	3103	1102	3455	1671	0.844	15.0	6.09	381	356 × 368 × 177
48590	17550	15.8	9.49	2684	948	2965	1435	0.844	17.0	5.11	251	356 × 368 × 153
40250	14610	15.6	9.43	2264	793	2479	1199	0.844	19.9	4.18	153	356 × 368 × 129
78870	24630	14.8	8.27	4318	1529	5105	2342	0.855	7.65	6.35	2034	305 × 305 × 283
64200	20310	14.5	8.15	3643	1276	4247	1951	0.854	8.74	5.03	1271	305 × 305 × 240
50900	16300	14.2	8.04	2995	1037	3440	1581	0.854	10.2	3.88	734	305 × 305 × 198
38750	12570	13.9	7.90	2369	808	2680	1230	0.851	12.5	2.87	378	305 × 305 × 158
32810	10700	13.7	7.83	2048	692	2297	1053	0.851	14.2	2.39	249	305 × 305 × 137
27670	9059	13.6	7.77	1760	589	1958	895	0.850	16.2	1.98	161	305 × 305 × 118
22250	7308	13.4	7.69	1445	479	1592	726	0.850	19.3	1.56	91.2	305 × 305 × 97
30000	9870	11.9	6.81	2075	744	2424	1137	0.851	8.49	1.63	626	254 × 254 × 167
22530	7531	11.6	6.69	1631	576	1869	878	0.850	10.3	1.19	319	254 × 254 × 132
17510	5928	11.3	6.59	1313	458	1484	697	0.848	12.4	0.898	172	254 × 254 × 107
14270	4857	11.2	6.55	1096	379	1224	575	0.850	14.5	0.717	102	254 × 254 × 89
11410	3908	11.1	6.48	898	307	992	465	0.849	17.3	0.562	57.6	254 × 254 × 73
9449	3127	9.28	5.34	850	299	977	456	0.850	10.2	0.318	137	203 × 203 × 86
7618	2537	9.18	5.30	706	246	799	374	0.853	11.9	0.250	80.2	203 × 203 × 71
6125	2065	8.96	5.20	584	201	656	305	0.846	14.1	0.197	47.2	203 × 203 × 60
5259	1778	8.91	5.18	510	174	567	264	0.848	15.8	0.167	31.8	203 × 203 × 52
4568	1548	8.82	5.13	450	152	497	231	0.847	17.7	0.143	22.2	203 × 203 × 46
2210	706	6.85	3.87	273	91.5	309	140	0.848	13.3	0.0399	19.2	152 × 152 × 37
1748	560	6.76	3.83	222	73.3	248	112	0.849	16.0	0.0308	10.5	152 × 152 × 30
1250	400	6.54	3.70	164	52.6	182	80.2	0.840	20.7	0.0212	4.63	152 × 152 × 23

For the purpose of the exercises in this chapter the values of p_{cc} may be taken as follows:

7.6 N/mm^2 for 1:1:2 mix (or C30 grade)

6.5 N/mm^2 for $1:1\frac{1}{2}:3$ mix (or C25 grade)

5.3 N/mm^2 for 1:2:4 mix (or C20 grade)

The value of p_{sc} for hot rolled steel bars complying with BS 4449 up to 40 mm dia. used as longitudinal reinforcement in columns was given as 125 N/mm^2.

BS 8110: Part 1: 1997 distinguishes between braced and unbraced as well as short and slender columns. A braced column is one which is part of a structure stabilized against lateral forces by walls, bracing or buttressing. The definition of a short braced (or unbraced) column is based on a ratio similar to that given above. (See Clause 3.8.1.3 for details.)

Clause 3.8.4.3 of BS 8110 states that the design ultimate axial load, N, supported by a short column which cannot be subjected to significant moments because of the nature of the structure, may be calculated from

$$N = 0.4 f_{cu} A_c + 0.75 A_{sc} f_y$$

where

f_{cu} = characteristic strength of concrete

f_y = characteristic strength of reinforcement

For short braced columns supporting an approximately symmetrical arrangement of beams, Clause 3.8.4.4 of BS 8110 permits the following version of the above equation to be used in the calculation of the design ultimate axial load:

$$N = 0.35 f_{cu} A_c + 0.7 A_{sc} f_y$$

provided that the beams are designed for uniformly distributed imposed loads and the beam spans do not differ by more than 15 per cent of the longest beam.

In both equations allowance is made for the partial safety factor for strength of material, γ_m.

Example 14.6

A 250 mm square braced reinforced concrete column with an effective length of 3 m contains four 25 mm dia. longitudinal bars. Calculate the safe axial load for the column if $p_{cc} = 5.3 \text{ N/mm}^2$ and $p_{sc} = 125 \text{ N/mm}^2$.

Solution

$$\text{Slenderness ratio} = \frac{\text{effective length}}{\text{least width}}$$

$$= 3000/250 = 12 \ (<15)$$

Therefore the column is a short one.

$$\text{Gross area of concrete} = 250 \times 250 = 62\,500 \text{ mm}^2$$

$$A_{sc} = 4 \times 491 \text{ mm}^2 = 1964 \text{ mm}^2$$

$$A_c = A_{gross} - A_{sc} = 60\,536 \text{ mm}^2$$

Therefore

$$P_0 = p_{cc}A_c + p_{sc}A_{sc}$$
$$= 5.3 \times 60\,536 + 125 \times 1964$$
$$= 320.8 + 245.5$$
$$= 566.3 \text{ kN}$$

Example 14.7

A short braced reinforced concrete column is required to carry an axial load of 900 kN. Design a square column containing 0.8 per cent of steel and 8.0 per cent of steel.

Solution

Assume $p_{cc} = 5.3 \text{ N/mm}^2$ and $p_{sc} = 125 \text{ N/mm}^2$.

For 0.8 per cent steel:

$$A_{sc} = \frac{0.8}{100} \times A_g$$

where A_g is gross cross-sectional area.

Therefore

$$P_0 = p_{cc}A_c + p_{sc}A_{sc}$$
$$= 5.3(A_g - 0.008A_g) + 125 \times 0.008A_g$$
$$= 5.3A_g - 0.04A_g + 1.0A_g$$
$$= 6.26A_g$$

Hence

$$A_g = 900\,000/6.26 = 143\,770 \text{ mm}^2$$

$$\text{Length of side} = \sqrt{143\,770} = \text{(say) } 380 \text{ mm}$$

$$A_{sc} = 0.008 \times 0.144 \times 10^6 = 1155 \text{ mm}^2$$

Make column 380 mm square with four No. 20 mm dia. reinforcing bars (1260 mm^2).

Now for 8.0 per cent of steel:

$$A_{sc} = \frac{8.0}{100} \times A_g$$

$$P_0 = 5.3(A_g - 0.08A_g) + 125 \times 0.08A_g$$
$$= 14.9A_g$$

Hence

$$A_g = 900\,000/14.9 = 60\,400 \text{ mm}^2$$

$$\text{Length of side} = \sqrt{60\,400} = \text{(say) } 250 \text{ mm}$$

$$A_{sc} = 0.08 \times 60\,400 = 4832 \text{ mm}^2$$

Make column 250 mm square with four No. 32 mm dia. and four No. 25mm dia. reinforcing bars (5180 mm^2).

Summary

Timber columns   Permissible stresses for various slenderness ratios are given in Table 14.1.

If the column is of solid rectangular cross-section, the slenderness ratio may be taken as effective length divided by the least lateral dimension. If not of rectangular cross-section, the slenderness ratio is effective length divided by least radius of gyration.

The effective length should be assumed in accordance with Table 14.3. The safe axial load for the column is obtained by multiplying the permissible stress by the area of the cross-section of the column.

Steel columns

$$\text{Slenderness ratio} = \frac{\text{effective length}}{\text{least radius of gyration}} = \frac{l}{r}$$

See Table 14.3 and Appendix D of BS 5950 for a guide to estimating effective lengths. Permissible stresses for various slenderness ratios are given in Table 14.2.

The safe axial load is obtained by multiplying the permissible stress by the area of the cross-section of the column.

Least r for solid rectangular section $= 0.289b$

$$\text{Least } r \text{ for hollow square section} = \tfrac{1}{4}\sqrt{\left(\frac{B^2 + b^2}{3}\right)}$$

$$\text{Least } r \text{ for hollow circular section} = \tfrac{1}{4}\sqrt{(D^2 + d^2)}$$

Least r for rolled sections, e.g. I-sections, channels and angles, can be obtained from tables in BS 4: Part 1: 1993.

Reinforced concrete columns   A short braced column is one where the ratio of the effective column length to least lateral dimension does not exceed 15. A large number of columns in buildings are therefore 'short'.

For permissible stress design:

$$\text{safe axial load } P_0 = p_{cc}A_c + p_{sc}A_{sc}$$

where

p_{cc} = permissible concrete stress

A_c = area of cross-section of concrete $= A_g - A_{sc}$

p_{sc} = permissible steel stress

A_{sc} = area of steel

For limit state design see Clause 3.8 of BS 8110: Part 1: 1997, or Part 3 of that code which contains design charts for rectangular columns.

Exercises

1 Calculate the safe axial loads for the following posts all of C14 Class timber: (a) 75 mm square, 2 m effective length; (b) 100 mm square, 2.5 m effective length.

2 Calculate the safe axial loads for the following posts of C14 Class timber all of 3 m effective length: (a) 150 mm × 50 mm; (b) 100 mm × 75 mm; (c) 85 mm square.

3 A timber post of 3.6 m effective length is required to support an axial load of 270 kN. Determine the length of side of a square section post of C24 Class timber.

4 A timber post of C14 Class timber has an effective length of 3 m and is 300 mm diameter (solid circular cross-section). Calculate the value of the axial load.

5 A timber post of solid circular cross-section is of timber C24 Class and 3 m effective length. It has to support an axial load of 300 kN. Determine the diameter of the post.

6 Two timber posts 200 mm × 74 mm in cross-section of C24 Class timber are placed side by side without being connected together to form a post 200 mm × 150 mm. The effective length of the posts is 2.6 m. Calculate the safe axial load for the compound post and compare this load with the safe axial load of one solid post 200 mm × 150 mm in cross-section of the same effective length.

7 Calculate the compression resistance of a 254 × 254 UC89 stanchion for the following end fixings and effective lengths: (a) 12 m, both ends fully restrained; (b) 9 m, one end fully restrained the other held in position; (c) 6 m, both ends held in position only.

8 Determine the maximum permissible effective length of a 254 × 254 UC73 section. Calculate the compression resistance for such an effective length.

9 Choose a suitable stanchion of I-section to support a factored axial load of 2000 kN, the effective length = 3 m.

10 Make calculations to determine which of the following stanchions has the greater compression resistance: (a) a stanchion consisting of two 152 × 152 UC23 sections placed side by side without being connected together; effective length = 2.6 m; (b) one 203 × 203 UC46 section of 2.6 m effective length.

11 Calculate the safe axial load for a 400 mm square 'short' reinforced concrete column containing eight 25 mm diameter bars. Use 1:2:4 mix of concrete.

12 Calculate the safe axial load for a 450 mm diameter 'short' circular cross-section reinforced concrete column containing six 32 mm diameter bars.

$$p_{cc} = 6 \text{ N/mm}^2 \qquad p_{sc} = 125 \text{ N/mm}^2$$

13 Design a square cross-section column to support a load of 2 MN assuming 4 per cent of steel and $1:1\frac{1}{2}:3$ concrete mix.

14 Design a short column to carry 2.5 MN, assuming (a) $p_{cc} = 7.6 \text{ N/mm}^2$, $p_{sc} = 140 \text{ N/mm}^2$, 8 per cent steel; (b) $p_{cc} = 4.3 \text{ N/mm}^2$, $p_{sc} = 125 \text{ N/mm}^2$, 0.8 per cent steel.

15 Compare the load-carrying capacities of a timber and a reinforced concrete column both 3 m effective length and 300 mm square. The timber is C24 Class and the reinforced concrete column contains eight 25 mm diameter bars.

$$p_{cc} = 6.0 \text{ N/mm}^2 \qquad p_{sc} = 125 \text{ N/mm}^2$$

16 One dimension of a 'short' reinforced concrete column must not exceed 250 mm. The column has to support an axial load of 730 kN. Assuming 4 per cent steel, design the column using 1:2:4 concrete mix.

17 A column is 400 mm square and 6 m effective length and it contains eight 25 mm diameter bars. Calculate the safe axial load assuming

$$p_{cc} = 5.7 \text{ N/mm}^2 \qquad p_{sc} = 125 \text{ N/mm}^2$$

18 A 'short' column is 450 mm square and it has to support an axial load of 2 MN. Calculate the area of steel required.

$$p_{cc} = 6.0 \text{ N/mm}^2 \qquad p_{sc} = 125 \text{ N/mm}^2$$

19 A square concrete column with 5 per cent of reinforcement is to carry an axial load of 554 kN. Given that $p_{cc} = 8 \text{ N/mm}^2$ and $p_{sc} = 125 \text{ N/mm}^2$ determine (a) a suitable size for the column and the cross-sectional area of the reinforcement; (b) the maximum length of the column so that it may be treated as a 'short' column, assuming full restraint at both ends.

Chapter fifteen Connections

The treatment of connections in this chapter is confined to direct shear connections in steelwork. The behaviour of rivets and bolts is presented in some detail. The criterion value of rivets or bolts is explained and its use in design calculations is demonstrated. High strength friction grip bolts are dealt with briefly to give an introduction to this type of connection. Finally, welding is also discussed because of its undoubted versatility and the resulting popularity.

Note: All the loads used in the design of connections in this chapter are **factored loads** (i.e. specified loads multiplied by the relevant partial factor) in accordance with Clause 6.1.1 of BS 5950: Part 1: 2000.

Riveting and bolting

A rivet or bolt may be considered simply as a *peg* inserted in holes drilled in two or more thicknesses of steel in order to prevent relative movement. For example, the two steel plates in Fig. 15.1(a) tend to slide over each other, but could be prevented from doing so by a suitable steel pin inserted in the holes in each plate, as shown. In order to prevent the *steel pin* from slipping out of holes, bolts with heads and nuts are used or rivet heads are formed, and these produce an effective connection (Fig. 15.1(b)).

The rivet heads (or bolt heads and nuts) do, in fact, strengthen the connection by pressing the two thicknesses of plate together, but this strength cannot be determined easily, and so the rivet or bolt strength is calculated on the assumption that its shank (shown shaded) only is used in building up its strength.

Single shear

If the loads W in Fig. 15.1(b) are large enough, the rivet or bolt could fail, as in Fig. 15.2, in *shear*, i.e. breaking by the sliding of its fibres along line A–A.

Fig. 15.1 Riveting

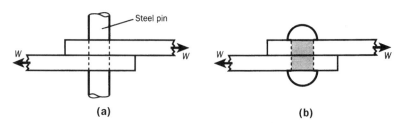

Steel pin

(a)   (b)

Fig. 15.2 Failure in single shear

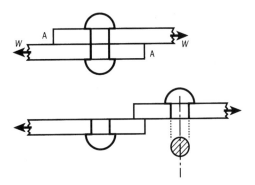

This type of rivet or bolt failure is known as **failure in single shear**. The area of steel rivet resisting this failure is the circular area of the rivet shank, shown hatched in Fig. 15.2, i.e.

$$\tfrac{1}{4}\pi \times (\text{diameter of river})^2 \quad \text{or} \quad 0.7854d^2 = A$$

The shear strength, p_s, for rivets may be assumed to be as follows:

110 N/mm^2 for hand-driven rivets

135 N/mm^2 for power-driven rivets

For ordinary bolts of grade 4.6 the value of p_s is given in Table 30 of BS 5950: Part 1: 2000 as 160 N/mm^2.

Power-driven rivets are usually driven by a special machine. The rivets and the rivet heads are formed more accurately than is possible in the case of hand-driven rivets and they are therefore permitted a higher stress. The holes are drilled 2 mm larger in diameter than the specified sizes of the rivets.

Since rivets are driven while hot and, therefore, their material fills the hole completely, it is necessary to distinguish between the nominal and the gross diameter of the rivets. The nominal diameter refers to the specified size of the rivet shank, i.e. the diameter of the rivet when it is cold, whilst the gross diameter is 2 mm larger than the specified (i.e. nominal) diameter of the rivet. The strength of a rivet is normally estimated on its *gross* diameter.

For example, the safe load in single shear (safe stress × area) of a 16 mm diameter power-driven rivet is

$$135 \times 0.7854 \times (16 + 2)^2 = 34.3 \text{ kN}$$

For bolts, the gross diameter is, of course, equal to the nominal diameter. Therefore the safe load in single shear, or single shear value (SSV) of a 16 mm diameter ordinary bolt of grade 4.6 is

$$160 \times 0.7854 \times 16^2 = 32 \text{ kN}$$

Double shear

In the type of connection shown, for example, in Fig. 15.3 (a double cover butt joint), the rivets or bolts on one side of the joint would have to shear across *two* planes, as shown. This is known as **failure in double shear**.

A rivet or bolt under these circumstances will need twice as much load to break it compared with a rivet or bolt in single shear, so the safe load on a

Fig. 15.3 Failure in double shear

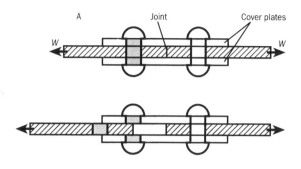

rivet in double shear is twice that on the same rivet in single shear. Thus, using the circular area of the rivet shank as before,

Safe load in double shear

$$= \text{area} \times 2 \times \text{single shear safe stress}$$

Therefore,

Safe load for the above ordinary bolt

$$= 2 \times \text{shear strength} \times \text{area}$$

$$= 2 \times 160 \times 201$$

$$= 64\,\text{kN}$$

or, simply,

$$2 \times \text{SSV} = 2 \times 32 = 64\,\text{kN}$$

Bearing

The two main ways in which the rivet or bolt itself may fail have been discussed. This type of failure assumes, however, fairly thick steel plates capable of generating sufficient stress to shear the rivet.

Consider Fig. 15.4(a). The heavy load of 120 kN taken by the 25 mm steel plates would certainly shear the 12 mm diameter rivet (single shear).

Now consider the opposite type of case, as in Fig. 15.4(b), where a thick steel rivet (24 mm diameter) is seen connecting two very thin steel plates. The steel plates in this case are much more likely to be torn by the rivet than the rivet is to be sheared by the weaker steel plates.

This type is known as **failure in bearing** (or *tearing*), and note should again be taken of the area which is effective in resisting this type of stress (Fig. 15.4(c)). The area of contact of the rivet with the plate on one side of it is actually semicylindrical, but since the bearing stress is not uniform, it is assumed that the area of contact is the plate thickness times the rivet diameter. This area is shown shaded in Fig. 15.4(d).

For bearing purposes, as for shear, the gross diameter of the rivet can be taken as the nominal diameter plus 2 mm.

When two plates of the same thickness are being connected, then *either* plate could tear, and the area resisting bearing would be the thickness of one plate times the diameter of the rivet (Fig. 15.5(a)). Where plates of different

Fig. 15.4 Failure in bearing: (a) plates would shear rivet; (b) rivet would tear plates; (c) bearing force; (d) effective area (section A–A)

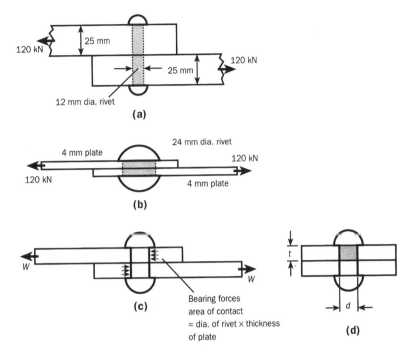

Fig. 15.5 Area resisting bearing: (a) plates of equal thickness; (b) plates of unequal thickness; (c) three plates

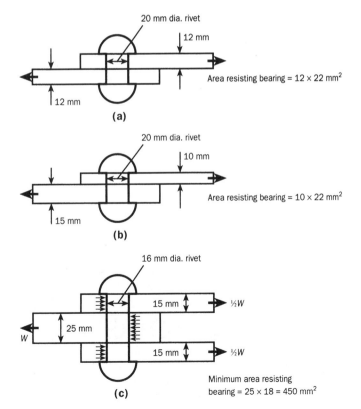

thicknesses are used, then the thinner of the two plates would tear first, so the area resisting bearing or tearing would be the thickness of the thinner plate times the diameter of the rivet (Fig. 15.5(b)). Where three thicknesses are concerned, as in Fig. 15.5(c), the two 15 mm plates are acting together and the 25 mm plate would tear before the two 15 mm plates, so the area resisting tearing would be $25 \times 18 = 450$ mm^2.

The bearing strength, p_{bb}, for ordinary bolts of grade 4.6 and for connected parts using ordinary bolts in clearance holes, p_{bb}, are given in BS 5950 as 460 N/mm^2 for Grade S275 steel.

The value of p_{bb} for hand-driven rivets may be taken as 350 N/mm^2 and for power-driven rivets as 405 N/mm^2.

Criterion value

It will be seen that rivets or bolts may be designed on the basis of their strength in shear, or their strength in bearing.

In actual design, the lesser of these two values will, of course, have to be used. This is called the **criterion value** of the rivet or bolt.

When designing this type of connection, the following questions should be asked:

• Is the connection in single or double shear?
• What is the safe appropriate shear load on one rivet or bolt?
• What is the safe bearing load on one rivet or bolt?

The criterion value, that is the lower of the values given by the last two questions, will then be used in determining the safe load on the connection.

Example 15.1

Calculate the safe load W on the lap joint shown in Fig. 15.6 in terms of the shear or bearing values.

Solution

There are only two plates, and therefore the connection is in single shear.

Single shear value of one rivet is

$$135 \times 0.7854 \times 22^2 = 135 \times 380 = 51.3 \text{ kN}$$

Fig. 15.6 Example 15.1

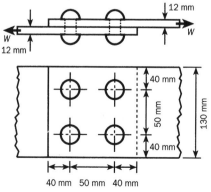

Four 20 mm diameter power-driven rivets

Bearing value of the rivet in 12 mm plate is

$$405 \times 22 \times 12 = 405 \times 264 = 106.9 \text{ kN}$$

The shear value is, therefore, the criterion value in this case, and the safe load for the connection is

$$4 \times 51.3 = 205 \text{ kN}$$

Example 15.2

What is the safe load W in kN on the tie shown in Fig. 15.7 with respect to rivets A and B?

Solution

Rivets A are in double shear and rivets B are in single shear.

Rivets A
DSV of one 20 mm diameter rivet is

$$2 \times 135 \times 380 = 102.6 \text{ kN}$$

Bearing value (BV) of that rivet in 15 mm plate is

$$405 \times 22 \times 15 = 133.6 \text{ kN}$$

$$\text{Criterion value} = 102.6 \text{ kN}$$

Therefore, total value of rivets A is

$$3 \times 102.6 = 308 \text{ kN}$$

Rivets B
SSV of one 20 mm diameter rivet is 51.3 kN
Bearing value of that rivet in 12 mm plate is

$$405 \times 22 \times 12 = 106.9 \text{ kN}$$

$$\text{Criterion value} = 51.3 \text{ kN}$$

Therefore, total value of rivets B is

$$4 \times 51.3 = 205 \text{ kN}$$

Fig. 15.7 Example 15.2

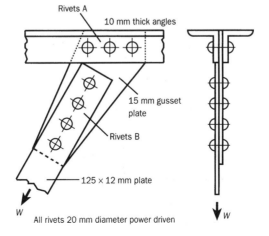

Rivets A
10 mm thick angles
15 mm gusset plate
Rivets B
125 × 12 mm plate
W
All rivets 20 mm diameter power driven
W

It appears, therefore, that the safe load W is decided by rivets B and is 205 kN. The strength of the plate in tension, however, should be investigated before the 205 kN load is accepted as the safe load.

The 125 mm × 12 mm plate is weakened by one 22 mm diameter rivet hole. Hence the net cross-sectional area of the plate is

$$(125 - 22) \times 12 = 1236 \text{ mm}^2$$

and the permissible tensile stress for Grade S275 steel is 275 N/mm² for plates up to 16 mm thick.

Tension value of the plate is

$$275 \times 1236 = 340 \text{ kN}$$

In this case, therefore, the strength of the connection is decided by rivets B. It should be noted, however, that the angles have not been checked, although they are usually satisfactory.

Example 15.3

A compound bracket, shown in Fig. 15.8, is connected to the 13 mm thick web of a stanchion by six 16 mm diameter ordinary bolts of grade 4.6. The bracket carries a reaction of 225 kN from a beam. Is the connection strong enough in terms of the bolts?

Solution

There are three thicknesses – the web, the angle and the cover plate – but the bolts are in single shear because the angle and the cover plate act as one.

SSV of one 16 mm diameter bolt is

$$160 \times 0.7854 \times 16^2 = 160 \times 201 = 32.2 \text{ kN}$$

Bearing value of the bolt in 13 mm plate is

$$460 \times 16 \times 13 = 460 \times 208 = 95.7 \text{ kN}$$

$$\text{Criterion value} = \text{shear value} = 32.2 \text{ kN}$$

$$\text{Safe load} = 6 \times 32.2 = 193.2 \text{ kN}$$

Fig. 15.8 Example 15.3

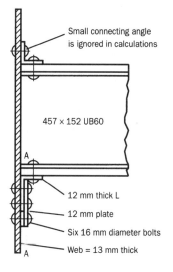

This is less than the applied load (reaction). Therefore, either the number of the 16 mm diameter bolts should be increased or larger diameter bolts will have to be used.

In practice the bolts in this type of connection would be also investigated for direct tension since, according to Clause 4.7.6 of BS 5950: Part 1: 2000, the reaction must be assumed to be applied at least 100 mm from line A–A (Fig. 15.8), thus creating an eccentricity of loading.

Double cover butt connections

In designing butt and other similar types of connections, it should always be borne in mind that not only can failure occur through an insufficient number of rivets or bolts being provided, but that the member itself may fail in tension.

Consider, for example, Fig. 15.9 noting, in particular, the layout of the rivets in what is called a 'leading rivet' arrangement.

One possible chance of failure is that the plate being connected would fail by tearing across face A–A or B–B under a heavy load. Therefore, no matter how many rivets are employed, the safe strength in tension across this and other faces could never be exceeded.

The strength of the rivets must be approximately equal to the strength of the member in tension for the connection to be considered economical.

Fig. 15.9 Double cover butt connections

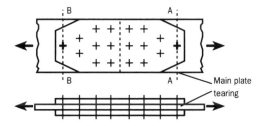

Example 15.4

A 125 mm × 18 mm steel plate used as a tension member in a structural frame has to be connected using a double cover butt connection with two 12 mm cover plates and 20 mm diameter power-driven rivets.

Design a suitable connection assuming that the permissible stress in tension for the steel plate is 250 N/mm^2.

Solution

However the rivets are arranged, the section will be weakened by having at least one rivet hole so the net cross-sectional area of the plate is

$$(125 - 22) \times 18 = 1854 \, \text{mm}^2$$

and the safe load carried by the plate must not exceed

$$250 \times 1854 = 463 \, \text{kN}$$

The rivets will be in double shear. DSV of one 20 mm diameter rivet is

$$2 \times 135 \times 380 = 102.6 \, \text{kN}$$

Bearing value of that rivet in 18 mm plate is

$$405 \times 22 \times 18 = 160.4 \, \text{kN}$$

$$\text{Criterion value} = 102.6 \, \text{kN}$$

Therefore, number of rivets required on each side of joint is

$$\frac{\text{Total load}}{\text{Value of one rivet}} = \frac{463}{102.6} = 4.5, \text{ say 5 rivets}$$

The arrangement of the rivets is shown in Fig. 15.10(a).

Check the strength of the plate:

At section A–A, the strength, as calculated, is 463 kN.

At Section B–B, the plate is weakened by two rivet holes, but, in the event of tearing of the plate across B–B, the connection would not fail until the rivet marked X also had failed.

Thus the strength across B–B is

$$250 \times (125 - 2 \times 22) \times 18 + 102\,600$$

$$= 250 \times 1458 + 102\,600 = 467 \, \text{kN}$$

and the strength at C–C is

$$250 \times 1458 + 3 \times 102\,600 = 672 \, \text{kN}$$

since in this case the rivets X, Y and Z have to be considered.

Finally, the cover plates have to be designed. The critical section here is section C–C. Assume 12 mm thick plates then the strength at C–C is

$$250 \times (125 - 2 \times 22) \times 24 = 486 \, \text{kN}$$

Therefore the above connection would carry 463 kN.

It is sometimes useful to check the 'efficiency' of the connection. This is given by

$$\text{Efficiency} = \frac{\text{safe load for the connection}}{\text{original value of the undrilled plate}} \times 100$$

In the above case, the efficiency would be

$$\frac{463\,000}{250 \times 125 \times 18} \times 100 = 82.3\%$$

Increasing the number of rivets above that which is required may, in some cases, actually weaken the connection.

Fig. 15.10 Example 15.4: (a) five rivets on each side of the joint; (b) six rivets on each side

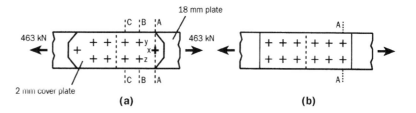

Consider the connection in Example 15.4. Had six rivets been used as in Fig. 15.10(b), instead of the required five, the value of the plate at section A–A would now be

$$250 \times (125 - 2 \times 22) \times 18 = 364\,kN$$

as against 463 kN for the leading rivet arrangement.

High strength friction grip bolts

The rivets and bolts discussed so far in this chapter relied on their shear and bearing strength to produce an effective connection capable of transmitting a load from one member to another, e.g. from beam to column.

The performance of high strength friction grip (HSFG) bolts is based on the principle that the transfer of the load may be effected by means of friction between the contact surfaces (interfaces) of the two members. To produce the necessary friction a sufficiently high clamping force must be developed, and this is achieved by tightening the bolts to a predetermined tension. In this way the bolts are subjected to a direct (axial) tensile force and do not rely on their shear and bearing strength.

The substantial forces needed to produce the necessary 'friction grip' require the bolts to be of high tensile strength and the interfaces to be meticulously prepared. Various methods are used to ensure that the bolts are tightened to the required tension in the shank. These include torque control by means of calibrated torque wrenches and special load–indicating devices.

HSFG bolts and their use are specified in BS 4395 and BS 4604, respectively, and further details may be obtained from manufacturers' literature.

Considering connections subject only to shear between the friction faces, the slip resistance, P_{SL}, may be determined from the following:

$$P_{SL} = 1.1 \times K_S \times \mu \times P_0 \text{ (connection to be non-slip in service)}$$

$$P_{SL} = 0.9 \times K_S \times \mu \times P_0 \text{ for waisted shank fasteners and for non-slip connection under factored loads}$$

The slip resistance is the limit of shear (i.e. the load) that can be applied before slip occurs. For parallel shank fasteners, however, it may be necessary to check the bearing capacity of the connection (see the note to Clause 6.4.1 of BS 5950: Part 1: 2000).

The value of the factor K_S is 1.0 for fasteners in clearance holes (as in most cases). It is less for oversized and slotted holes, see Clause 6.4.2 of BS 5950: 2000.

μ is the slip factor (i.e. the coefficient of friction between the surfaces) and may be taken as 0.45 for general grade fasteners. Its value is limited to 0.55, see Table 35, BS 5950: 2000.

P_0 is the minimum shank tension as given in Table 15.1.

Table 15.1 High strength friction grip bolts (general grade)

Nominal size and thread diameter	Proof load, (minimum shank tension) (kN)
M16	92.1
M20	144
M22	177
M24	207
M27	234
M30	286
M36	418

(*Source*: Adapted from BS 4395: Part 1)

Example 15.5

Consider the connection in Example 15.3 (Fig. 15.8). Assume that the six bolts are 16 mm diameter HSFG bolts (general grade). Is the connection strong enough now?

Solution

$$\text{Safe load} = 1.1 \times 1.0 \times 0.45 \times 92.1 \times 6 = 273\,kN > 225\,kN$$

i.e. the connection is now satisfactory.

It must be pointed out again that here, as in the case of Example 15.3, the bolts would also be subject to tension caused by the eccentricity of loading. This tension reduces the effective clamping action of the bolts and, therefore, the safe load would have to be suitably decreased. (See Clause 6.4.5 of BS 5950: Part 1: 2000.)

Welding

Welding for structural purposes is governed by the requirements of BS 5135, *Process of welding of carbon and carbon–manganese steels*, and the design of welds is covered by Clause 6.7 of BS 5950: Part 1: 2000.

The two types of weld used are butt welds and fillet welds.

Butt welds

These require the edges of the plates to be prepared by bevelling or gouging as shown in Fig. 15.11(a). This preparation and the need for careful alignment when welding make the butt weld generally the more expensive of the two.

For the purpose of strength calculations, butt welds are treated as the parent metal, i.e. the allowable stresses for the weld are the same as those for the connected plates.

Fillet welds

No special preparations are needed and the strength of the weld is calculated on the throat thickness (see Fig. 15.11(b)).

The design strength, p_w, depends on the grade of the steel of the connected parts and is 220 N/mm² for Grade S275 steel.

The size of the weld is specified by the minimum leg length of the weld, e.g. the strength of an 8 mm fillet weld for Grade S275 steel is

$$8 \times 0.7 \times 220 = 1232 \text{ N/mm}$$

i.e. each millimetre length of this weld is capable of carrying a load of 1232 N.

Fig. 15.11 Welding: (a) butt welds; (b) fillet welds

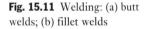

Single V   Single U   Single J   Single bevel

(a)

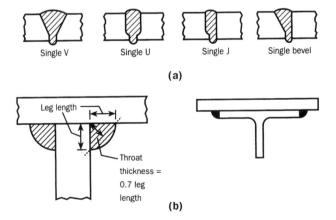

Leg length

Throat thickness = 0.7 leg length

(b)

When deciding on the size of a weld it is wise to consider that the amount of weld metal increases faster than the strength of the weld, e.g. compare 6 mm and 8 mm welds:

Increase in strength 33%

Increase in weld metal 78%

Example 15.6

A tension member in a framework consists of an 80 mm × 10 mm flat and is subject to a direct force of 206 kN. Design a suitable fillet weld connection using a gusset plate, as shown in Fig. 15.12.

Solution

Welding along the two edges of the flat requires a minimum length of weld of 80 mm on each side (Clause 6.7.2.4 of BS 5950), i.e. minimum length of weld is 160 mm.

Use 6 mm weld:

$$\text{Required length} = \frac{206\,000}{6 \times 0.7 \times 220} = 222.94 \text{ mm}$$

The weld should be returned continuously around the corner for a distance not less than 2 × weld size to comply with Clause 6.7.2.2 of BS 5950 and an allowance of one weld size should be made at the open end of the weld.

The overall length of the welds should be

$$\tfrac{1}{2} \times 222.94 + 1 \times 6 + \text{return end}$$

Fig. 15.12 Example 15.6

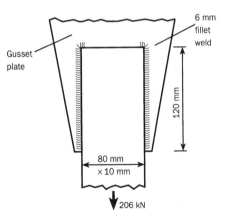

Gusset plate

6 mm fillet weld

120 mm

80 mm × 10 mm

206 kN

Summary

Rivets and bolts

SSV of one rivet or bolt = Ap_s

DSV of one rivet or bolt = $2Ap_s$

BV of one rivet or bolt in a plate of thickness t mm = $dt_p p_{bb}$

A is the area of cross-section of the rivet shank or bolt shank.

For rivets, A may be taken as the area of a circle 2 mm greater in diameter than the specified (nominal) diameter.

For bolts, A is the area calculated from the nominal diameter.

d is the diameter of the rivet or bolt.

For rivets, d = nominal diameter plus 2 mm

For bolts, d = nominal diameter

p_s = shear strength

p_{bb} = bearing strength

t_p = thickness of connected plate (see Clause 6.3.3.1 of BS 5950)

In certain problems, the strength of the plate in tension may have to be investigated. The permissible tension stress should not exceed $0.84 \times$ minimum ultimate tensile strength.

When deducting the areas of rivet or bolt holes to determine the strength of a plate, the diameter of the hole is taken as 2 mm greater than the nominal diameter of the rivet or bolt.

HSFG bolts rely on their tensile strength to induce friction between the connected parts.

A *butt weld* is considered to be as strong as the parent metal.

The strength of a *fillet weld* per millimetre of its length is calculated as

$$0.7 \times \text{size of weld} \times \text{design strength}, p_w$$

Exercises

Note: Permissible tension stress for Grade S275 steel may be assumed to be 250 N/mm². These examples may be adapted for HSFG bolts and fillet welds.

1 The size of each plate in a simple lap joint is 100 mm × 12 mm and there are six 20 mm diameter ordinary bolts in a single line. Calculate the safe load in tension.

2 In a double cover butt connection, the joined plate is 125 mm × 12 mm and the cover plates are 125 mm × 8 mm. There are two 20 mm diameter power-driven rivets each side of the joint (four rivets in all). Calculate the maximum safe tension for the plates.

3 A simple lap joint with five 24 mm diameter hand-driven rivets is shown in Fig. 15.Q3. Calculate the maximum safe pull, W.

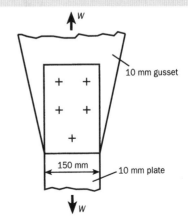

Fig. 15.Q3

4 Eight bolts, 24 mm diameter, connect a flat bar to a gusset plate as shown in Fig. 15.Q4. Calculate the safe load, *W*.

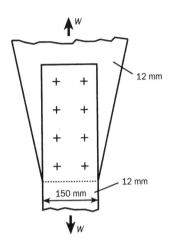

Fig. 15.Q4

5 Figure 15.Q5 gives two different bolted connections (a) and (b). In each case, calculate the safe load, *W*.

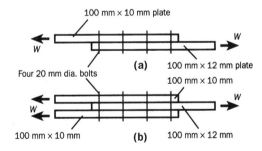

Fig. 15.Q5

6 A 100 mm wide plate is connected, by means of a 10 mm gusset plate to the flange of a 254 × 254 UC107

stanchion, as shown in Fig. 15.Q6. Calculate the required thickness of the plate and the number of 20 mm diameter bolts.

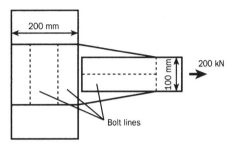

Fig. 15.Q6

7 A 10 mm thick tie member is to be connected to a 10 mm thick gusset plate, as indicated in Fig. 15.Q7. Calculate the necessary width, *x*, of the tie and calculate the number of 16 mm diameter bolts to connect the tie to the gusset.

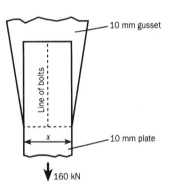

Fig. 15.Q7

8 Figure 15.Q8 shows a joint in a tension member. Determine the safe load, *W*. (Calculations are required for the strength of the middle plate at sections A–A and B–B; the strength of the cover plate at C–C, and the strengths of the rivets in shear and bearing.)

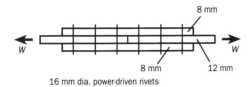

16 mm dia. power-driven rivets

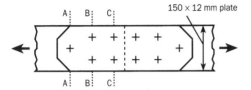

Fig. 15.Q8

10 Calculate the value of the maximum permissible load, W, for the connection shown in Fig. 15.Q10. The rivets are 20 mm diameter, hand-driven.

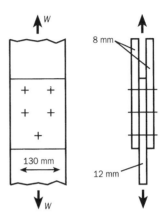

Fig. 15.Q10

9 Referring to Fig. 15.Q9, calculate the maximum safe load, W, confining the calculations to that part of the connection which lies to the right of line A–A. Then, using this load, find the required thickness of the 75 mm wide plate and the necessary number of 16 mm diameter hand-driven rivets.

All rivets 16 mm dia. hand-driven rivets

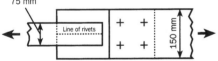

Fig. 15.Q9

Previous chapters have dealt largely with two main types of stress:

- *Direct or axial stress*   This occurs when a load (tensile or compressive) is spread evenly across a section, as in the case of a short column axially loaded. Here, the unit stress or stress per mm² is found by dividing the total load by the sectional area of the member, i.e. direct stress = $f_d = W/A$.
- *Bending stress*   In this case, the stress has been seen to vary – having different values at different distances from the neutral axis of the section. The stress at a distance y from the neutral axis is $f_b = M \times y/I$, or where the stress at the extreme fibres is required $f_{bmax} = M/Z$.

Cases frequently arise, however, where *both* these types of stress occur at the same time, and the combined stress due to the addition of the two stresses is to be determined.

Consider, for example, the short timber post shown in plan in Fig. 16.1. The load passes through the centroid of the section, and the resulting stress is pure axial stress

$$W/A = 90\ 000/22\ 500 = 4\ \text{N mm}^2$$

The stress is the same at all points of the section, and this is shown by the rectangular shape of the 'stress distribution diagram'.

Figure 16.2(a)(i) shows the same section with a load of 90 kN as before, but this time the load, whilst still on the X–X axis, lies 20 mm away from the Y–Y axis.

As before, there will still be a direct stress everywhere on the section of $W/A = 4\ \text{N/mm}^2$, but the eccentricity of the load with regard to the Y–Y axis causes a moment of $90\ 000 \times 20 = 1.8 \times 10^6$ N mm, which has

Fig. 16.1 Direct stress:
(a) loading of a short timber post;
(b) stress distribution

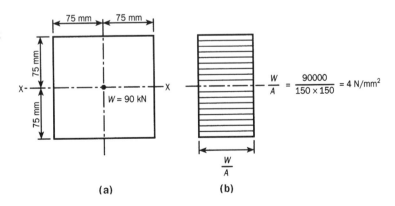

(a)   (b)

Fig. 16.2 Direct stress plus bending stress: (a) Y – Y axis eccentricity = 20 mm; (b) Y – Y axis eccentricity = 25 mm; (i) loading diagram, (ii) direct stress, (iii) bending stress, (iv) combined stress distribution

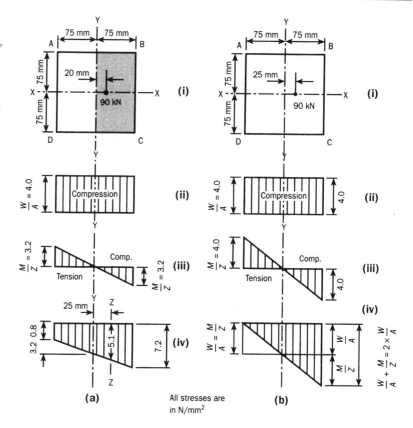

All stresses are in N/mm²

the effect of increasing the compression on the area to the right of Y–Y (shaded) and decreasing the compression on the portion to the left of Y–Y.

Figure 16.2(a)(ii) shows the direct stress W/A in compression.

Figure 16.2(a)(iii) shows the bending stress of

$$\frac{M}{Z} = \frac{1\,800\,000 \times 6}{150^3} = 3.2 \text{ N/mm}^2 \text{ compression at BC}$$

and 3.2 N/mm² tension at AD

Figure 16.2(a)(iv) shows how these two stress distribution diagrams, (ii) and (iii), are superimposed, giving the final diagram as shown. The compressive stress on the edge BC is

$$\frac{W}{A} + \frac{M}{Z} = 4.0 + 3.2 = 7.2 \text{ N/mm}^2$$

whilst the stress on edge AD is

$$\frac{W}{A} - \frac{M}{Z} = 4.0 - 3.2 = 0.8 \text{ N/mm}^2$$

Note: The stress is still compressive at any point, but it varies from face AD to face BC.

The stress at section Z–Z (25 mm from Y–Y) is

$$\frac{W}{A} + \frac{My}{I_{yy}} \quad \text{where } y = 25 \text{ mm}$$

$$= 4.0 + \frac{1\,800\,000 \times 25 \times 12}{150^4}$$

$$= 4.0 + 1.1 = 5.1 \text{ N/mm}^2$$

Figure 16.2(b)(i) shows the same short column as in the previous example, but this time the eccentricity about the Y–Y axis has been increased to 25 mm. As before,

$$\text{Direct stress} = W/A = 90\,000/22\,500 = 4 \text{ N/mm}^2$$

but the M is now 90 000 × 25 = 2.25 × 10^6 N mm. Thus

$$\frac{M}{Z} = \frac{2.25 \times 10^6}{562\,500} = 4 \text{ N/mm}^2$$

Figure 16.2(b)(ii) and (iii) show the direct and bending stresses respectively, whilst in Fig. 16.2(b)(iv) the two diagrams have been superimposed, showing that:

- The stress on face BC is

$$\frac{W}{A} + \frac{M}{Z} = 4.0 + 4.0 = 8.0 \text{ N/mm}^2$$

- The stress on face AD is

$$\frac{W}{A} - \frac{M}{Z} = 4.0 - 4.0 = 0$$

The original compression of W/A on face AD has been cancelled out by the tension at AD caused by the eccentricity of the loading away from that face, and the stress intensity at AD = 0, whilst that on face BC = 8.0 N/mm^2.

This interesting example demonstrates a general rule that, in the case of square or rectangular sections, the load may be eccentric from one axis by no more than $\frac{1}{6}$th of the width unless tension on one face is permissible.

In the case of a steel, timber or reinforced concrete column, tension would be allowable within reasonable limits, but, in a brick wall, tension would be undesirable and the resultant load or thrust in a masonry structure should not depart more than $d/6$ (or $b/6$) from the axis.

This is usually called the **law of the middle third**, and it states that if the load or the resultant thrust cuts a *rectangular* section at a point not further than $d/6$ from the centre line (i.e. within the middle third), then the stress everywhere across the section will be compressive.

Consider the masonry retaining wall shown in Fig. 16.3(a). The earth pressure P and the weight W of the wall combine to form a resultant R which cuts the base, as shown, at a distance e from the neutral axis (which, in this case, is also the centre line) of the base.

Fig. 16.3 The law of the middle third

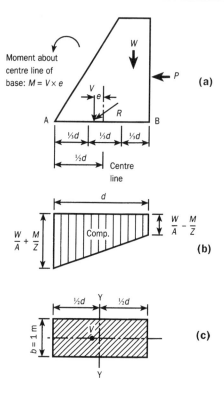

The vertical component V of the resultant causes a moment of $V \times e$, which adds compressive stress between the wall and the earth below at A, and reduces the pressure between the surfaces at B.

If V cuts the base within the middle third of its width d, then the stress everywhere between A and B will be compressive, as in Fig. 16.3(b).

Note: As the wall itself may be of any length, it is customary in design to consider only one metre length of wall, i.e. to estimate the weight of one metre of wall and the earth pressure on one metre of length also.

Thus, if the rectangle shown in Fig. 16.3(c) represents the area of wall resting on the base AB (as seen in a plan), then the eccentricity of the load V will tend to rotate this area about the axis Y–Y, so that, in calculating Z, the value $bd^2/6$ will be in effect $1 \times d^2/6 = \frac{1}{6}$th of d^2, and will normally be in m³.

Eccentricity about both axes

The short pier shown in Fig. 16.4 carries one load only, eccentric to both X–X and Y–Y axes. Consider the stresses at A, B, C and D.

At C, the stress is

$$\frac{W}{A} + \frac{M_{xx}}{Z_{xx}} + \frac{M_{yy}}{Z_{yy}}$$

because the load has moved towards C from both axes.

Fig. 16.4 Eccentricity about both axes

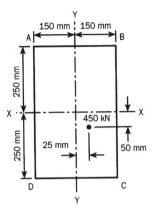

Stress at corner B will be

$$\frac{W}{A} - \frac{M_{xx}}{Z_{xx}} + \frac{M_{yy}}{Z_{yy}}$$

for the load is away from the Y–Y axis on the same side as B, but is away from the X–X axis on the side furthest from B.

Similarly,

$$\text{Stress at corner A} = \frac{W}{A} - \frac{M_{xx}}{Z_{xx}} - \frac{M_{yy}}{Z_{yy}}$$

$$\text{Stress at corner D} = \frac{W}{A} + \frac{M_{xx}}{Z_{xx}} - \frac{M_{yy}}{Z_{yy}}$$

The solution is outlined below:

$$\frac{W}{A} = \frac{450\,000}{300 \times 500} = 3.0\,\text{N/mm}^2$$

$$\frac{M_{xx}}{Z_{xx}} = \frac{450\,000 \times 50 \times 6}{300 \times 500^2} = 1.8\,\text{N/mm}^2$$

$$\frac{M_{yy}}{Z_{yy}} = \frac{450\,000 \times 25 \times 6}{400 \times 300^2} = 1.5\,\text{N/mm}^2$$

Hence,

$$\text{Stress at corner A} = 3.0 - 1.8 - 1.5$$
$$= -0.3\,\text{N/mm}^2 \text{ tension}$$

$$\text{Stress at corner B} = 3.0 - 1.8 + 1.5$$
$$= +2.7\,\text{N/mm}^2 \text{ compression}$$

$$\text{Stress at corner C} = 3.0 + 1.8 + 1.5$$
$$= +6.3\,\text{N/mm}^2 \text{ compression}$$

$$\text{Stress at corner D} = 3.0 + 1.8 - 1.5$$
$$= +3.3\,\text{N/mm}^2 \text{ compression}$$

Example 16.1

A short stanchion carries three loads, as shown in Fig. 16.5(a). Calculate the intensity of stress at corners A, B, C and D.

$$\text{Area} = 7640 \text{ mm}^2 \qquad Z_{xx} = 584\,000 \text{ mm}^3 \qquad Z_{yy} = 201\,000 \text{ mm}$$

Fig. 16.5 Example 16.1:
(a) method 1; (b) method 2

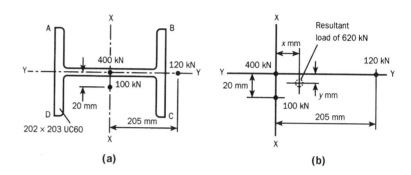

(a)   (b)

Solution

Method 1

$$\text{Direct stress } \frac{W}{A} = \frac{400 + 120 + 100}{7640} = 81.2 \text{ N/mm}^2$$

$$\text{Bending stress } \frac{M_{xx}}{Z_{xx}} = \frac{120\,000 \times 205}{584\,000} = 42.1 \text{ N/mm}^2$$

$$\text{Bending stress } \frac{M_{yy}}{Z_{yy}} = \frac{100\,000 \times 20}{201\,000} = 9.9 \text{ N/mm}^2$$

Therefore, the combined stresses are

At A $81.2 - 42.1 - 9.9 = +29.2 \text{ N/mm}^2$

At B $81.2 + 42.1 - 9.9 = +113.4 \text{ N/mm}^2$

At C $81.2 + 42.1 + 9.9 = +133.2 \text{ N/mm}^2$

At D $81.2 - 42.1 + 9.9 = +49.0 \text{ N/mm}^2$

Method 2

The value and position of the resultant of the three loads are calculated and the resultant load applied to the column in place of the three loads.

To find distance x (see Fig. 16.5(b)) take moments of loads about the X–X axis:

$$(400 + 120 + 100) \times x = 120 \times 205$$

$$x = 24\,600/620 = 39.7 \text{ mm}$$

For distance y take moments about the Y–Y axis:

$$620 \times y = 100 \times 20$$

$$y = 2000/600 = 3.22 \text{ mm}$$

Now, the direct stress W/A remains the same and the bending stresses are

$$\frac{M_{xx}}{Z_{xx}} = \frac{620\,000 \times 39.7}{584\,000} = 42.1 \text{ N/mm}^2$$

$$\frac{M_{yy}}{Z_{yy}} = \frac{620\,000 \times 3.22}{201\,000} = 9.9 \text{ N/mm}^2$$

Both are the same as in Method 1.

Application to prestressed concrete beams

The addition of direct and bending stresses is not confined to columns only. It is used extensively in the design calculations for prestressed concrete beams. The prestressing force is generally applied in the direction of the longitudinal axis of the beam, thus producing direct stresses in the beam. The bending stresses result mainly from the loads the beam has to support, but there is usually also bending caused by an eccentricity of the prestressing force.

Example 16.2

A plain concrete beam, spanning 6 m, is to be prestressed by a force of 960 kN applied axially to the 240 mm × 640 mm beam section (see Fig. 16.6).

Calculate the stresses in the top and bottom fibres of the beam if the total inclusive UDL carried by the beam is 136 kN.

Fig. 16.6 Example 16.2

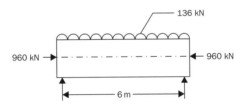

Solution

The direct stress caused by the prestressing force P is

$$\frac{P}{A} = \frac{960\,000}{240 \times 640} = 6.25 \text{ N/mm}^2$$

The bending stresses, on the other hand, are produced by the inclusive UDL and these are M/Z where

$$M = \tfrac{1}{8}Wl = \tfrac{1}{8} \times 136 \times 6 \times 10^6 = 102 \times 10^6 \text{ N mm}$$

$$Z = \tfrac{1}{6}bd^2 = \tfrac{1}{6} \times 240 \times 640^2 = 16\,384\,000 \text{ mm}^3$$

Therefore,

$$\frac{M}{Z} = \frac{102\,000\,000}{16\,384\,000} = 6.23 \text{ N/mm}^2$$

Fig. 16.7 Example 16.2: (a) axial prestressing; (b) prestressing below the centre line; (i) beam details, (ii) stress diagrams

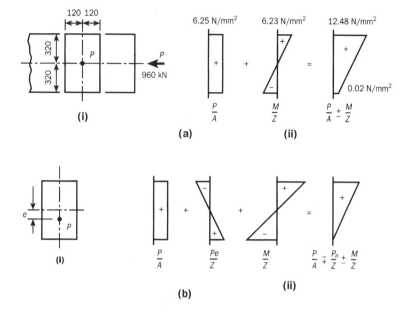

and the combined stresses are

At the top $6.25 + 6.23 = +12.48\,\text{N/mm}^2$

At the bottom $6.25 - 6.23 = +0.02\,\text{N/mm}^2$

as shown diagrammatically in Fig. 16.7(a).

In order to increase the beam's load bearing capacity it is usual to apply the prestressing force below the centre line of the section. This induces tensile bending stresses in the top fibres and compressive bending stresses in the bottom fibres of the beam. The applied loads produce bending stresses of the opposite sign, i.e. tension in bottom and compression in top fibres.

The combination of the stresses is shown in Fig. 16.7(b).

Example 16.3

Consider the beam in Example 16.2 with the prestressing force applied on the vertical axis of the section but 120 mm below its intersection with the horizontal axis. Calculate the stresses in top and bottom fibres.

Solution

Direct stress

$$\frac{P}{A} = +6.25\,\text{N/mm}^2 \text{ (as before)}$$

Bending stress due to prestressing force

$$\frac{P \times c}{Z} = \frac{960\,000 \times 120}{16\,384\,000} = \pm 7.03\,\text{N/mm}^2$$

Bending stress due to applied loads

$$\frac{M}{Z} = \pm 6.23\,\text{N/mm}^2 \text{ (as before)}$$

The combined stresses are

At the top $+6.25 - 7.03 + 6.23 = +5.45\,\text{N/mm}^2$

At the bottom $+6.25 + 7.03 - 6.23 = +7.05\,\text{N/mm}^2$

It may be concluded from this example that the UDL could be doubled without indicating tensile stresses in the concrete when the beam is fully loaded. There are, however, various problems which have to be taken into account when designing prestressed concrete beams and which are outside the scope of this textbook.

Summary

Maximum combined stress = direct stress + bending stress. When load W is eccentric to axis X–X only:

$$\text{Maximum stress} = \frac{W}{A} + \frac{We_x}{Z_{xx}}$$

When load W is eccentric to axis Y–Y only:

$$\text{Maximum stress} = \frac{W}{A} + \frac{We_y}{Z_{yy}}$$

When load W is eccentric to both axes:

$$\text{Maximum stress} = \frac{W}{A} + \frac{We_x}{Z_{xx}} + \frac{We_y}{Z_{yy}}$$

In prestressed concrete beams, where opposing bending stresses are produced by the prestressing force P and the applied loads W,

$$\text{Maximum stress} = \frac{P}{A} \mp \frac{P \times e}{Z} \mp \frac{M_w}{Z}$$

Exercises

1 A short timber post carries a load of 40 kN eccentric from one axis only as shown in Fig. 16.Q1. Calculate the intensity of stress (a) at face BC (f_{bc}), (b) at face AD (f_{ad}).

2 Calculate the compression stress on the faces BC and AD of the short timber post, as shown in Fig. 16.Q2.

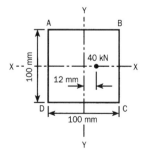

Fig. 16.Q1

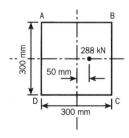

Fig. 16.Q2

3 Repeat Question 2, but increasing the eccentricity to 60 mm.

4 Calculate the stress at the corner C of the short timber post shown in Fig. 16.Q4.

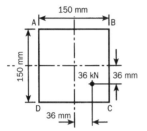

Fig. 16.Q4

5 A short steel stanchion consists of a 203 × 203 UC60 and carries loads as shown in Fig. 16.Q5. Calculate the stress at faces AD and BC.

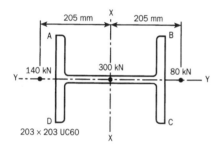

Fig. 16.Q5

6 A short steel post consisting of a 152 × 152 UC30 carries three loads, as shown in Fig. 16.Q6. Calculate the stress at each of the corners A, B, C and D.

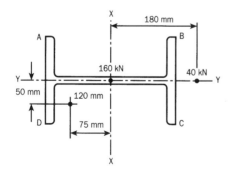

Fig. 16.Q6

7 The short steel post, as shown in Fig. 16.Q7, is a 152 × 152 UC37. There is a central point load of 120 kN and an eccentric load of W kN as shown. The maximum stress on the face BC is 80 N/mm². What is the value of the load W?

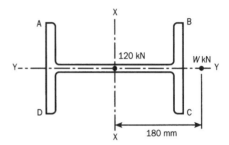

Fig. 16.Q7

8 The vertical resultant load on the earth from a metre run of retaining wall is 40 kN applied as shown in Fig. 16.Q8. Calculate the intensity of vertical stress under the wall at A and B.

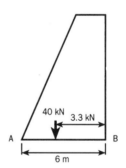

Fig. 16.Q8

9 A triangular mass wall, as shown in section in Fig. 16.Q9, weighs 20 kN/m³ and rests on a flat base AB. What is the intensity of vertical bearing stress at A and B per metre length of wall?

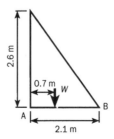

Fig. 16.Q9

10 The concrete beam shown in Fig. 16.Q10 is prestressed by the application of two loads of 300 kN applied at points 40 mm below the neutral axis of the section. (All dimensions are in mm.) (a) What are the stresses at the upper and lower faces of the beam? (b) At what distance below the neutral axis would the loads have to be applied for there to be no stress at all on the top face of the beam? (All dimensions are in mm.)
Note: Neglect self-weight of beam.

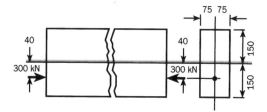

Fig. 16.Q10

11 A double angle rafter member in a steel truss is sub-jected to an axial compression of 60 kN and also to bending moment from a point load of 10 kN, as shown in Fig. 16.Q11. The member is composed of two

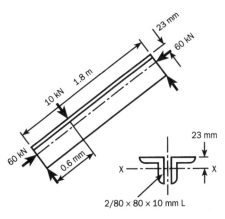

Fig. 16.Q11

angles, as shown, for which the properties (given for the double angle section) are

$$A = 3000 \text{ mm}^2 \quad I_{xx} = 1.75 \times 10^6 \text{ mm}^4$$

Calculate the maximum compressive stress in the member.

12 A timber beam subjected to horizontal thrusts at each end, shown in Fig. 16.Q12, is loaded with a central concentrated load of 5 kN (all dimensions are in mm). The beam itself weighs 150 kN. What is (a) the maxi-mum compressive stress and (b) the maximum tensile stress?

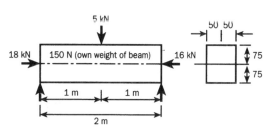

Fig. 16.Q12

13 A plain concrete beam, 240 mm × 600 mm, spans 6 m and weighs 3.2 kN/m. A horizontal compressive force of 900 kN is applied to the beam at 180 mm above its underside.

Calculate (a) the stresses in the top and bottom fibres of the beam; (b) the additional UDL the beam could carry without developing tension in its bottom fibres.

Chapter
seventeen   Portal frames and arches

Portal frames

Portal frames are designed to carry lightweight roof coverings. They are often constructed of steel, aluminium, concrete, timber or composite materials. They are commonly used for single- or multi-span buildings with flat, pitched, arch north light and monitor roofs (see Fig. 17.1).

The three basic forms of portal frames are as follows:

Rigid portal frame

Figure 17.1(d). These are, *normally*, frames of height and span less than or equal to 5 m and 15 m respectively. They are connected directly to their foundation structure through rigid joints. Rigid connection gives a more evenly distributed moment of lower magnitude acting on the supporting elements compared with the two-pin or three-pin portal frames (see Fig. 17.2).

Two-pin portal frames

Figure 17.1(a and c). A pin joint or hinged joint at the base connections is *normally* considered by the designer for portal frames of height and span more than 5 m and 15 m respectively. Hinges can also be introduced at the centre or apex of the spanning member. Structural engineers commonly considers two pin portal frames where high base moment and weak ground bearing conditions are encountered. No moment at the base eliminates the tendency of the base or foundation to rotate.

Fig. 17.1 Single-storey portal frames: (a) flat-pinned base; (b) three-hinged; (c) two-pinned; (d) fixed base; (e) north light; (f) monitor roof; (g) industrial carrying crane; (h) multi-portal frame building

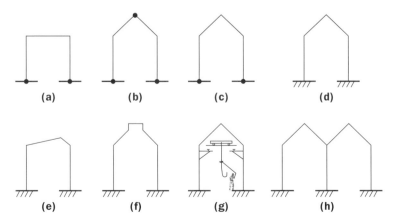

Fig. 17.2 Portal frames: comparison of moments (a) high sagging moment on the beam; (b) low moment on the beam, moment on the base not recommended on weak bearing ground; (c) high moment on supporting members compared with case (b), used to eliminate base rotation; (d) reduced moment in the centre of the spanning member but increased deflection

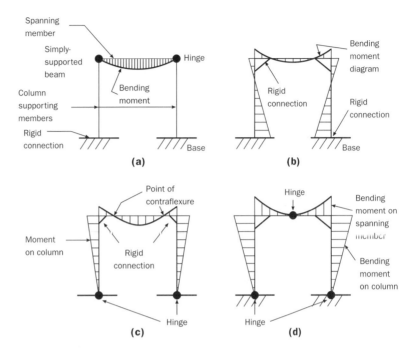

Three-pin portal frames

Figure 17.1(b). These are statically determinate structures, simple in design and analysis, and easier to erect. The hinge at the centre of the spanning member, or other location (if necessary), is introduced to reduce the bending moment in the spanning member. However, the introduction of the hinge will increase the deflection of the spanning member. The designer commonly uses a pitched spanning member or deeper section to overcome deflections of high magnitude.

The advantages of portal frames include but are not limited to the following:

- Provide an economic means of increasing the useful span. These types of buildings can achieve spans of up to 60 m.
- Eliminate the need for a lattice of struts and ties within the roof space, giving a more pleasing internal finish and greater usable headroom to the structure.
- Accurate fabrication/manufacturing under factory control production conditions.
- Ease of site assembly of components, since the criteria for design and quality of manufacturing recommended in the code of practice and building regulations can be accurately controlled by skilled labour under high quality factory conditions, and placed during the production process.
- The portal frame buildings are aesthetically pleasing.

Arches

Framed domes, arches, circular and elliptical tubes are structural forms which are visually some of the most attractive structures. They are commonly used to cover sports areas, exhibition pavilions, churches, etc. Fig. 17.3 shows four examples of arched structural forms.

Past use of arches can be found in brick arch floors, inverted brick arch foundations, tunnels, chimneys, etc. Structurally, they are the most efficient method of forming a structure with materials that have good compressive strength and low tensile strength, such as brick and concrete. The design engineer, *in order to achieve the most efficient structural solution*, can alter the structural configurations to produce an equilibrium condition made up of compression forces, as can be seen in Fig. 17.4. The figure shows typical masonry arches and the common design terms such as loads and forces acting on them. It is common practice that the line of thrust remains within the middle third of the depth of the arch (see Fig. 17.4).

For case (a):

$$\text{total stress,} \quad f = P/A$$

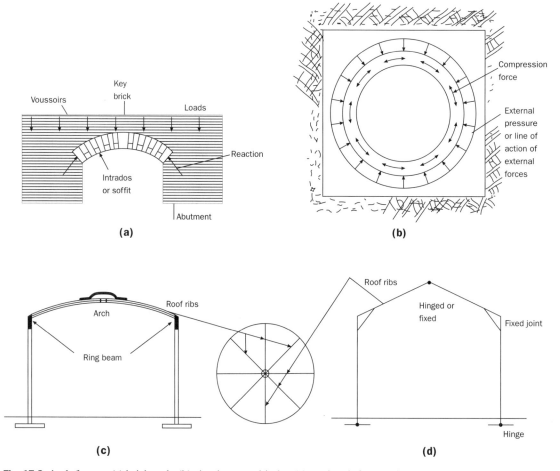

Fig. 17.3 Arch frames: (a) brick arch; (b) circular tunnel/tube; (c) steel arch frame; (d) steel portal arch or dome frame

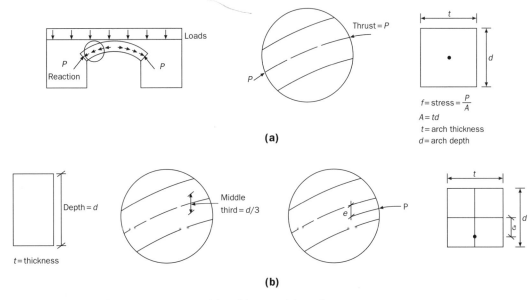

Fig. 17.4 The line of thrust: (a) no eccentricity; (b) eccentricity > 0

For case (a):

$$\text{total stress} = f = \frac{P}{A}$$

For case (b):

$$\text{total stress at top fibres,} \quad f = \frac{P}{A} - \frac{Pe}{Z}$$

$$\text{total stress at bottom fibres,} \quad f = \frac{P}{A} + \frac{Pe}{Z}$$

Always keep the value of $e \leq d/6$, which is the maximum value of e for the section to be kept in compression. This is derived as follows:

$$f = \frac{P}{A} - \frac{Pe}{Z}$$

At the limit for no tension, $f \geq 0$

$$\frac{P}{A} - \frac{Pe}{Z} \geq 0$$

$$\frac{P}{A} \geq \frac{Pe}{Z}$$

$$e \leq \frac{Z}{A}$$

Since $Z = \dfrac{td^2}{6}$ (for a rectangular section)

$$A = dt$$

$$e \leq \left(\frac{td^2}{6}\right)/dt \quad \text{or} \quad e \leq \frac{d}{6}$$

Provided that the line of the thrust does not pass outside a distance $d/6$ to each side of the centre line of the arch, no tension stress will develop. This is well known by designers as the **middle third rule**.

Portal frame analysis

A portal frame is analysed either by assuming linear elastic behaviour or by basing the analysis on the basic principles of plastic theory. The reactions, moments and shear forces are calculated either by hand calculation or by using any suitable computer software. Plastic designed frames are lighter than elastically designed frames, however, higher stability bracing and checks will be needed. The designer must make sure that stability is provided both for the frame as a whole and for its individual elements. This includes stability checks using the general rules (see the current code of practice BS 5950) and checks that movement of the frame under all loading cases is not sufficient to cause damage to either the frame itself, or the finishes such as cladding and brick walls adjacent to constructions.

The following example shows the basic procedures both for the analyses of three-pinned portal frames and for selection of the main frame elements such as the rafters and columns.

Example 17.1

Portal frame analysis and design
The portal frame shown in Fig. 17.5 spans 18 m, 6.34° pitch and 6 m frame centres. The loading details are as follows:

Dead loads including self-weight   = 0.2 kN/m²
Services   = 0.1 kN/m²
Imposed loads (from access and snow) = 0.6 kN/m ²
Wind loads:

    on the windward wall   = 0.4 kN/m²
    on the leeward wall   = 0.125 kN/m²

(i)   draw the shear force, bending moment and thrust diagrams for the columns and the rafters

(ii) select suitable hot rolled sections for the columns and rafters

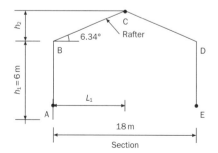
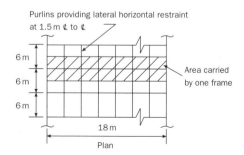

Fig. 17.5 Plan and a section of a portal frame structure

Solution

The three critical loading combinations that are often considered by structural engineers are:

Load case 1: design dead loads + design imposed loads (both at ultimate limit state values)

$$W = 1.4G_k + 1.6Q_k \quad \text{(this gives the maximum vertical design load)}$$

Load case 2: dead loads + imposed loads + wind load

$$W = 1.2G_k + 1.2Q_k + 1.2W_k$$

where $1.2G_k$ and $1.2Q_k$ are vertical loads and $1.2W_k$ is wind load

Load case 3: dead loads + wind loads

$$W = 1.0G_k + 1.4W_k$$

where $1.0G_k$ is vertical loads and $1.4W_k$ is wind load

$$\frac{h_2}{L_1} = \tan 6.34°$$

therefore $h_2 = L_1 \tan 6.34°$
$$= 9 \times 0.111 = 1 \text{ m}$$

Load case 1: design dead loads + design imposed loads

Design loads, $W = 1.4G_k + 1.6Q_k = 1.4 (0.2 + 0.1) + 1.6 \times 0.6$
$$= 1.38 \text{ kN/m}^2$$
Load carried by one frame $= 1.38 \times 6 \times (1 \text{ m length}) = 8.28 \text{ kN/m}$
From Fig. 17.6(a), and since the load is symmetrical:

$$\sum V = 0, \text{ therefore } R_{AV} = R_{EV} = \frac{149.04}{2} = 74.52 \text{ kN}$$

From Fig. 17.6(b),
$$M_C = 0$$

Fig. 17.6 Load case 1

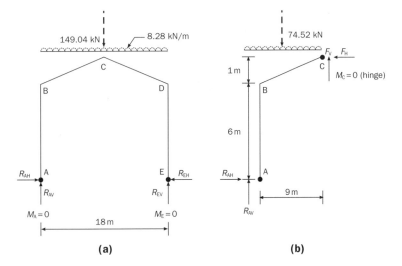

(a)

(b)

Therefore $R_{AV} \times 9 = R_{AH}(6 + 1) + 74.52 \times (9/2)$

$74.52 \times 9 - 335.34 = 7R_{AH}$

$R_{AH} = \dfrac{335.34}{7} = 47.906 \text{ kN}$

$M_B = R_{AH} \times 6 = 47.906 \times 6 = 287.4 \text{ kN m}$

Thrust force in member AB = 74.52 kN
Shear force in member AB = 47.906 kN

Thrust force in member BC is calculated using the method of resolution of forces at joint B.
At point B (see Figs 17.7 and 17.8):

$F_{BC} = R_{AH} \cos 6.34° + R_{AV} \sin 6.34°$

$= 47.613 + 8.229$

$= 55.842$

Thrust force in member BC = 55.842 kN (at B)
Thrust at C = 47.906/cos 6.34 = 48.2 kN

Fig. 17.7 Load case 1: point B

(at point B)   (at point C)

Fig. 17.8 Load case 1: thrust and bending moment diagrams

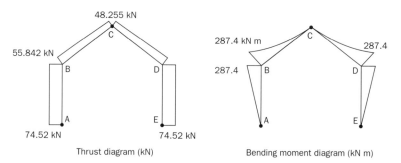

Thrust diagram (kN)   Bending moment diagram (kN m)

Note: Useful information. If the designer decided to replace the hinge at point C with a rigid connection, the frame becomes statically indeterminate and its analysis can be carried out using a suitable computer analysis program or simply by hand calculations using the formula for rigid frames, see *Steel Designers' Manual*, Fifth Edition, eds Graham W. Owens and Peter R. Knowles, Blackwell Scientific Publications, 1994, pages 1080–97, extracted from Kleinlogel, *Rahmenformeln* 11. Auflage Berlin-Verlag von Wilhelm Ernst & Sohn.

For example, the moment at B can be calculated as follows (see Fig. 17.9):

$$M_B = M_D = -\dfrac{(wL^2(3 + 5m))}{16N}$$

$$M_C = \dfrac{wL^2}{8} + mM_B$$

$$N = B + mC$$

Fig. 17.9 Calculation of moment at B and D

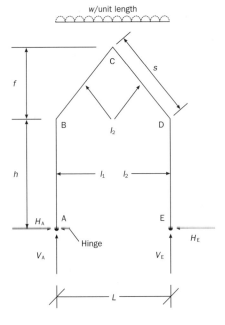

$$K = \frac{I_2 h}{I_1 S}$$

$$\phi = \frac{f}{h}$$

$$m = 1 + \phi$$

$$B = 2(K + 1) + m$$

$$C = 1 + 2m$$

$$H_A = H_E = -\frac{M_B}{h}$$

$$V_A = V_E = \frac{wL}{2}$$

$I_1 = I_2$ (if the same cross-section is used for both the rafter and the column),

$$K = \frac{6}{9.06} = 0.662$$

$$\phi = \frac{f}{h} = \frac{1}{6} = 0.17$$

$$m = 1 + \phi = 1.17$$

$$C = 1 + 2m = 1 + 2(1.17) = 3.34$$

$$B = 2(K + 1) + m = 2(0.662 + 1) + 1.17 = 4.494$$

$$N = 4.494 + (1.17 \times 3.34) = 8.402$$

$$M_B = M_D = -\frac{[8.28 \times 18^2(3 + (5 \times 1.17))]}{(16 \times 8.402)}$$

$$= -176.610 \text{ kN m}$$

This is a reduction of $((287.4 - 176.6)/287.4) \times 100\% = 38.6\%$ less moment, thus the size of the column section can be made smaller (see Fig. 17.10).

Fig. 17.10 Comparison of bending moments

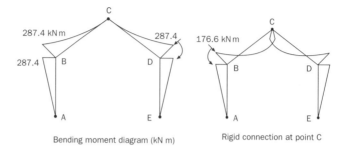

Bending moment diagram (kN m)   Rigid connection at point C

Load case 2: dead loads + imposed loads + wind loads

(See Fig. 17.11)

Design load, $W = 1.2G_k + 1.2Q_k + 1.2W_k$

where $1.2G_k + 1.2Q_k$ are vertical loads and $1.2W_k$ is wind load

Design vertical loads of $1.2G_k + 1.2Q_k = 1.2(0.2 + 0.1) + 1.2 \times 0.6$
$$= 1.08 \text{ kN/m}^2$$

Design vertical load per metre $= 1.08 \times 6 = 6.48$ kN/m
Design wind loads on:

(a)   windward wall $= 1.2 \times 0.4 \times 6 = 2.88$ kN/m
(b)   suction on leeward wall $= 1.2 \times 0.125 \times 6 = 0.9$ kN/m

Calculation of R_{AV} and R_{EV}:

$$\sum M \text{ (at A)} = 0, \text{ therefore}$$

$$R_{EV} \times 18 = (0.9 \times 6 \times 3) + (6.48 \times 18 \times 9) + (2.88 \times 6 \times 3)$$
$$= 16.2 + 1049.76 + 51.84$$

$$R_{EV} = \frac{1117.8}{18} = 62.1 \text{ kN} \uparrow$$

Fig. 17.11 Load case 2

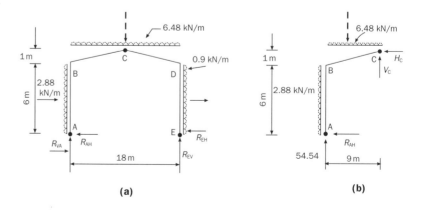

(a)   (b)

$$\sum V = 0,$$

$$R_{AV} + R_{EV} = 6.48 \times 18$$

Therefore, $R_{AV} = 6.48 \times 18 - 62.1 = 54.54 \text{ kN}$

Calculation of R_{AH} and R_{EH}:

$$\sum M \text{ (at C)} = 0$$

$$(R_{AH} \times 7) + (54.54 \times 9) = (6.48 \times 9 \times (9/2))$$

$$+ (2.88 \times 6 \times (3 + 1))$$

$$R_{AH} = -22.76 \text{ kN} \rightarrow$$

$$R_{EH} = (2.88 \times 6) + (0.9 \times 6) + 22.76 = 45.44 \text{ kN} \leftarrow$$

$$M \text{ at B} = (2.88 \times 6 \times 3) + (22.76 \times 6) = 188.4 \text{ kN m}$$

$$M \text{ at D} = (45.44 \times 6) - (0.9 \times 6 \times 3) = 256.44 \text{ kN m}$$

Thrust in member AB = 54.54 kN

Thrust in member BC at point B can be calculated as follows (see Figs 17.12 and 17.13).

At B:

$$\sum H = 0$$

$$F_{BC} \cos 6.34 = R_{AH} + (2.88 \times 6) = 40.04$$

$$F_{BC} = \frac{40.04}{(\cos 6.34)} = 40.286 \text{ kN}$$

Thrust in member DE = R_{EV} = 62.1 kN

Fig. 17.12 Load case 2: point B

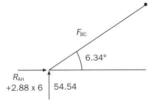

Fig. 17.13 Load case 2: thrust and bending moment diagrams

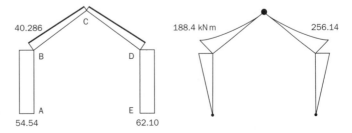

Thrust diagram (kN)

Max. bending moment (kN m)

Thrust in member CD:

At D,

$$F_{CD} \cos 6.34 = 45.44 - (0.9 \times 6) = 40.04 \text{ kN}$$

$$F_{CD} = \frac{40.04}{(\cos 6.34)} = 40.286 \text{ kN}$$

Design values

From load cases 1 and 2, it can be seen that load case 1 produces the critical design values.

For the column use the following load combinations:

design axial load = 74.52 kN

design moment = 287.4 kN m

For the rafter, use:

axial loads = 55.842 kN

design moment = 287.4 kN m at B and 0 kN m at A

Design of the column

Design the column section to carry a design axial load of 74.52 kN and design moment = 287.4 kN m. As the column carries a significantly large moment, try a 533 × 210 × 122 UB, steel grade S275 (new code), steel grade 43 (old code). This section has greater moment capacity than an equivalent H universal column section.

$T = 21.3$ mm, therefore $P_y = 265$ N/mm^2, see Table 9, BS 5950: Part 1: 2000.

Check the adequacy of the column to carry the loads and the moment.

Local capacity check

$$\frac{F}{(Ap_c)} + \frac{M_x}{(S_x p_y)} + \frac{M_y}{(Z_y p_y)} \leq 1.0$$

F = axial design loads
A = cross-sectional area of the steel section
p_c = compressive strength, see Table 14.2
M_x = maximum design moment about the X–X axis
M_y = maximum design moment about the Y–Y axis
S_x = the plastic modulus of the section at the X–X and Z_y = the section modulus about the Y–Y axis respectively

At B, axial design loads = 74.52 kN   and M_x = 281.4 kN m

At A, axial design loads = 74.52 kN   and $M_x = M_y = 0$ kN m

Therefore the local capacity check is more critical at position B than at position A.

Check if the section carries light shear force:

$$\text{Acting shear force} = F_{AH} = 47.906 \text{ kN}$$

$$\text{Section shear capacity } p_v = 0.6 p_y t D$$

$$= \frac{0.6 \times 275 \times 12.7 \times 544.5}{1000}$$

$$= 1140.99 \text{ kN}$$

The thickness of the web, $t = 12.7$ mm, $D = 544.5$ mm, and therefore p_y(web) = 275 N/mm^2 (see Table 11.3 or Table 9 of BS 5950: Part 1: 2000).

$$\text{Therefore } 0.6 \times p_v = 0.6 \times 1140.99 > F_{AH}$$
$$= 684.594 \text{ kN} > 47.906 \text{ kN}$$

Therefore the section carries a light shear force, and

$$\text{Section moment resistance } M_x = p_y S_x = \frac{265 \times 3196}{1000}$$

$$= 846.94 \text{ kN m}$$

For p_c, calculate λ and λ_y, and use the highest value to determine the axis of buckling (in this case, λ_y has the greatest value, hence the axis of buckling will be the Y–Y axis).

$$\lambda_y = L_E/r_y = \frac{3000}{46.7} = 64.23, \text{ say } 65$$

(L_E = effective length, see Clause 4.3.5 of BS 5950: Part 1: 2000) therefore from Table 12.2, $p_c = 205$ N/mm^2

$$\frac{M_y}{(p_y Z_y)} = \frac{0.0}{(p_y Z_y)} = 0$$

$$\frac{74.52}{3177.5} + \frac{287.4}{846.94} + 0 < 0.0234 + 0.339 = 0.363 < 1.0 \text{ (OK)}$$

Overall capacity check (see Fig. 17.14):

$$\frac{F}{A_g p_c} + \frac{m_{LT} M_x}{M_b} + \frac{m_{LT} M_y}{(S_y p_y)} \leq 1.0$$

$$M_b = S_x p_b$$

$$\beta = \frac{M_{small}}{M_{large}} = \frac{0}{287.4} = 0$$

m_{LT} = 0.6, see Table 18 of BS 5950: Part 1: 2000

L_E = 3 m

For p_b, see Table 11 of the BS 5950: Part 1: 2000

$$\lambda_{LT} = uv\lambda(\sqrt{\beta w}), \beta w = 1 \text{ as the section is plastic}$$

$$\lambda = \frac{L_E}{r_y} = \frac{3000}{46.7} = 64.23$$

Fig. 17.14 Column bending
moment diagram

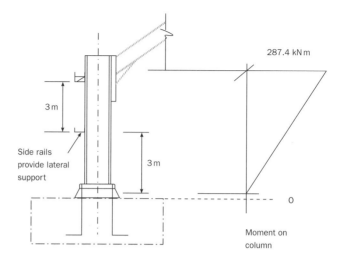

where u = a buckling parameter, see Clause 4.3.6.8 of BS 5950: Part 1: 2000

v = a slenderness factor, see Table 19 of BS 5950: Part 1: 2000

m_{LT} = equivalent moment uniform factor, see Clause 4.3.6.6 of BS 5950: Part 1: 2000

βw = ratio defined in Clause 4.3.6.9 of BS 5950: Part 1: 2000

βw = 1 for class 1 plastic or class 2 compact cross-section

x = 27.6 (from table of section properties)

$\dfrac{\lambda}{x}$ = 2.3

v = 0.951, see Table 19 of BS 5950: Part 1: 2000

βw = 1, see Clause 4.3.6.9 of BS 5950: Part 1: 2000

$\lambda_{LT} = uv\lambda(\sqrt{\beta_w}) = 0.877 \times 0.95 \times 64.23 \times 1 = 53.5$

From Table 16 of BS 5950: Part 1, $p_b = 230$ N/mm^2

Therefore $M_b = S_x p_b = \dfrac{3196 \times 230}{1000} = 735$ kN m

$\dfrac{74.52}{3177.5} + \dfrac{(0.6 \times 287.4)}{735} + 0.0 = 0.246 < 1.0$ (OK)

Design of the rafter

Design the rafter for the following loading combinations:
    Axial loads = 55.84 kN and max. design moment = 287.4 kN m.
    As the size of the section of the rafter is not yet known, and for economical reasons as well as choosing a lightweight rolled section, assume that the thickness of the flange for the rafter section is less than 16 mm. This means that $p_y = 275$ N/mm^2, see Table 9 of the BS 5950: Part 1 or Table 11.2 of this book (p. 220).

$$S_x \text{ (required)} = \frac{M_{max}}{p_y} = \frac{(287.4 \times 1000 \times 1000)}{((275) \times 1000)}$$

Try $457 \times 152 \times 52$ UB, with $S_x = 1096 \text{ cm}^3$
See, the table of section properties, Table 11.3 (p. 224).
Therefore S_x (provided) $> S_x$ (required) (OK)

Check that the section is adequate to carry the axial load and the action of the loads in terms of bending moment.

Overall capacity check

Students also need to check local capacity.

$$\frac{F}{(A_g p_c)} + \frac{m_{LT} M_x}{M_b} + \frac{m_y M_y}{(Z_y p_y)} \leqslant 1.0$$

Calculate p_c;

$$\lambda = \frac{L_E}{r} = \frac{1.5 \times 1000}{31.1} = 48.231 \text{ (rafter is restrained in direc-}$$

tion at B and held effectively in position at both ends and at 1.5 m centre to centre, see Fig. 17.5)

$F = 55.84 \text{ kN}$

$A_g = 66.6 \text{ cm}^2$

$m_{LT} = M_{max} = 287.4 \text{ kN m}$

$M_y = 0.0$

$p_c = 238 \text{ N/mm}^2$, see Table 14.2

$M_b = S_x p_b$

p_b from Table 16 of BS 5950: Part 1: 2000

$\lambda_{LT} = uv\lambda\sqrt{\beta_w}$

$\beta_w = 1$, for plastic section

$u = 0.859$, see Table 11.3

$x = 43.9$, see table of section properties, Table 11.3

$\lambda/x = 48.231/43.9 = 1.098$

$v = 0.99$, Table 19, BS 5950: Part 1: 2000

$u = 0.859$, Table of section properties

$\lambda_{LT} = uv\lambda(\sqrt{\beta_w}) = 0.95 \times 0.859 \times 0.94 \times 48.231$

$$= 36.997$$

$p_b = 271 \text{ N/mm}^2$, see Table 16 of BS 5950: Part 1: 2000

$$M_b = S_x p_b = 1096 \text{ (S_x provided by the section)} \times \frac{271}{1000}$$

$$= 297.016 \text{ kN m}$$

m_{LT}, from Table 18: BS 5950

$$\beta = \frac{M_{\mathrm{small}}}{M_{\mathrm{max}}} = \frac{0.0}{287.4} = 0.0$$

$$m_{\mathrm{LT}} = 0.60$$

$$\frac{55.842}{(66.6 \times 100 \times 238)/1000} + \frac{0.6 \times 287.4}{297.016} + 0.0$$

$$= 0.035 + 0.58 + 0 = 0.615 < 1.0 \text{ (OK)}$$

Analysis of ribbed dome structure

A framed dome or arch is one of the most spectacular structures. The analysis of ribbed dome structures is explained in this section, as such structures have recently become attractive to use.

Example 17.2

Ribbed dome, analyses and design

A ribbed dome structure is shown in Fig. 17.15, is to be octagonal in plan on a 20 m diameter base, 3 m high at the eves and 5.68 m at the crown. There are to be four braced bays with masonry walls. The roof is constructed from felt on timber on purlins with a ceiling. The details of loadings are as follows:

$$G_{\mathrm{k}} = 1.0 \text{ kN/m}^2 \text{ on slope}$$

$$Q_{\mathrm{k}} = 0.75 \text{ kN/m}^2 \text{ on plan}$$

Design the column sections for the four three-pinned arches as shown in Fig. 17.15, use steel grade S355.

For further information, see Lee G.C., Ketter R.L. and Hus T.L., *The Design of Single Story Rigid Frames*, Metal Manufacturers Association, Cleveland, Ohio, 1981.

Fig. 17.15 Ribbed dome structure: (a) plan; (b) section

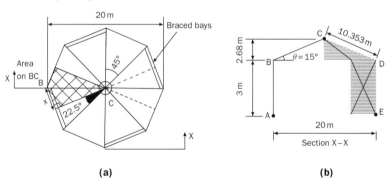

(a)   **(b)**

Solution

$$x/10 = \sin 22.5°$$

$$x = 10 \sin 22.5 = 3.827 \text{ m}$$

$$\text{design loads} = 1.4 G_{\mathrm{k}} + 1.6 Q_{\mathrm{k}}$$

$$= 1.4\left(\frac{1.0}{\cos 15}\right) + 1.6 \times 0.75$$

$$= 2.649 \text{ kN/m}^2$$

design loads on half portal $= 2.649 \times$ shaded area

$$= 2.649 \times 2\left(\frac{10 \times \cos 22.5 \times 3.827}{2}\right)$$

$$= 93.66 \text{ kN}$$

From Fig. 17.17,

$$\sum V = 0$$

$$R_{AV} = R_{EV} = 93.66$$

$$\sum M \text{ (at C)} = 0$$

$$R_{AH} \times 5.68 + 93.66 \times (10 \times 2/3) - 93.66 \times 10 = 0$$

$$R_{AH} = 54.964 \text{ kN} \rightarrow$$

$$M_{B} - R_{AH} \times 3 - 54.964 \times 3 - 164.893 \text{ kN m}$$

(See Figs 17.16 and 17.17.)

Fig. 17.16 Ribbed dome structure: point B

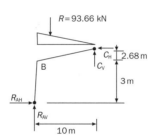

Fig. 17.17 Ribbed dome structure: (a) design loads (kN); (b) moment diagram (kN m)

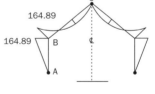

Design loads (kN)   Moment diagram (kN M)

For member AB, calculate S_x as follows;

$$S_x \text{ (required)} = \frac{M_{max}}{p_y}$$

$$= \frac{164.88 \times 1000}{(355)}$$

$$= 464.45 \text{ cm}^3$$

$305 \times 127 \times 37$ UB for which $S_x = 539 \text{ cm}^3$

Exercises

1 Discuss the factors that have led to an increase in the use of portal frame and ribbed dome structures over recent years.

2 Discuss the factors to be considered in the selection of one of the three basic forms of portal frames for a particular application.

3 The portal frame showing in Fig. 17.Q3 spans 21 m, and has 10° pitch and 6 m frame centres. The loading details are as follows:

Dead loads including self-weight $= 0.2 \text{ kN/m}^2$
Services $= 0.1 \text{ kN/m}^2$
Imposed loads (from access and snow) $= 0.6 \text{ kN/m}^2$

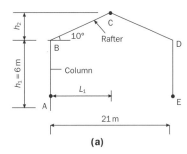

(a)

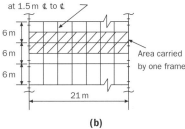

(b)

Fig. 17.Q3 Exercise 3: (a) section; (b) plan

Wind loads:
    On the windward wall = 0.4 kN/m^2
    On leeward wall = 0.125 kN/m^2

(i)   Draw the shear force, bending moment and thrust diagrams for the columns and the rafters.

(ii) Select suitable hot rolled sections for the columns and rafters.

(iii) Compare your results with those of Example 17.1, and comment on your findings.

4 Use a suitable computer software, recalculate the shear force, bending moment and thrusts for the columns and the rafters in Example 17.1 assuming 10°, 15°, 20° and 30° pitch. Consider the three critical loading conditions, i.e.

Load case 1: design dead loads + design imposed loads (both at ultimate limit state values)

$$W = 1.4G_k + 1.6Q_k \text{ (vertical design loads)}$$

Load case 2: design dead loads + design imposed loads + design wind loads

$$1.2G_k + 1.2Q_k \text{ (design vertical loads)}$$

$$1.2W_k \text{ (design wind load)}$$

Load case 3: design dead loads + design wind load

$$W = 1.0G_k + 1.4W_k$$

$$1.0G_k \text{ (vertical loads)}$$

$$1.4W_k \text{ (design wind load)}$$

Comment on your results.

5 A ribbed dome structure, shown in Fig. 17.Q5, is to be hexagonal in plan on an 18 m diameter base, 3 m high at the eaves and 5 m at the crown. There are to be three braced bays with masonry walls. The roof is constructed from felt on timber on purlins with a ceiling. The loading details are as follows:

$$G_k = 2.5 \text{ kN/m}^2 \text{ on slope}$$

$$Q_k = 1.75 \text{ kN/m}^2 \text{ on plan}$$

Design the column and rafter sections for the three three-pinned arches shown in Fig. 17.Q5, use steel grade S275.

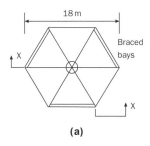

(a)

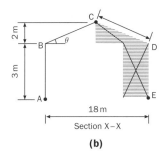

(b)

Fig. 17.Q5 Exercise 5: (a) plan; (b) section

6 In Question 5 above, replace the hinge at C with a rigid connection.

(i)   Recalculate the shear forces, bending moments and thrusts for the columns and the rafters. Use the formula for rigid frames; see the *Steel Designers' Manual*, 5th edition, The Steel Construction Institute, 1994 and any suitable computer software.

(ii) Design the column and rafter sections.

Comment on your results.

The previous chapter has been concerned mainly with the addition of direct and bending stresses when these two types of stress occur within a material, e.g. the variation of stress across the face of a column section which arises due to eccentricity of loading.

The principles involved in those cases are used in much the same way on occasions where, for example, a wall, resting on soil or on a concrete footing and acted upon by horizontal forces, is transmitting to the soil or footing stresses which consist of:

- direct stress from the wall's weight
- stress due to the overturning moment.

Before these resultant stresses are considered, it will be necessary to study the effect of the combined action of the vertical and horizontal (or inclined) forces on the overall behaviour of the wall.

As a result of that action the wall may fail in three ways:

- sliding
- overturning
- overstressing.

But since the main purpose of a retaining wall is to provide resistance to the horizontal (or inclined) forces caused by the retained material, the nature of the pressures these forces exert on the wall will be investigated first.

Horizontal forces – wind pressure

This is the simplest case, because wind pressure is assumed to be uniform. Therefore, the total resultant force acts at the centre of the area over which the pressure is applied and is given by

$$P = p \times \text{area newtons}$$

Or since, in the case of a wall, 1 m length of wall is generally considered, P is given by

$$P = p \times 1 \times H = p \times H \text{ newtons}$$

where p = unit wind pressure in N/m^2

H = the height of the part of the wall subject to that wind pressure (see Fig 18.1(a)) in meters

The determination of the unit wind pressure p is detailed in BS 6399–2: 1997 Part 2: Wind loads. It is based on several factors, including not only the basic speed of the wind and the topography in the locality of the wall, but also a

Fig. 18.1 Horizontal forces:
(a) wind pressure; (b) water
pressure (i) vertical retaining
wall, (ii) pressure variation;
(c) action of resultant force;
(d) action of resultant force
where the liquid does not reach
the top of the dam

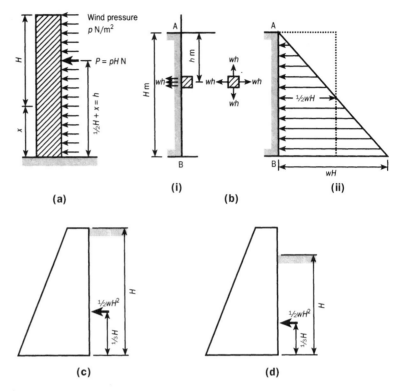

statistical factor which takes into account the probability of the basic wind
speed being exceeded within the projected life span of the wall.

Horizontal forces – liquid pressure

Consider the vertical surface AB, shown in Fig. 18.1(b)(i), to be the face of
a wall which is retaining a liquid. It can be shown that a cubic metre of
liquid, situated at a depth h metres below the surface, exerts a pressure of
$w \times h$ kN outwards on all its six side surfaces. w in this case is the equivalent density or the unit weight of the liquid in kN/m^3.

Thus the intensity of outward pressure varies directly with the depth and
will have a maximum value of $w \times H$ kN/m^2 at H m, the maximum depth
as indicated in Fig. 18.1(b)(ii).

At the surface of the liquid (where $h = 0$), the pressure will be zero. So,
as the maximum is wH kN/m^2, the average pressure between A and B is

$$\tfrac{1}{2}wH \text{ kN/m}^2$$

In dealing with retaining walls generally, it is convenient, as was said
earlier, to consider the forces acting on one metre length of wall, that is, an
area of wall H m high and 1 m measured perpendicular to the plane of the
diagrams (Fig. 18.1(b)). Thus, as the 'wetted area' concerned is H m^2, the
total force caused by water pressure on a one metre strip of wall is

$$\text{'wetted' area} \times \text{average rate of pressure} = H \times \tfrac{1}{2}wH$$
$$= \tfrac{1}{2}wH^2 \text{ kN}$$

This total resultant force on the wall's vertical surface from the liquid is (as will be seen from Fig. 18.1(b)(ii)) the resultant of a large number of forces, which range from zero at the top to wH at the base. The resultant will therefore act at a point $\frac{1}{3}H$ from the base, as shown in Fig. 18.1(c).

Note: If the liquid does not reach the top of the wall, as, for example, in Fig. 18.1(d), then the resultant force is calculated with H as the depth of the liquid and not as the height of the wall. The force is again $\frac{1}{2}wH^2$ and it acts at a point $\frac{1}{3}H$ (one-third the depth of the liquid) from the wall's base.

Example 18.1

A masonry dam retains water on its vertical face. The wall is, as shown in Fig. 18.2(a), 3.7 m, but the water level reaches only 0.7 m from the top of the wall. What is the resultant water pressure per metre run of wall?

Fig. 18.2 Example 18.1:
(a) vertical dam;
(b) non-vertical dam

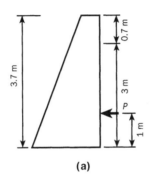
(a)

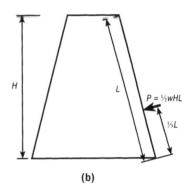
(b)

Solution

The equivalent density w of water $= 10$ kN/m³. Therefore

$$P = \tfrac{1}{2}wH^2 = \tfrac{1}{2} \times 10 \times 3 \times 3 = 45 \text{ kN}$$

acting at 1 m above base.

In cases where the wall in contact with the water is not vertical, as in Fig. 18.2(b), the wetted area will be larger than in the case of a vertical back, and the resultant pressure will thus be increased to $\frac{1}{2}wHL$ (i.e. the wetted area will be L m² instead of H m² considering one metre run of wall).

Horizontal forces – soil pressure

It is obvious that pressures on walls from retained soil or other granular materials cannot be determined with quite the same accuracy as with water. Soils vary in weight and character; they behave quite differently under varying conditions of moisture, etc., and, in general, the resultant pressures on vertical and non-vertical surfaces from soils are obtained from various soil pressure theories. Numerous theories exist for the calculation of soil pressures, and these theories vary in the assumptions which they make, and the estimated pressures which they determine. A great deal of research is still being carried out on this subject, but most of it is beyond the scope of a volume of this type, and therefore only the well-tried Rankine theory will be dealt with in detail.

Rankine's theory of soil pressure

It has been seen that a cubic metre of liquid at a depth h below the surface presses outwards horizontally by an amount wh kN/m^2 (w being the equivalent density of liquid). In the case of soil weighing w kN/m^3, the outward pressure at a depth of h m below the surface will be less than wh kN/m^2, since some of the soil is 'self- supporting'.

Consider, for example, the soil retained by the vertical face AB in Fig. 18.3(a). If the retaining face AB were removed, then some of the soil would probably collapse at once, and in the course of time the soil would assume a line BC, as shown. The angle ϕ made between the horizontal and the line BC varies with different types of soil, and is called the **angle of repose** or *the angle of internal friction* of the soil. It can be said, therefore, that only part of the soil was in fact being retained by the wall and was exerting pressure on the wall. Thus, it follows that the amount of pressure on the wall from the soil depends upon the angle of repose for the type of soil concerned, and Rankine's theory states in general terms that the outward pressure per square metre at a depth of h m due to a level fill of soil is

$$wh\left(\frac{1 - \sin\phi}{1 + \sin\phi}\right) \text{kN/m}^2$$

as compared with (wh) kN/m^2 in the case of liquids. Thus, by similar reasoning as used in the case of liquid pressure, the maximum pressure at the bottom of the wall is given by

$$\text{Maximum pressure} = wH\left(\frac{1 - \sin\phi}{1 + \sin\phi}\right) \text{kN/m}^2$$

$$\text{Average pressure} = \tfrac{1}{2}wH\left(\frac{1 - \sin\phi}{1 + \sin\phi}\right) \text{kN/m}^2$$

The soil acts at this average rate on an area of H m^2 of wall, so the total resultant force per metre run of wall is

$$P = \tfrac{1}{2}wH^2\left(\frac{1 - \sin\phi}{1 + \sin\phi}\right) \text{kN}$$

and this acts, as shown in Fig. 18.3(b), at $\tfrac{1}{3}H$ above the base of the wall.

Fig. 18.3 Rankine's theory of soil pressure: (a) angle of repose; (b) action of resultant force

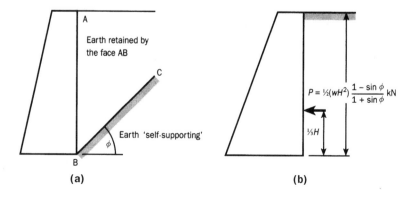

Earth retained by the face AB

C

Earth 'self-supporting'

ϕ

B

(a)

$P = \tfrac{1}{2}(wH^2)\dfrac{1 - \sin\phi}{1 + \sin\phi}$ kN

$\tfrac{1}{3}H$

(b)

Example 18.2

Soil weighing 15 kN/m^3 and having an angle of repose $\phi = 30°$ exerts pressure on a 4.5 m high vertical face. What is the resultant horizontal force per metre run of wall?

Solution

$\sin \phi = \sin 30° = 0.5$ and

$$P = \tfrac{1}{2}wH^2\left(\frac{1 - \sin \phi}{1 + \sin \phi}\right)$$

$$= \tfrac{1}{2} \times 15 \times 4.5 \times 4.5 \times \frac{1 - 0.5}{1 + 0.5}$$

$$= 151.875 \times \tfrac{1}{3} = 50.625 \text{ kN}$$

Modes of failure – sliding

It was stated at the beginning of the chapter that a retaining wall may fail by sliding. The possibility that the wall may slide along its base exists unless the weight of the wall is sufficient to prevent such movement.

The resistance to sliding depends upon the interaction of the weight of the wall and the friction between the material of the wall and the soil directly in contact with the base of the wall.

The basis of this interaction may be explained as follows. Consider a wooden block of weight W resting on a steel surface, as in Fig. 18.4(a). The steel surface presses upwards in reaction to the downward weight of the block, as shown.

Fig. 18.4 Failure by sliding: (a) normal loading; (b) addition of an angled force; (c) angle and coefficient of friction

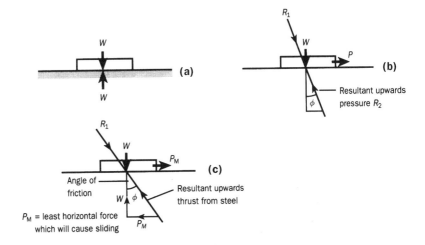

If, however, a small load P (not sufficient to move the block) is also placed on the block, as in Fig. 18.4(b), then the resultant of the forces W and P will be R_1, and the resultant upward pressure R_2 from the steel surface will also be inclined as shown. This resultant upward pressure will be inclined at an angle ϕ, as indicated.

If the force P is now gradually increased, the angle ϕ will also increase, until at a certain load (which depends on the nature of the two surfaces in contact and on the weight W) the block will move horizontally.

The angle ϕ which the resultant upward thrust makes with the vertical at the stage where the block starts to slide is known as the *angle of friction* between the two surfaces.

The tangent of ϕ is P_M/W (see Fig. 18.4(c)), so that

$$\tan \phi = \frac{\text{least force which will cause sliding}}{\text{weight of block}} \qquad (1)$$

and $\tan \phi$ is known as the *coefficient of friction* for the two materials and is denoted by the letter μ (mu).

For most materials this coefficient of friction will vary between 0.4 and 0.7, and it will be appreciated from equation (1) that

$$P_M = \text{the leastforce that will cause sliding}$$

$$= W \times \text{coefficient of friction } \mu$$

In the case of retaining walls, P_M, the force which would cause sliding, can be calculated as $W \times$ coefficient of friction, but the horizontal force P of the retained material should not exceed approximately half of the force P_M.

Example 18.3

The masonry dam, shown in Fig. 18.5, retains water to the full depth, as shown. The coefficient of friction between the base of the wall and the earth underneath is 0.7. Check if the wall is safe against sliding.

Solution

$$P = \text{actual horizontal pressure on side of wall}$$

$$= \tfrac{1}{2}wH^2 = \tfrac{1}{2} \times 10 \times 4^2 = 80 \text{ kN}$$

$$P_M = \text{horizontal force which would just cause sliding}$$

$$= 0.7 \times W = 0.7 \times \tfrac{1}{2}(1 + 3) \times 4 \times 18$$

$$= 0.7(144) = 100.8 \text{ kN}$$

The actual pressure (80 kN) exceeds half the value of P_M, and the factor of safety against sliding is

$$\frac{\mu W}{P} = \frac{100.8}{80} = 1.26 < 2$$

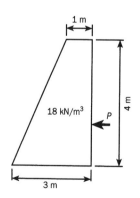

Fig. 18.5 Example 18.3

which is undesirable.

Modes of failure – overturning

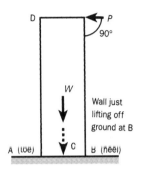

Fig. 18.6 Failure by overturning

A retaining wall may have quite a satisfactory resistance to sliding, but the positive action of the horizontal forces may tend to overturn it about its toe (Fig. 18.6).

Equilibrium will be upset when the wall just rises off the ground at the heel, B, the turning point being the toe, A. (It is assumed that sliding of the wall will not occur.) When the wall is just lifting off the ground, the overturning moment due to the force P is just balanced by the restoring or balancing moment due to the weight of the wall, the wall being balanced on the edge A. Since the inclination of AB to the horizontal is very slight, distances can be taken assuming the wall is vertical.

Taking moments about A,

$$P \times \text{distance AD} = W \times \text{distance AC}$$

or

$$\text{Overturning moment} = \text{restoring (balancing) moment}$$

In practice it is usual to apply a factor of safety of 2, where

$$\text{Factor of safety against overturning} = \frac{\text{restoring (balancing) moment}}{\text{overturning moment}}$$

and, therefore,

$$\text{Overturning moment} = \tfrac{1}{2} \times \text{restoring (balancing) moment}$$

Example 18.4

A long boundary wall 2.7 m high and 0.4 m thick is constructed of brickwork weighing 18 kN/m^3 (Fig. 18.7(a)). If the maximum wind pressure uniformly distributed over the whole height of the wall is 500 N/m^2, calculate the factor of safety against overturning, neglecting any small adhesive strength between the brickwork and its base.

Solution

$$\text{Weight of 1 m run of wall} = 2.7 \times 1.0 \times 0.4 \times 18$$

$$= 19.44 \, \text{kN}$$

Fig. 18.7 Example 18.4

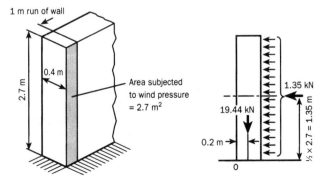

$$\text{Wind pressure on 1 m run of wall} = 2.7 \times 1.0 \times 0.5$$
$$= 1.35 \text{ kN}$$

and this can be taken as acting at the centre of height of the wall for the purposes of taking moments about O (Fig. 18.7(b)).

$$\text{Restoring moment} = 19.44 \times \tfrac{1}{2} \times 0.4$$
$$= 3.888 \text{ kN m}$$

$$\text{Overrunning moment} = 1.35 \times \tfrac{1}{2} \times 2.7$$
$$= 1.823 \text{ kN m}$$

Therefore

$$\text{FS against overturning} = 3.888/1.823 = 2.13$$

It will be found that a satisfactory factor of safety against overturning is achieved if the resultant of the horizontal and vertical forces crosses the base of the wall within its 'middle third', i.e. when no tensile stresses are allowed to develop in the wall.

This comes within the scope of the third mode of failure of retaining walls and is the topic of the following section of this chapter.

Modes of failure – overstressing

When the weight of a wall per metre, W, and the resultant pressure from the soil or liquid, P, have been calculated, these two forces may be compounded to a resultant, as in Fig. 18.8(a). It will be shown that the position along the base at which this resultant cuts (i.e. at S) has an important bearing on the stability of the wall and on the pressures exerted by the wall upon the earth beneath.

Consider now the wall shown in Fig. 18.8(b)(i). The values in kN of W and P are as shown, and W acts through the centroid of the wall section.

From the similar triangles ASC and ADE

$$\frac{y}{1.4} = \frac{10}{28} \quad \text{so} \quad y = \frac{10 \times 1.4}{28} = 0.5 \text{ m}$$

Thus the resultant force cuts the base at $1.0 + 0.5 = 1.5$ m from the point G, the heel of the wall.

It can be seen that the resultant force cuts the base at point S, as shown in Fig. 18.8(b)(i) and (ii).

In considering the effect of this resultant on the soil or concrete under the base FG, it is normally convenient to resolve this force into the vertical and horizontal components W and P from which this resultant was compounded. (Note that the vertical component of the resultant is equal to the weight W of the wall only when the soil pressure is horizontal.)

These two component forces, W and P, are shown again in Fig. 18.8(b)(ii). Component force P tends to *slide* the wall along the plane FG, so need not be considered when calculating the pressure on the soil beneath

Fig. 18.8 Failure by overstressing: (a) resultant of weight and pressure; (b)(i) loading diagram, (ii) resultant, (iii) stress or pressure distribution

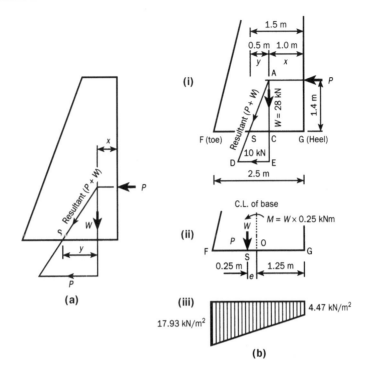

FG. Component force W, however, does exert pressure on the soil or concrete under FG, and as the point S is to the left of the centre line of the base, the soil under the left-hand part of the base will have greater pressure exerted upon it than the soil under the right-hand half. Here

$$\text{Direct stress} = \frac{W}{A} = \frac{\text{weight of wall}}{\text{area of base}}$$

$$= \frac{28}{2.5 \times 1.0} = 11.2 \text{ kN/m}^2$$

The moment due to the eccentricity of the resultant is

$W \times$ distance of S from the centre of the base

$$W \times 0.25 = 28 \times 0.25 = 7 \text{ kN m}$$

and the section modulus about an axis through the centre line of the base is

$$Z = \tfrac{1}{6}bd^2 = \tfrac{1}{6} \times 1 \times (2.5)^2 = 1.04 \text{ m}^2$$

$$\text{Bending stress} = \frac{M}{Z} = \frac{7}{1.04} = 6.73 \text{ kN/m}^2$$

Thus, the pressure under wall at F is

$$\frac{W}{A} + \frac{M}{Z} = 11.20 + 6.73 = 17.93 \text{ kN/m}^2$$

and the pressure under wall at G is

$$\frac{W}{A} - \frac{M}{Z} = 11.20 - 6.73 = 4.47 \text{ kN/m}^2$$

as shown in the stress or pressure distribution diagram of Fig. 18.8(b)(iii).

Example 18.5

A masonry wall is shown in Fig. 18.9(a) and weighs 20 kN/m³. It retains on its vertical face water weighing 10 kN/m³. The water reaches the top of the wall. Calculate the pressures under the wall at the heel and the toe.

Solution

$$W = \tfrac{1}{2}(1.0 + 3.0) \times 4.5 \times 20 = 180.0 \text{ kN}$$

$$P = \tfrac{1}{2}wH^2 = \tfrac{1}{2} \times 10 \times (4.5)^2 = 101.25 \text{ kN}$$

To determine the centroid of the wall,

$$x = \frac{1.0 \times 4.5 \times 0.5 + \tfrac{1}{2} \times 2.0 \times 4.5 \times 1.7}{1.0 \times 4.5 + \tfrac{1}{2} \times 2.0 \times 4.5}$$

$$= \frac{2.25 + 7.50}{4.50 + 4.50} = 1.08 \text{ m}$$

By similar triangles

$$\frac{y}{1.5} = \frac{101.25}{180.0}$$

$$y = \frac{101.25 \times 1.5}{180.0} = 0.84 \text{ m}$$

Fig. 18.9 Example 18.5: (a) loading diagram; (b) the equilibrium about the base's centroid

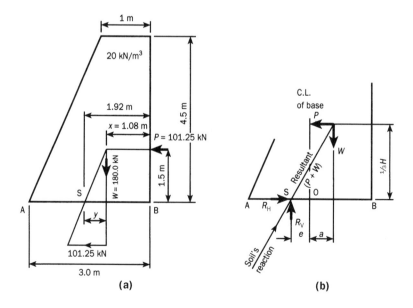

Thus, the resultant force cuts at $1.08 + 0.84 = 1.92$ m from B at S or 0.42 m to the centre line of the base.

Hence, the pressure at A is

$$\frac{W}{A} + \frac{M}{Z} = \frac{180.0}{3.0 \times 1.0} + \frac{180.0 \times 0.42 \times 6}{1.0 \times 3.0 \times 3.0}$$

$$= 60.0 + 50.4 = +110.4 \, \text{kN/m}^2$$

and the pressure at B is

$$\frac{W}{A} + \frac{M}{Z} = 60.0 - 50.4 = +9.6 \, \text{kN/m}^2$$

The purpose of calculating distance y was to find the eccentricity of W, i.e. distance e, so that $M = We$ could be determined. Distance e, however, can be obtained directly by considering the equilibrium about the centroid of the base of all the forces acting on the wall. This is illustrated in Fig. 18.9(b).

Let R_H and R_V be, respectively, the horizontal and vertical components of the soil's reaction acting at the intersection of the resultant of P and W with the base of the wall.

Then, for equilibrium, $R_H = P$, and $R_V = W$, both of which have been calculated previously.

Hence, taking moments about O, the centre of area (or centroid) of the base:

$$W \times a - P \times \tfrac{1}{3}H + W \times e = 0$$

$$e = \frac{P \times \tfrac{1}{3}H - W \times a}{W}$$

$$= \frac{101.25 \times 1.5 - 180.0 \times (1.50 - 1.08)}{180.0}$$

$$= 0.42 \, \text{m as before}$$

Example 18.6

The trapezoidal retaining wall shown in Fig. 18.10(a) weighs 22 kN/m³ and retains on its vertical face soil with an equivalent density of 16 kN/m³ and an angle of repose (or internal friction) ϕ of 30°. The retained soil carries a superimposed vertical load of 9.6 kN/m². Determine the pressure under the base AB.

Solution

The imposed load is converted to an equivalent additional height of soil as shown in Fig. 18.10(b)(i), where

$$h_1 = \frac{\text{intensity of imposed load (kN/m}^2\text{)}}{\text{equivalent density of soil (kN/m}^3\text{)}}$$

$$= \frac{9.6}{16.0} = 0.6 \, \text{m}$$

Fig. 18.10 Example 18.6:
(a) loading of a trapezoidal
retaining wall; (b)(i) equivalent
additional height of soil,
(ii) determining the eccentricity,
(iii) pressure distribution
diagram

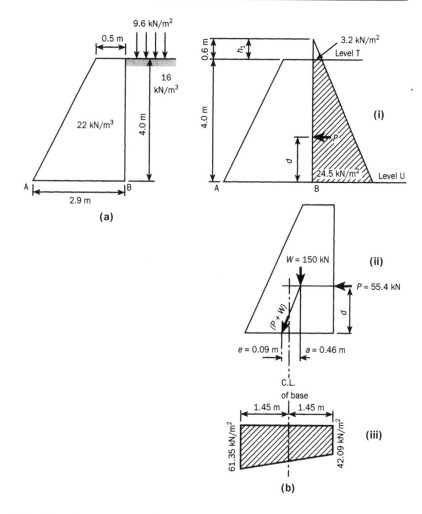

Therefore, the pressure at the top of the wall, level T, is

$$w \times h_1 \times \frac{1 - \sin\phi}{1 + \sin\phi} = 16 \times 0.6 \times \tfrac{1}{3} = 3.2 \ \text{kN/m}^2$$

and that on the underside of the wall, at level U, is

$$w \times (h_1 + H) \times \frac{1 - \sin\phi}{1 + \sin\phi} = 16 \times 4.6 \times \tfrac{1}{3} = 24.5 \ \text{kN/m}^2$$

The force P, acting on the vertical face of the wall, is found from the shaded trapezoid of pressure and it acts at the centroid of the trapezoid, i.e.

$$P = \tfrac{1}{2}(3.2 + 24.5) \times 4 = 55.4 \ \text{kN}$$

The distance d between the base of the wall and the line of action of force P is

$$d = \tfrac{1}{55.4}\left[3.2 \times 4 \times 2 + (24.5 - 3.2) \times \tfrac{4}{2} \times \tfrac{4}{3}\right]$$

$$= \frac{25.6 + 56.8}{55.4} = 1.49 \text{ m}$$

Now, the vertical force W due to the weight of the wall is

$$W = \tfrac{1}{2}(2.9 + 0.5) \times 4 \times 22 = 150 \text{ kN}$$

To find distance a between the line of action of force W and the centre line of the base, take moments about that centre line, then

$$a = \frac{0.5 \times 4 \times (1.45 - 0.25) + 2.4 \times \tfrac{1}{2} \times 4 \times (1.45 - \tfrac{1}{3} \times 2.4 - 0.5)}{0.5 \times 4 + 2.4 \times 2}$$

$$= \frac{2.4 + 0.72}{2.0 + 4.8} = \frac{3.12}{6.80}$$

$$= 0.46 \text{ m}$$

Finally, to determine the eccentricity, e, consider the moments of *all* the forces (i.e. the applied loads, P and W, and the reaction of the soil, R) about the centre line of the base, as demonstrated above. Therefore,

$$e = \frac{150 \times 0.46 - 55.4 \times 1.49}{150} = \frac{-13.5}{150}$$

$$= -0.09 \text{ m (i.e. to the left of centre line. Fig. 18.10(b)(ii))}$$

Thus, pressure at A is

$$\frac{W}{A} + \frac{We}{Z} = \frac{150}{1.0 \times 2.9} + \frac{150 \times 0.09 \times 6}{1.0 \times 2.9 \times 2.9}$$

$$= 51.72 + 9.63 = 61.35 \text{ kN/m}^2$$

and pressure at B is

$$51.72 - 9.63 = 42.09 \text{ kN/m}^2$$

The pressure distribution diagram is shown in Fig. 18.10(b)(iii).

The previous examples have been chosen so that the resultant force has cut the base within the middle third. Hence, although the pressures at B and A (the heel and toe of the base) have been of different amounts, the stresses have been compressive at both points. Consider, however, the case shown in Fig. 18.11(a)(i). The resultant here cuts outside the middle third and thus there will be a tendency to tension at B, i.e. M/Z will be greater than W/A.

If the joint at B is capable of resisting tensile stress then the stress at A will be

$$\frac{W}{A} + \frac{M}{Z} = \frac{42}{1.0 \times 3.0} + \frac{42 \times (1.5 - 0.75) \times 6}{1.0 \times 3.0 \times 3.0}$$

$$= 14.0 + 21.0 = 35 \text{ kN/m}^2$$

Fig. 18.11 Overstressing where the resultant does not cut within the middle third: (a) resultant cuts outside middle third; (b) joint B cannot resist tension; (i) loading diagrams, (ii) pressure distribution diagrams

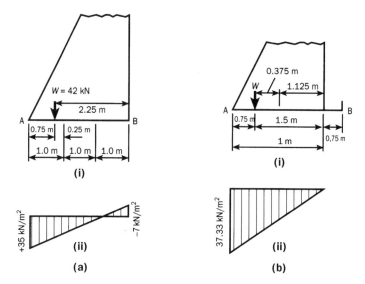

and the stress at B will be

$$\frac{W}{A} - \frac{M}{Z} = 14.0 - 21.0$$

$$= -7.0 \text{ kN/m}^2 \quad \text{or} \quad 7.0 \text{ kN/m}^2 \text{ tension}$$

The diagram of pressure distribution under the base will be as in Fig. 18.11(a)(ii).

If the joint at B is not capable of resisting tension, then there will be a tendency for the base to lift where tension would occur. In this case, the point at which the resultant cuts the base should be considered the middle-third point, as in Fig. 18.11(b)(i), so that the effective width of the base becomes three times the distance AS, i.e.

$$3 \times 0.75 = 2.25 \text{ m}$$

The stress at A (the toe) then becomes

$$\frac{W}{A} + \frac{M}{Z} = \frac{42}{1.0 \times 2.25} + \frac{42 \times 0.375 \times 6}{1.0 \times 2.25 \times 2.25}$$

and the pressure distribution diagram is as shown in Fig. 18.11(b)(ii).

Example 18.7

For the retaining wall shown in Fig. 18.12 determine:

- The position of the intersection point of the resultant of forces P and W with the base AB when soil is level with top of wall.
- The reduced height of the soil which will cause the point of intersection to coincide with the edge of the middle third of the base.

Assume:

$$w \text{ for wall} = 20 \text{ kN/m}^3$$

$$w \text{ for soil} = 15 \text{ kN/m}^3$$

$$\phi \text{ for soil} = 30°$$

Solution

$$W = \tfrac{1}{2}(1.0 + 2.5) \times 6 \times 20 = 210 \text{ kN}$$

$$P = \tfrac{1}{2}(15 + 6^2) \times \tfrac{1}{3} = 90 \text{ kN}$$

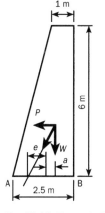

Fig. 18.12 Example 18.7

To determine the values of a and e take moments about centre line of base:

$$a = \frac{90 \times \tfrac{1}{3} \times 6.0 \times 0.75 - 1.5 \times \tfrac{1}{2} \times 6.0 \times 0.25}{\tfrac{1}{2}(1.0 + 2.5) \times 6.0}$$

$$= \frac{4.5 - 1.125}{10.5} = 0.32 \text{ m to right of centre}$$

$$e = \frac{90 \times \tfrac{1}{3} \times 6.0 - 210 \times 0.32}{210} = \frac{180 - 67.5}{210}$$

$$= 0.54 \text{ m} > 2.5/6, \text{ i.e. tension will develop}$$

Now to find the reduced height of the soil:

$$\tfrac{1}{2}(15 \times H^2) \times \tfrac{1}{3}H \times \tfrac{1}{3} - 210 \times (0.32 + \tfrac{1}{6} \times 2.5) = 0$$

$$H^3 = \frac{(67.5 + 87.5) \times 18}{15} = 186$$

$$H = 5.71 \text{ m}$$

The above example may be used quite conveniently to verify the final statement in the previous section on overturning. Check the stability of the above wall in both cases.

Moments are taken about the toe of the wall at A.

Soil level with the top

$$\frac{\text{Restoring moment}}{\text{Overturning moment}} = \frac{210 \times (1.25 + 0.32)}{180}$$

$$= 1.83 < 2 \text{ unsatisfactory}$$

Soil height reduced

$$\frac{\text{Restoring moment}}{\text{Overturning moment}} = \frac{210 \times 1.57 \times 18}{15 \times (5.71)^3}$$

$$= 2.13 \text{ satisfactory}$$

Finally, it should be noted that the main point of calculating the pressures under the base of the wall is to ensure that the soil on which the wall is founded is not overstressed. It is, therefore, essential to keep the actual pressures within the limits of the safe bearing capacities of the soil.

The safe bearing capacities vary over a very wide range of values from, say, 50 kN/m² for made-up ground to 650 kN/m² for compact, well-graded sands or stiff boulder clays. The values for rocks are, of course, higher.

Summary

The three *modes of failure* of a retaining wall are:

* sliding
* overturning
* overstressing

The *horizontal force*, P, acting on 1 m length of the vertical face of a retaining wall is

* wind: $p \times H$
* liquid: $\frac{1}{2}wH^2$

* soil (or granular material), using Rankine's formula

$$\frac{1}{2}wH^2 \times \frac{1 - \sin \phi}{1 + \sin \phi}$$

where p = unit wind pressure in N/m²

w = equivalent density (or unit weight) of retained material in kN/m³

ϕ = angle of repose (or internal friction) of soil (or granular material)

H = height (or depth) of retained material (wind included) in metres

Resistance to *sliding*

$$\frac{W \times \mu}{P_{\mathrm{M}}} = \text{factor of safety (usually 2)}$$

where W = weight of wall in kN

μ = coefficient of friction between wall and supporting soil (0.4–0.7)

P_{M} = the least force that will cause sliding in kN

Resistance to *overturning*

$$\frac{\text{Restoring (balancing) moment}}{\text{Overturning moment}} = \text{factor of safety (usually 2)}$$

(Moments are taken about an outer edge – toe or heel – of the base of the wall.)

Resistance to *overstressing*

The maximum stress resulting from the combination of direct and bending stresses must be kept within limits of the safe bearing capacity of the soil supporting the wall.

For 'no tension', the resultant of the applied loads P and W must cross the base wit hin its middle third or

$$e \not> \tfrac{1}{6} \times \text{width of base}$$

Exercises

1 Calculate the resultant horizontal force P acting on the wall shown in Fig. 18.Q1 for (a) a liquid of equivalent density $w = 10$ kN/m³; (b) a soil of equivalent density $w = 16$ kN/m³ and angle of repose $\phi = 30°$, given that $H =$ (i) 2 m, (ii) 3 m, (iii) 7 m.

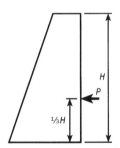

Fig. 18.Q1

2 The walls shown in Fig. 18.Q2(a) and (b) weigh 22 kN/m³. Calculate the value of P in kN per m run of wall, if there is to be a factor of safety of 2 against sliding; $\mu = 0.6$.

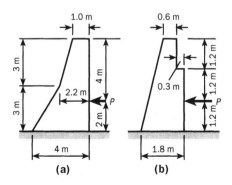

Fig. 18.Q2

3 The horizontal forces on the walls shown in Fig. 18.Q3(a), (b) and (c) represent the forces on a 1 m run of wall. (All dimensions are in m.) All the walls weigh 22 kN/m³. Calculate the factor of safety against overturning of the walls. (Only the position of the vertical line in which the c.g. is situated need be calculated.)

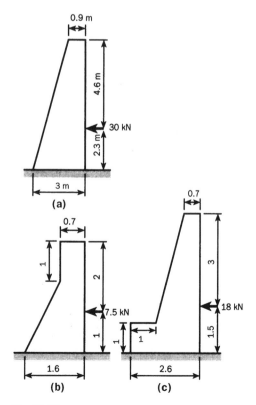

Fig. 18.Q3

4   The wall, shown in Fig. 18.Q4 has a resultant vertical force of 57 kN/m length, due to the weight of the wall W and the pressure P. Calculate the pressure at the toe (point A).

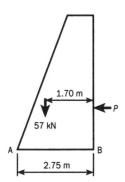

Fig. 18.Q4

5   A masonry retaining wall is trapezoidal, being 4 m high, 1.4 m wide at the top and 2.0 m at the base. The masonry weighs 20 kN/m^3 and the wall retains on its vertical face earth weighing 15 kN/m^3 at an angle of repose of 40°. Calculate the maximum pressure under the wall.

6   A brick pier H m high is 450 mm $\times$ 300 mm in section and weighs 18 kN/m^3. It has a uniform wind pressure of 720 N/m^2 on one side, as shown in Fig. 18.Q6 and rests on a concrete block to which it is not connected. How high is the wall if the resultant force cuts through the point A?

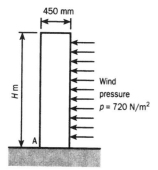

Fig. 18.Q6

7   A concrete wall of trapezoidal section, 4.8 m high, has a top width of 1.0 m and a base width of 2.5 m with one face vertical and a uniform slope on the other face. There is water pressure on the vertical face, with top water level 1 m below the top of the wall. If the water weighs 10 kN/m^3 and the masonry 24 kN/m^3, calculate the maximum and minimum pressures on a horizontal plane 2 m above the base of the wall.

8   A brick wall 450 m thick and 3.0 m high is built on a level solid concrete base slab. The wall weighs 20 kN/m^3. Assuming no tension at the bed joint, calculate the intensity of the uniformly distributed horizontal wind force in N/m^2 over the full height of the wall that would just cause overturning of the wall.

9   A concrete retaining wall is as shown in Fig. 18.Q9. Using the Rankine formula, calculate the pressure under the wall at A and B. Assume the soil to weigh 16 kN/m^3, and the natural angle of repose of the soil to be 30°. The concrete weighs 24 kN/m^3.

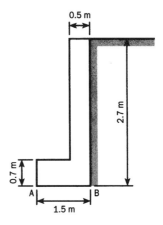

Fig. 18.Q9

10   Figure 18.Q10 shows the section of a mass concrete retaining wall, weighing 24 kN/m^3. The position and amount of the resultant horizontal force are as shown. Calculate the vertical pressures under the wall at A and B, assuming (a) tension stresses are permitted; (b) no tension stresses can develop.

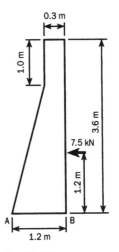

Fig. 18.Q10

11 A small mass concrete retaining wall, 2 m high, is trapezoidal in section, 0.3 m wide at the top and 1.2 m wide at the base. The wall weighs 24 kN/m³ and it retains on its vertical face soil weighing 14 kN/m³ at an angle of repose of 30°. Calculate the pressure under the base of the wall at the heel and at the toe.

12 A wall 0.23 m thick, weighing 19 kN/m³, rests on a solid foundation. It has a uniform horizontal pressure of 0.7 kN/m² on one face. If the resultant force at the base just cuts at the edge of the wall (i.e. if overturning is just about to take place), how high is the wall?

13 How high will be a wall, similar to that mentioned in Question 12, if the resultant pressure at the base cuts at the middle third, i.e. if tension is about to occur?

14 An L-shaped retaining wall is shown in Fig. 18.Q14. The masonry weighs 22 kN/m³. Calculate the max-

imum and minimum pressures under the wall in kN/m².

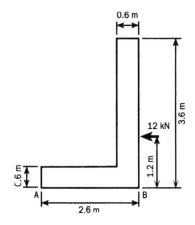

Fig. 18.Q14

Computer software

At the present there is much computer software available for structural analysis and design. Some of it is provided with integrated word processing, spreadsheets and mathematical calculations which enable users to produce calculation sheets, graphs and sketches of the type and format that meet their requirements.

Sketches are sometimes linked to equations and spreadsheet cells. Some of the software can even carry out optimisation techniques on a certain element of a structure or on the structure as a whole. They are easy to use and powerful in two-dimensional or three-dimensional analysis. The reader is encouraged to use this new technology for 'making the best of valuable talent'. Possessing such a skill is important for continuous and professional development. The following is a list of new education packages supplied by EdSoft Ltd. You are also encouraged to see their web site at www.Reel.co.uk.

QSE – Space 2/3D analysis
Staad – Pro (FE, 2nd order, seismic + other advanced feature with QSE interface)
QSE Steel Designer BS 5950+EC3
QSE – Concrete Designer BS 8110 + CE2
QSE – Timber Designer BS 5628 + EC5
QSE – Section Wizard (all properties, stress)

Answers to exercises

Chapter 2

2   Joist, 5.74 kN; Beam A, 51.92 kN; Beam B, 21.63 kN

3   20.4 kN/m = 163.208 kN

4   A1–A2 roof: 16.4 kN, parapet cavity wall 0.5 m high
    and steel 0.5 kN/m
    A1–A2 floor: 30.8 kN, assumed that 0.5 kN/m is the
    self-weight of the external beam
    Lower length 1-4-7: 671.32 kN

Chapter 3

1   (a) 51.8 N, 73.2 N; (b) 183.0 N, 224.2 N; (c) 577.4 N, 288.7 N

2   14.5°

3   L = 0.87 kN, R = 0.50 kN, Reaction = 1.36 kN

4   L = 2.04 kN, R = 0.75 kN, W = 2.88 kN, M = 1.06 kN

5   19.9 kN at 31° to the vertical

6   7.3 kN at 25° to the vertical

7   9.6 kN at 81° to the vertical

8   A = 32°, resultant = 17.2 kN

9   7.3 kN at 70° to the vertical

10 559 N at 23° to the vertical

11 677 N at 17° to the vertical

12 774 N at 64°, 554 N at 22.5°, 793 N at 37.5° to the vertical

13 (a) 35 kN, 40 kN; (b) 8 kN, 8 kN; (c) 17 kN, 10 kN

14 9.2 kN, x = 22.5°, 8.5 kN, 3.5 kN

15 L = 2.5 kN, L_v = 1.3 kN, L_h = 2.2 kN; R = 4.3 kN, R_v = 3.7 kN, R_h = 2.2 kN

16 5.5 kN, 6.7 kN, 4.8 kN, R_L = 4.8 kN, R_R = 2.8 kN

17 $T = S$ = 5 kN, $A_v = B_v$ = 4.3 kN, $A_h = B_h$ = 2.5 kN

18 AB = 173 N, BC = 100 N, A_v = 150 N, A_h = 87 N

19 5.0 kN, 6.1 kN, 6.8 kN

20 82.5°, 6.1 kN

21 2.9 kN

22 A = 9.0 kN, B = 7.3 kN

23 0.29 kN

24 4.0 kN in inclined members, 2.3 kN in horizontal member and in cable, vertical reactions = 2 kN

25 BC = CD = 3.0 kN, BD = 3.5 kN, AB = AD = 1.7 kN, R_A = 3.0 kN

26 (a) X = 64.6 N down, Y = 82.0 N up; (b) X = 4.6 N down, Y = 22.0 N up; (c) X = 1.6 kN up, Y = 3.2 kN down; (d) X = 0.4 kN down, Y = 1.2 kN down; (e) 38.6 kN at 43° to vertical; (f) 41.4 kN at 40° to the vertical

27 13.7 kN, 11.8 kN.

Chapter 4

1   A = 5.30 kN, B = 3.95 kN

2   R_A = 17.3 kN, R_B = 20.0 kN, AC = 20.0 kN, BC = 17.3 kN, AB = 10.0 kN, R_{Bv} = 10.0 kN, R_{Bh} = 17.3 kN

3   Top = 385 N, bottom = 631 N (385 N horizontal, 500 N vertical)

4   Cable 1.9 kN, hinge 1.9 kN

5   Cable 1.4 kN, hinge 1.4 kN

6   A = 0.50 kN, B = 0.87 kN

7   A = 0.67 kN, B = 1.20 kN (1.00 kN vertical, 0.67 kN horizontal)

8 String 1000 N, hinge 900 N

9 Rope 6.1 kN, hinge 6.0 kN

10 500 N, 500 N

11 Cable 1.15 kN, hinge 1.53 kN

12 Rope 14.1 kN, hinge 22.4 kN

13 Rope 3.5 kN, hinge 7.9 kN

14 3.87 kN, 8.24 kN, 8.87 kN

15 A = 7.64 kN, B = 5.77 kN, AB = 10.0 kN, AC = 2.75 kN, BC = 5.77 kN

16 x = 20°

17 354 N, 250 N

18 A = 334 N, B = 213 N

19 A = 315 N, B = 192 N

20 A = 0.37 kN, B = 0.78 kN

21 433 N, 578 N, 807 N

22 4.72 kN, 4.25 kN, 2.51 kN

23 A = 6.3 kN, B = 2.9 kN

24 (a) 62.5 N at 70° to the vertical, 100 mm from bottom r.h. corner; (b) 60.0 N at 65° to the vertical, 600 mm from bottom l.h. corner; (c) 75.0 N at 48° to the vertical, 1.1 m from bottom l.h. corner

25 3.8 kN at 72° to the vertical, 1.7 m from bottom

26 6.0 kN, 6.0 m from bottom

27 38.5 kN, 41.0 kN at 32.5° to the vertical

28 60.0 kN, 118.6 kN at 37.5° to the vertical

29 X = 11.4 kN, Y = 6.5 kN at 6.6° to the vertical

30 50.0 kN, 22.4 kN at 63.4° to the vertical

31 55.0 kN, 65.0 kN

32 4.4 kN, 11.5 kN at 41° to the vertical

33 R = 5.7 kN at 41° to the vertical, L = 9.4 kN at 23° to the vertical

34 8.2 kN, 16.8 kN at 80° to the vertical

35 11.6 kN at 45° to the vertical, 8.7 kN at 71° to the vertical

36 43.3 kN, 49.8 kN at 20.5° to the vertical

37 R = 44.2 kN at 11.0° to the vertical, L = 47.5 kN at 10.5° to the vertical

38 13.2 kN, 15.4 kN at 27° to the vertical

39 R = 40.5 kN at 7° to the vertical, L = 57.0 kN at 5° to the vertical.

Chapter 5

1 110 N, 205 N

2 1.5 m

3 375 N, 225 N (upwards)

4 35 mm, 265 N

5 0.24 m

6 1.64 kN, 1.06 kN

7 4.23 kN, 1.53 kN (downwards)

8 W = 22.5 N, A = 425.0 N, B = 237.5 N

9 See Answers to Exercise 3

10 1.44 kN, 2.36 kN at 17.7° to the vertical

11 2.5 kN, 1.8 kN at 65.3° to the vertical

12 30.6 kN, 34.2 kN, 25.0 kN, 43.3 kN

13 6.0 kN, 5.34 kN downwards at 76.5° to the vertical

14 3.9 kN, 6.9 kN

15 (a) 0.41 m from hinge, 70 N, 176 N at 20° to the vertical; (b) 2.41 m from hinge, 4.12 kN, 6.10 kN at 42.5° to the vertical; (c) 2.27 m from hinge, 3.0 kN, 5.78 kN at 21.5° to the vertical

16 (a) 2.33 m from hinge, 3.5 kN, 5.7 kN at 38° to the vertical; (b) 1.8 m from the hinge, 30 kN, 39 kN at 40° to the vertical; (c) 3.17 m from the hinge, 15.8 kN, 21.8 kN at 46.5° to the vertical; (d) 3.0 m from the hinge, 10.0 kN, 22.4 kN at 27° to the vertical

17 15.0 kN, 9.92 kN at 41° to the vertical, 6.50 kN

18 300 mm, 45 000 N mm

19 (a) 50 kN, 70 kN; (b) 110 kN, 130 kN; (c) 120 kN, 10 kN; (d) 55 kN, 245 kN; (e) 70 kN, 150 kN

20 (a) 225 kN, 135 kN; (b) 199 kN, 249 kN; (c) 324 kN, 108 kN; (d) 172.8 kN, 115.2 kN; (e) 221.5 kN, 60.5 kN; (f) 242.8 kN, 107.2 kN; (g) 523.1 kN, 231.9 kN

21 A = 40 kN, B = 50 kN, C = 40 kN, D = 10 kN, E = 26.7 kN, F = 13.3 kN

22 14.07 kN

Chapter 6

For solutions to Questions 1 to 31 see Figs A.1–A.31. All forces are in kN; ' + ' indicates compression and ' − ' indicates tension.

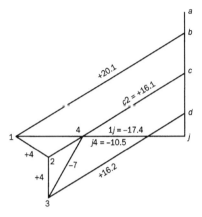

Fig. A.1

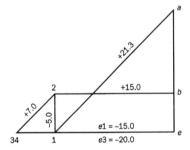

Fig. A.2

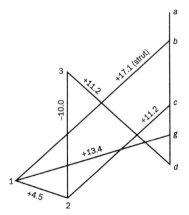

Fig. A.3

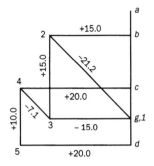

Fig. A.4

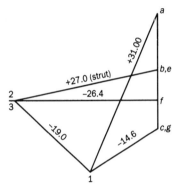

Fig. A.5

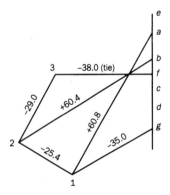

Fig. A.6

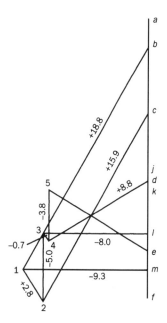

Fig. A.7

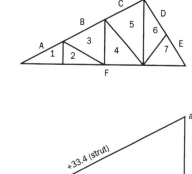

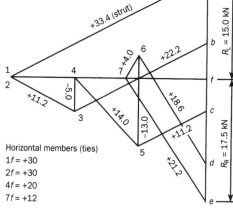

Horizontal members (ties)
1f = +30
2f = +30
4f = +20
7f = +12

Fig. A.8

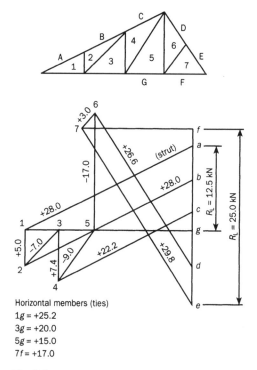

Horizontal members (ties)
1g = +25.2
3g = +20.0
5g = +15.0
7f = +17.0

Fig. A.9

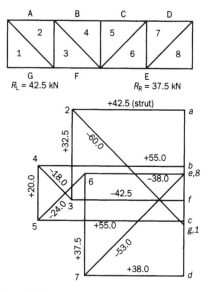

Fig. A.10

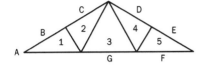

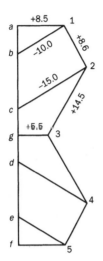

Fig. A.11

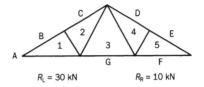

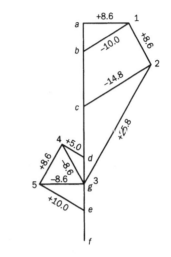

Fig. A.13

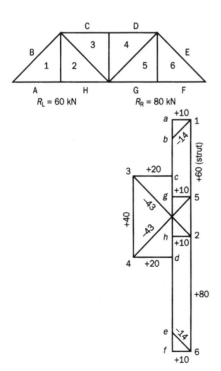

Fig. A.12

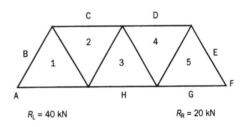

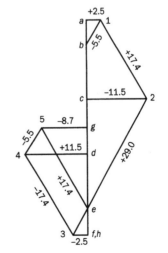

Fig. A.14

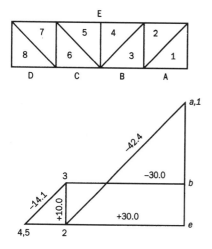

Fig. A.15

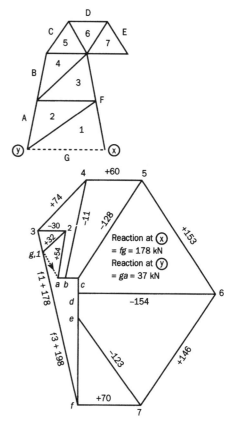

Fig. A.17

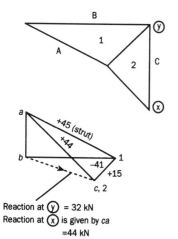

Reaction at ⓨ = 32 kN
Reaction at ⓧ is given by *ca*
=44 kN

Fig. A.16

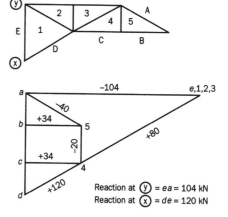

Reaction at ⓨ = *ea* = 104 kN
Reaction at ⓧ = *de* = 120 kN

Fig. A.18

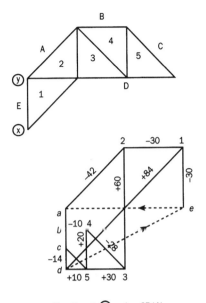

Reaction at (x) = de = 67 kN
Reaction at (y) = ea = 60 kN

Fig. A.19

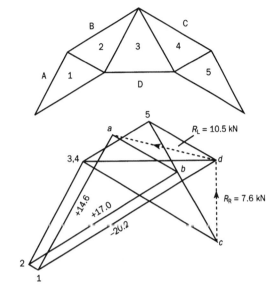

Fig. A.21

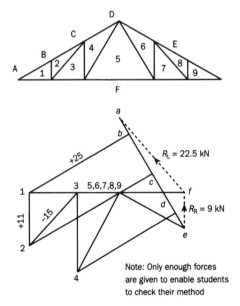

Note: Only enough forces
are given to enable students
to check their method

Fig. A.20

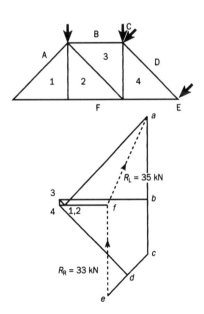

Fig. A.22

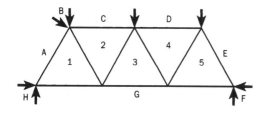

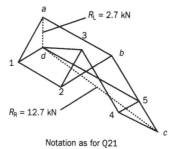

Notation as for Q21

Fig. A.25

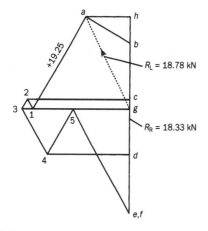

Fig. A.23

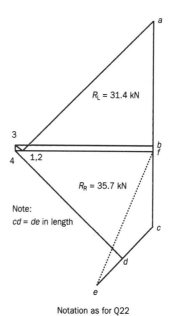

Note:
cd = de in length

Notation as for Q22

Fig. A.26

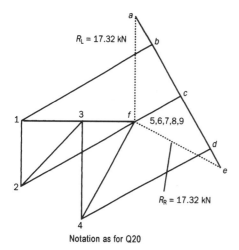

Notation as for Q20

Fig. A.24

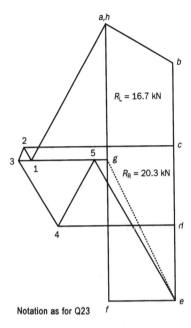

Notation as for Q23

Fig. A.27

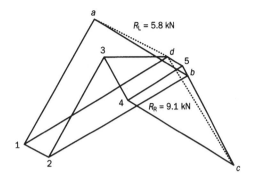

Notation as for Q21

Fig. A.29

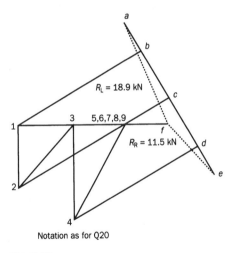

Notation as for Q20

Fig. A.28

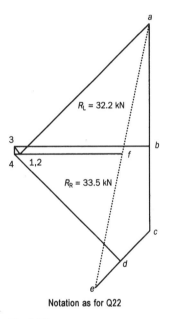

Notation as for Q22

Fig. A.30

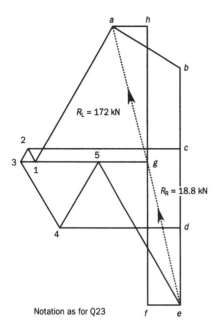

Notation as for Q23

Fig. A.31

32 AB = 2.39 kN
    AC = 0.98 kN } all struts
    AD = 2.25 kN

33 AB = 16.62 kN (strut)
    AC = 9.82 kN (tie)
    AD = 7.25 kN (tie)

34 AB = AC = 0.7 kN (tie)
    AD = AF = 2.43 kN (struts)
        AE = 3.33 kN (tie)
        BC = 0.5 kN (strut)
    BF = DC = zero

Chapter 7

1   135 N/mm²

2   69.75 kN

3   8.9 mm

4   122.2 N/mm²

5   157.8 kN

6   18.5 mm

7   80 kN, 26 mm

8   55 mm, 99 kN

9   9.3 mm

10 460 mm

11 5

12 442 mm

13 (a) 68 N/mm², (b) 516 mm, (c) 2.45 m

14 (a) 251 N/mm², (b) 404 N/mm², (c) 176 839 N/mm²

15 1.5 mm

16 141 N/mm², 2 mm

17 8727 N/mm², 1.6 mm

18 146.25 kN, 1.6 mm

19 Copper 0.27 mm, brass 0.28 mm

20 15.5 N/mm², 0.001, 15 500 N/mm²

21 202.5 kN, 2 mm

22 2 × 6.6 mm

23 Timber 5.1 N/mm², steel 129.8 N/mm², 1.9 mm

24 69.4 kN, steel 106.6 N/mm², brass 41.6 N/mm²

Chapter 8

1   $L = 91$ kN, $R = 59$ kN, $M_{max} = 58$ kN m 1.03 m from L

2   $L = 145$ kN, $R = 135$ kN, $M_{max} = 151.9$ kN m 1.75 m from L

3   $L = 100$ kN, $R = 60$ kN, $M_{max} = 83.3$ kN m 1.67 m from L

4   $L = 24$ kN, $R = 36$ kN, $M_{max} = 57.6$ kN m 2.8 m from L

5   $L = 31.5$ kN, $R = 73.5$ kN, $M_{max} = +16.5$ kN m 1.05 m from L and -15.0 kN m at R

6   $L = 140.4$ kN, $R = 114.6$ kN, $M_{max} = 219.1$ kN m 3.18 m from L

7   $L = 99.6$ kN, $R = 104.4$ kN, $M_{max} = +14.3$ kN m 1.79 m from L and -35.0 kN m at R

8   $L = R = 60$ kN, $M_{max} = 96$ kN m at centre of span

9   $L = 112.1$ kN, $R = 27.9$ kN, $M_{max} = +7.38$ kN m
    0.87 m from L and -30.0 kN m at L

10 $L = 120$ kN, $R = 200$ kN, $M_{max} = +120$ kN m
    between 1 m and 5 m from L and -120 kN m at R

11 $L = 97.5$ kN, $R = 187.5$ kN, $M_{max} = +35.1$ kN m
    0.25 m from L and -120.0 kN m at R

12 $L = 75$ kN, $R = 175$ kN, $M_{max} = +5$ kN m 1 m from
    L and -140 kN m at R

13 $L = 169$ kN, $R = 201$ kN, $M_{max} = 282.95$ kN m 2.04 m
    from L

14 $L = 230$ kN, $R = 140$ kN, $M_{max} = 512.88$ kN m 4.5 m
    from L

15 $M_{max} = +28.9$ kN m and -80.0 kN m

16 37.63 kN m

17 2.83 m ($\sqrt{8}$ m)

18 (a) $L = 150$ kN, $R = 145$ kN; (b) $+158.75$ kN m 4.5 m
    from L, -70 kN at L

19 (a) $L = 142$ kN, $R = 148$ kN; (b) 181.4 kN m 2.4 m
    from L

20 $R = 10$ kN, $M_{max} = -30$ kN m at support

21 Max. SF = 146.7 kN at R, $M_{max} = 136.67$ kN m 2.5 m
    from L

22 (a) $L = 215.5$ kN, $R = 235.5$ kN; (b) $M_{max} = 708.25$
    kN m 5.5 m from L

23 (a) 15.0 kN m; (b) 245.0 kN m; (c) 357.5 kN m

24 400 kN m

25 (a) $R_L = 20$ kN, $R_R = 40$ kN; (b) $M_{max} = 46.20$ kN m;
    (c) $M_C = 31.11$ kN m

26 (a) $+20$ kN m; (b) -10 kN m; (c) $+5$ kN m

27 120 kN

28 6 m

29 60 kN

30 (a) $\frac{1}{2}Wl$; (b) $\frac{3}{8}Wl$

Chapter 9

1   444 mm east and 444 mm south of A

2   (a) 22.3 mm; (b) 55.9 mm; (c) 82.1 mm; (d) 104.2 mm

3   (a) $\bar{x} = 83$ mm, $\bar{y} = 325$ mm;
    (b) $\bar{x} = 134.7$ mm, $\bar{x} = 62.9$ mm;
    (c) $\bar{x} = 76.3$ mm, $\bar{y} = 49.6$ mm;
    (d) $\bar{x} = 83.8$ mm, $\bar{y} = 148.5$ mm;
    (e) $\bar{x} = 55.0$ mm, $\bar{y} = 41.7$ mm;

4   $I_{xx} = 381.24 \times 10^6$ mm^4, $I_{yy} = 47.50 \times 10^6$ mm^4

5   $I_{xx} = 247.30 \times 10^6$ mm^4, $I_{yy} = 126.25 \times 10^6$ mm^4

6   $a = 29.4$ mm, $I_{xx} = 4.20 \times 10^6$ mm^4, $I_{yy} = 9.54 \times$
    10^6 mm^4

7   $a = 53.8$ mm, $I_{xx} = 11.6 \times 10^6$ mm^4, $I_{yy} = 2.15 \times$
    10^6 mm^4

8   $a = 49.5$ mm, $b = 24.5$ mm, $I_{xx} = 6.56 \times 10^6$ mm^4,
    $I_{yy} = 2.37 \times 10^6$ mm^4

9   $a = 398$ mm, $I_{xx} = 998.79 \times 10^6$ mm^4, $I_{yy} = 76.00 \times$
    10^6 mm^4

10 $a = 71$ mm

11 $I_{xx} = 12.13 \times 10^9$ mm^4, $I_{yy} = 3.19 \times 10^9$ mm^4

12 $x = 127.7$ mm, $I_{xx} = 199.34 \times 10^6$ mm^4

13 $x = 90.5$ mm, $I_{xx} = 117.51 \times 10^6$ mm^4

14 $x = 49.5$ mm, $I_{xx} = 6.56 \times 10^6$ mm^4

15 $I_{xx} = 37.83 \times 10^6$ mm^4, $I_{yy} = 9.02 \times 10^6$ mm^4

16 $I_{xx} = I_{yy} = 3.54 \times 10^6$ mm^4

17 $x = 37$ mm, $I_{xx} = 161.97 \times 10^6$ mm^4, $I_{yy} = 74.73 \times$
    10^6 mm^4

18 $I_{xx} = 159.31 \times 10^6$ mm^4, $I_{yy} = 44.95 \times 10^6$ mm^4

19 24 mm

20 160 mm

21 24 mm

22 221 mm

23 $I_{xx} = 2.69 \times 10^9$ mm^4

Chapter 11

1   25 kN

2   258 mm

3   409 kN

4   169.5 kN

5   150 N/mm^2

6   96 kN

7   6.8 kN

8   8.8 kN

9   8.88 kN/mm^2

10   (a) 4.86 N/m^2; (b) 0.156 $\times$ 10^6 mm^3

11   (a) 137 mm; (b) I_{xx} = 280.9 $\times$ 10^6 mm^4, Z_{xxC} = 1.26 $\times$ 10^6 mm^3, Z_{xxT} = 2.05 $\times$ 10^6 mm^3; (c) 25.3 kN/m

12   453 mm

13   (a) 305 $\times$ 165 UB40; (b) and (c) 305 $\times$ 102 UB83; (d) 610 $\times$ 229 UB101; (e) 457 $\times$ 191 UB82; (f) 254 $\times$ 146 UB37

14   (a) 51.0 N/mm^2; (b) 17.8 N/mm^2

15   152 $\times$ 89 UB16

16   Joists: 142 mm $\times$ 75 mm, 133 mm $\times$ 48 mm; A = 305 $\times$ 102 UB28, B = 254 $\times$ 102 UB25

17   l = 6.4 m, beam = 356 $\times$ 127 UB39

18   22 m

19   406 $\times$ 178 UB74

20   305 $\times$ 102 UB33

21   10.5 kN

Chapter 12

1   (a) Steel: 120 N/mm^2, timber: 8 N/mm^2; (b) 81.6 $\times$ 10^6 N mm

2   99.5 kN

3   74.8 $\times$ 10^6 N mm

4   (a) 44.9 kN; (b) 20 mm

5   592 mm

6   d_1 = 127 mm, A_{st} = 475 mm^2

7   (a) 590 mm; (b) 1520 mm^2

8   d_1 = 289 mm, A_{st} = 776 mm^2

9   d_1 = 380 mm, A_{st} = 1246 mm^2, W = 12.5 kN

10   d_1 = 318 mm, A_{st} = 856 mm^2

11   d_1 = 200 mm, A_{st} = 844 mm^2

12   Beam A: d_1 = 323 mm, A_{st} = 865 mm^2; beam B: d_1 = 167 mm, A_{st} = 224 mm^2

13   d_1 = 430 mm, A_{st} = 1879 mm^2

Chapter 13

1   18 mm

2   14 mm

3   16 mm

4   105 kN

5   5.8 mm

6   1.5 mm

7   1.4 mm

8   6.0 mm

9   7.83 mm

10   11.5 mm

11   (a) 7.6 kN; (b) 3.91 kN

12   (a) 205 mm; (b) 247 mm

13   9.68 mm

14   7.57 mm

15   75 $\times$ 160 mm

16   7.5

17   127 mm

18   14.9 mm

19   200 mm $\times$ 600 mm

20   347.6 mm, 45.31 kN

21   10 mm

Chapter 14

1   (a) 11.64 kN; (b) 25.3 kN

2   (a) 4.3 kN; (b) 8.7 kN; (c) 8.8 kN

3   227.7 mm

4   443 kN

5   220 mm

6   49.8 kN; 106.5 kN

7   (a) 1096 kN; (b) 1266 kN; (c) 1797 kN

8   11.6 m, 502 kN

9   203 $\times$ 203 UC71

10 (a) 1180 kN; (b) 1385 kN

11 1.3 MN

12 1.6 MN

13 422 mm square, 7117 mm^2

14 (a) 370 mm square, 10 994 mm^2; (b) 690 mm square, 3800 mm^2

15 587.1 kN, 1007.2 kN

16 250 mm × 290 mm, 2900 mm^2

17 980 kN

18 6597 mm^2

19 (a) 200 mm square, 2000 mm^2; (b) 4.0 m

Chapter 15

1 264 kN (plate)

2 205 kN (rivets)

3 292 kN (rivets)

4 323 kN (plate)

5 (a) and (b) 201 kN (bolts)

6 9.3 mm, 4 bolts each in plate and column

7 $x = 76$ mm, 5 bolts

8 AA, BB 408 kN, CC 488 kN, rivets 208 kN

9 224 kN, 16 mm, 8 rivets

10 324 kN (plate)

Chapter 16

1 (a) +6.88 N/mm^2; (b) +1.12 N/mm^2

2 $f_{bc} = 6.4$ N/m^2, $f_{ad} = 0$

3 $f_{bc} = +7.04$ N/mm^2, $f_{ad} = -0.64$ N/mm^2

4 7.81 N/mm^2

5 AD = +89 N/mm^2, BC = +47 N/mm^2

6 $A = +8.8$ N/mm^2, $B = -7.4$ N/mm^2, $C = +158.4$ N/mm^2, $D = +174.6$ N/mm^2

7 160.6 kN

8 $A = +8.67$ kN/m^2, $B = +4.67$ kN/m^2

9 $A = 52.0$ kN/m^2, $B = 0$

10 (a) +12 N/mm^2(bottom), +1.33 N/mm^2(top); (b) 50 mm

11 72 N/mm^2

12 (a) +8.0 N/mm^2; (b) −5.6 N/mm^2

13 (a) −0.25 N/mm^2(top), +12.75 N/mm^2(bottom); (b) $W = 244.8$ kN

Chapter 17

3 (i)

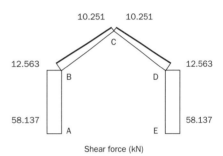

Shear force (kN)

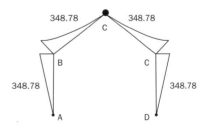

Moment (kN m)

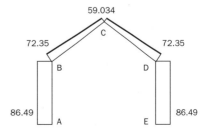

Thrust (kN)

(ii) Columns S33 × 210 UB122; Rafters 457 × 152 UB60 or 457 × 152 UB67

5 Columns 254 × 254 UC 89; Rafters 457 × 152 UB 60

6 (ii) Columns 203 × 203 UC 71; Rafters 305 × 165 UB 46

Chapter 18

1 (a) (i) 20 kN/m², (ii) 45 kN/m², (iii) 245 kN/m²; (b) (i) 10.7 kN/m², (ii) 24 kN/m², (iii) 131 kN/m²

2 (a) 93.1 kN; (b) 29.7 kN

3 (a) 8.3; (b) 9.3; (c) 9.3

4 35.42 kN/m²

5 80 kN/m²

6 5 m

7 66.1 kN/m², 36.9 kN/m²

8 1350 N/m²

9 $A = 47.5$ kN/m², $B = 18.1$ kN/m²

10 (a) 101.5 kN/m², −11.5 kN/m²; (b) 102.9 kN/m²

11 31.1 kN/m², 28.9 kN/m²

12 1.44 m

13 0.48 m

14 $A = 6.13$ kN/m², B = 50.73 kN/m²

Bibliography

1 British Standards and Eurocodes

BS 6399–1:1996 Loading for buildings – Part 1: Code of practice for dead and imposed loads

BS 6399–2:1997 Loading for buildings – Part 2: Code of practice for wind loads

BS 6399–3: 1988 Loading for buildings – Code of practice for imposed roof loads

BS 5950–1: 2000 Structural use of steelwork in building – Part 1: Code of practice for design – rolled and welded sections

BS 8110–1: 1997 Structural use of concrete – Part 1: Code of practice for design and construction

BS 8110–2: 1985 Structural use of concrete – Part 2: Code of practice for special circumstances

BS 8110–3: 1985 Structural use of concrete – Part 3: Design charts for singly reinforced beams, doubly reinforced beams and rectangular columns

BS 5628–1: 1992 Code of practice for use of masonry – Part 1: Structural use of unreinforced masonry

BS 5268–2: 1996 Structural use of timber – Part 2: Code of practice for permissible stress design, materials and workmanship

BS EN 338: 1995 Structural timber – strength classes

BS 8004: 1986 Code of practice for foundations

Extracts from British Standards for students of structural design p. 7312: 1998.

For further information, contact: British Standards Institute, 289 Chiswick High Road, London, W4 4AL, UK Customer Services Tel: + 44 (0)208 996 7000, Fax: + 44 (0) 208 996 7001

Building Regulations. Part A. Structure of the building regulations, Part B. Fire safety, HMSO. London

DD ENV 1993-1-1, Eurocode 3, 1992, Design of steel structures-Part 1.1, General rules for buildings

DD ENV 1992–1–1, Eurocode 2, Design of concrete structures – Part 1

DD ENV 206, Concrete performance, production, placing and compliance criteria

2 Textbooks and other publications

D. Beckett and A. Alexandrou, *Introduction to Euro-Code 2, Design of concrete structures*, E & FN Spon 1997

W.H. Mosley, R. Hulse and J.H. Bungy, *Reinforced Concrete Design to Euro-Code 2*, Macmillan Press Ltd 1996

T.J. MacGinley, *Steel Structures Practical Design Studies*, 2nd edition, E & FN Spon 1998

W.G. Curtin, G. Shaw, J.K. Beck, W.A. Bray, *Structural Masonry Designers' Manual*, 2nd edition, BSP Professional Books 1991

G.W. Owen, P.R. Knowles, Steel Construction Institute, *Steel Designers' Manual*, 5th edition. Blackwell Scientific Publications 1994

J.M. Dinwoodie, *Timber: Its Nature and Behaviour*, 2nd edition, E & FN Spon 2000

Standing Committee on Structural Safety, *Structural Safety 1997–1999: Review and Recommendations*, 12th report of SCOSS, February 1999

R. Cather, How to get better curing, *Concrete*, September 1992

N. Jackson and R.K. Dhir, *Civil Engineering Materials*, 5th edition, Macmillan 1996

J.M. Illston, *Construction Materials*, E & FN Spon 1994

3 Structural design and analysis by computer

R. Hulse and W.H. Mosley, *Reinforced Concrete Design by Computer*, Macmillan, Basingstoke 1986

R. Hulse and W.H. Mosley, *Prestressed Concrete Design by Computer*, Macmillan, Basingstoke 1987

Computer educational package supplied by EdSoft Ltd. Readers are encouraged to see their web site: www. Reel.co.uk

QSE – Space 2/3D analysis

Staad – Pro (FE, 2nd order, seismic + other advanced features with QSE interface)

QSE – Steel Designer BS 5950 + EC3

QSE – Concrete Designer BS 8110 + CE2

QSE – Timber Designer BS 5628 + EC5

QSE – Section Wizard (all properties, stress)